Activity-Based Statistics

Instructor Resources

Springer
New York
Berlin
Heidelberg
Barcelona
Budapest
Hong Kong
London
Milan
Paris
Santa Clara
Singapore
Tokyo

Activity-Based Statistics

Instructor Resources

Richard L. Scheaffer
University of Florida

Ann Watkins
California State University, Northridge

Mrudulla Gnanadesikan
Fairleigh Dickinson University

Jeffrey A. Witmer
Oberlin College

Springer

This project was funded by National Science Foundation Grant USE-9150836

Printed on acid-free paper.

Production coordinated by University Graphics, Inc. and managed by Terry Kornak; manufacturing supervised by Joe Quatela.
Photocomposed by University Graphics, Inc., York, PA.
Printed and bound by Hamilton Printing Co., Rensselaer, NY.
Printed in the United States of America.

9 8 7 6 5 4 3

ISBN 0-387-94597-0 Springer-Verlag New York Berlin Heidelberg SPIN 10635409

Advisory Committee

Donald Bentley
Pomona College

Gail Burrill
Greenfield, WI

George Cobb
Mt. Holyoke College

Elizabeth Eltinge
Texas A & M University

Joan Garfield
University of Minnesota

James Landwehr
AT&T Bell Laboratories

Donald Paxson
Florida Power and Light

Dennis Pearl
Ohio State University

Judith Singer
Harvard University

Frank Soler
De Anza College

W. Robert Stephenson
Iowa State University

Evaluator

Mary E. Huba
Iowa State University

Overview to the Student Guide

Their fast-paced world of action movies, rapid-fire TV commercials, and video games does not prepare today's students to sit and absorb a lecture, especially on a supposedly dull subject like statistics. To capture the interest of these students, teaching must move away from a lecture-and-listen approach toward innovative activities that engage students in the learning process. The goal of the Activity-Based Statistics Project (ABSP) is to develop a set of such activities that cover the statistical concepts essential to any introductory course. These activities can be used in a variety of class settings to allow students to "discover" concepts of statistics by working through a set of "laboratory" exercises. Whether the "laboratory" is the classroom, the student's usual place of study, or a more formal statistics laboratory, the traditional lectures in introductory statistics should be supplemented or supplanted by a program that requires the active participation of the students, working individually or in groups. Statistics, then, should be taught more as an experimental science and less as traditional mathematics.

The activities are organized around the major topics covered in most introductory courses. The overarching topic is exploring data. Statistical ideas begin with data, but data should be collected for a purpose. Relating data collection and analysis to the solving of a real problem, much as is done in statistical process improvement, allows exploratory techniques to be used extensively but with a purpose in mind.

Random behavior is fundamental to the decision-making process of statistics but is a difficult topic for students to grasp. Thus, a number of activities concentrate on developing an understanding of randomness, without going into the mathematical formalities of classical probability distributions. Simulation is key to this process. Related material on sampling distributions is included, to be introduced at the instructor's discretion. Since the amount of required probability is highly variable from course to course, the activities throughout the ABSP are designed to be completed with a minimum of probability.

Sampling, such as in the ubiquitous opinion poll, is the most common application of planned data collection in the "real" world and in the classroom. This topic allows the early introduction of the idea of random sampling and its implications for statistical inference. Thus, it serves as an excellent bridge between probability and inference. The difficulty and importance of collecting data that fairly represents a population are emphasized.

Estimation and hypothesis testing frame the basic approach to inference in most courses, and numerous activities deal with the conceptual understanding of the reasoning process used here. Again, simulation plays a key role as concepts of sampling error, confidence interval, and p-value are introduced without appeal to formulas.

Experiments are the second major technique for planned data collection. By appealing to activities that focus on optimization of factors, the ABSP emphasizes that experiments are designed to compare "treatments" rather than to estimate population parameters.

Modeling the relationship between two variables, especially through the use of least-squares regression, is widely used throughout statistics and should be part of an introductory course. Technology is required here in order to use time efficiently. Correlation, a topic much confused by students, is singled out for special study.

The ABSP contains far more activities than can be used efficiently in one course. It is hoped that instructors will choose a few activities from each of the topics so that students will be exposed to a range of concepts through hands-on learning.

Each activity has student pages and notes for the instructor. This allows the flexibility to use the activities as class demonstration, group work in class, or take-home assignments. In whatever setting they may be used, the activities should engage the students so that they become true participants in the teaching–learning process.

Overview to Instructor Resources

Incorporating Activities into a Statistics Course

The ABS collection of activities are many and varied. Some instructors will want to build their introductory course around these activities while others will incorporate selected ones within an existing course.

The following outline provides the instructor with examples of activities with which to teach a typical introductory statistics course. Activities shown in italics were used by one of the authors in such a course.

Exploring Data—One Variable

Getting to Know the Class and *The Shape of the Data* can be used on the first or second day of the course. *Measurement Bias* fits with a discussion of measurement and variation. After discussing box plots the *Living Box Plot* activity is a natural choice. *Matching Graphs to Variables* is a good activity for the student who needs to review reading a histogram and encourages the student to think about how the shape of a histogram is related to features in the data. Likewise, *Matching Statistics to Graphs* fits with early work on distributions.

Exploring Data—Two Variables

Getting Rid of the Jitters can be used when students learn about scatter plots and scatter plot smoothing. *Is Your Shirt Size Related to Your Shoe Size?* and *Models, Models, Models* fit with a discussion of regression. *Matching Descriptions to Scatter Plots* helps students develop an understanding of how regression and correlation results can depend on influential points. *The Regression Effect* presents an important lesson about regression. *Predictable Pairs* should be given with an exploration of categorical data.

Data Production

Gummy Bears in Space and *Jumping Frogs* give students a chance to work with designed experiments; they can be used when discussing data production or when discussing ANOVA. *Funnel Swirling* is a more complex activity that might work best as a review lesson late in the course. *How to Ask Questions* goes naturally with a discussion of survey design. *Random Rectangles* or *Stringing Students Along* provide good ways to introduce bias in sampling. *Spinning Pennies* gives an introduction to sampling distributions, *Capture/Recapture* fits in well when discussing how data are collected, and *Flick the Nick* can be used when discussing control charts.

Probability

Any of the random behavior activities (*What's the Chance?*, *What is Random Behavior?*, etc.) can be used when studying probability. The *Dueling Dice* activity can be presented as a way to review sampling distributions under the guise of wanting to study the sampling distribution of the larger of two rolls of a die. In fact, the activity shows that statistical analysis can reveal trends that are otherwise missed by most observers.

Distributions

Estimating Proportions: How Accurate Are the Polls? and *Streaky Behavior: Runs in Binomial Trials* fit with a discussion of the binomial distribution. *Cents and the Central Limit Theorem*, *Lets us Count*, and *The Central Limit Theorem and the Law of Large Numbers* should be paired with a discussion of sample means and the central limit theorem. *Sampling Error and Estimation* can be used here as well.

Introduction to Inference

What Is a Confidence Interval Anyway?, fits with an introduction to interval estimation. *Introduction to Hypothesis Testing* can be used to introduce the idea of examining evidence against a null hypothesis.

Inference for Means and for Proportions

The Bootstrap can be used to show an alternative to the t-test. *Confidence Intervals for the Percentage of Even Digits* fits with a discussion of proportion data and *Coins on Edge* with hypothesis testing for a single proportion, although it could also be used when introducing hypothesis testing. *Statistical Evidence of Discrimination* aligns with a discussion of equality of two proportions. *Estimating the Difference Between Two Proportions* can be used when discussing the difference between proportions.

Other Topics

Is Your Class Differently Aged?, uses the Chi-square goodness-of-fit test. *Relating to Correlation* correlates with the topic of inference in regression. Either *Gummy Bears in Space* or *Jumping Frogs* can be used as a one-way ANOVA example but they are really designed as two-way ANOVA examples with interactions.

Contents

ESTIMATION, 161

HYPOTHESIS TESTING EXPERIMENTS, 243

EXPERIMENTS, 275

MODELING, 305

ASSESSMENT, 337

EXPLORING DATA

A Living Box Plot

NOTES FOR THE INSTRUCTOR

Statistical Setting

This activity should be used when box plots are introduced.

Objectives

The goals of this lesson are to reinforce the ideas involved in constructing a box plot and to provide a vivid exercise to aid retention of these ideas.

Prerequisites for Students

Students should know how to find quartiles and how to construct a box plot using paper and pencil, although this activity can be used in the same class period in which box plots are first presented.

Procedure

1. This activity works best with 20 or fewer students. If you have a large class (say, more than 25), then ask for (or randomly select) a sample of 15 students. The rest of this description is based on a sample of $n = 15$; other sample sizes require simple modifications to these instructions.

2. Tell the students that you are going to construct a box plot of the times they went to sleep the previous night. (Other questions could be used; sleep time is but one of many possibilities.)

3. Find out who went to sleep the earliest.

4. Find out who went to sleep the latest.

5. Place marks on the floor representing the times given in steps 3 and 4.

6. Mark on the floor the location of (earliest + latest)/2, as a reference point. It is important that the students realize that the three marks form the basis of a scale, so that they place themselves correctly in step 7.

7. Tell the students to place themselves at the appropriate points along the line between earliest and latest, according to when they went to sleep the previous night. If two students went to sleep at the same time, then one should stand behind the other. As they arrange themselves, check that they are placing themselves accurately. For example, mark a few more places (times), and check that students near those spots went to sleep at the right times, given where they are standing. Have them resist the temptation to space themselves evenly.

8. After everyone is in an appropriate spot, ask "Who represents the median?" When the class tells you that this is the eighth person, count off to find that person and have him or her take a step forward and announce his or her time. One way to do this is to have students count off from either end so they see that person number 8 counting from the top of the distribution down is the same as person number 8 counting from the bottom of the distribution up. (*Note*: Don't let your students confuse the median, $x_{(8)}$, with the number "8"; make this distinction clear.)

9. Ask the class, "Who represents the first quartile?" When the class tells you that this is the fourth person, count off to find that person and have him or her take a step forward and announce his or her time. Again, you can do this by counting up from the minimum and down from the person just below the median. As before, don't let your students confuse the first quartile, $x_{(4)}$, with the number "4."

10. Ask the class, "Who represents the third quartile?" When the class tells you that this is the 12th person, count off to find that person and have him or her take a step forward and announce his or her time.

11. Compute the interquartile range. Point out to the class that the IQR equals the distance from person 4 to person 12.

12. Ask "Are there any outliers?" Compute 1.5*IQR and add this to the third quartile to find the upper limit of non-outliers. If there are any outliers at the upper end of the distribution, have the corresponding students step forward and turn sideways, so that they look different from the non-outliers.

13. Subtract 1.5*IQR from the first quartile to find the lower limit of non-outliers. If there are any outliers at the lower end of the distribution, have the corresponding students step forward and turn sideways.

14. You might want to draw the boxplot on the wall behind the students or on the floor in front of them.

15. Now that the living box plot is complete, have a class discussion of what the box plot tells you about this particular distribution; i.e., interpret the results.

Extensions

Have the median person sit down. Ask the class how this affects the box plot (now that $n = 14$). Have persons 7 and the new 8 step forward together. Indicate that the median when $n = 14$ is $(x_{(7)} + x_{(8)})/2$. Discuss how other parts of the box plot are affected by having $n = 14$ rather than $n = 15$.

Some instructors like to make a living histogram. To do this, you need to choose bins (i.e., subintervals) and have students stand in the appropriate group. The most likely issue to confuse students here is how to deal with the endpoints of the bins. For example, consider using sleep time as the variable. Suppose that the bins are 9–10, 10–11, 11–12, etc. The class needs to decide whether someone who went to sleep at 10 p.m. is in the first bin or the second bin.

Sample Assessment Questions

1. Here are the populations, as of April 1992, of 11 cities and towns in the United States that are named Springfield:

State	Springfield Population
FL	8,715
IL	105,227
MA	156,983
MI	5,582
MO	140,494
OH	70,487
OR	44,683
PA	24,160
TN	11,227
VT	9,579
VA	23,706

Construct a box plot of these data.

2. Describe the distribution shown by the box plot in step 1 here.

3. Based on the data in step 1 above, how large would a population have to be in order to be an outlier?

4. In New Jersey there is a township named Springfield with a population of 13,420. Add this township to the 11 data values listed above, and construct a box plot of the 12 data values.

Getting to Know the Class

NOTES FOR THE INSTRUCTOR

Statistical Setting

Much data that statisticians analyze are collected in surveys. This activity introduces issues of question wording, reliability, validity, and measurement process.

Prerequisites for Students

None. This activity can be used on the first day of the course.

Materials

The only materials needed are copies of a questionnaire.

Procedure

Prepare a questionnaire such as the following one—which is only an example—and distribute it to the students. Assure them that they will remain anonymous and that you are collecting data for educational purposes but that they should feel free to skip any question they feel uncomfortable answering. Have them scan the questions you have prepared and then *ask them if there are other questions they would like to add to the list.* Number any extra questions you choose to use, and write them on the board. Then have the students fill in the questionnaire.

1. What is your height *in inches*? __________

2. What is your pulse? __________

3. What is your gender? __________

4. How much did you spend on your last haircut (including the tip)? __________

5. How much money do you have with you right now, in change only? Report the combined value of your change, not the number of coins. __________

6. Choose a random number in the range 1 to 20. __________

7. Aside from class time, how many hours per week, on average, do you expect to spend studying for this course? __________

8. Do you have a job in which you work at least 10 hours per week? __________

9. How many CDs do you own? __________

10. Do you smoke? __________

11. What time did you go to bed last night? __________

12. __________

13. __________

One option is to give each student a note card and ask the class to submit questions to you. Then select questions from those the students suggest and prepare a questionnaire to hand out during the following class meeting.

The more questions you ask, the longer it will take for students to complete the questionnaire. Students are generally most interested in the questions they suggest in class, but you don't want the questionnaire to be too long. Thus you probably cannot use all questions that class members suggest.

As you choose questions and as the students suggest questions, be aware that some questions may be viewed as reinforcing stereotypes or bias concerning race, gender, religion, etc. Other questions may be viewed as unacceptably intrusive. Avoid questions that may embarrass students, such as asking about age or weight. Use judgment and care.

The questionnaire should include at least two nominal variables (such as gender and whether or not the person holds a job) and at least two interval variables (such as "How much did you spend on your last haircut?" and "How many CDs do you own?") so that, at the appropriate time during the course, you can discuss association using a 2×2 table for the nominal variables and a scatter plot for the interval variables. You will want to discuss with the class the difference between a nominal variable and an interval variable. You will also want to include at least one question that will generate a skewed distribution (such as "How much did you spend on your last haircut?").

It is good to include at least one question that involves measurement. For example, the question "What is your pulse?" can lead to a discussion of how one measures pulse. Let the students suggest various methods (e.g., count for 30 seconds and multiply by 2), and then have the class choose a single method and have everyone use the same method when collecting the data. Discuss why this is important.

Here are some other questions you might use:

How old do you think the professor is? ________
What is the population of your home town? ________
What is your shoe size? ________
How many cups of coffee did you drink yesterday? ________
How much did you spend on textbooks this semester? ________
How many credits are you taking this semester? ________
How many different people have you dated in the last 30 days? ________
How did you get to class today? By car? By walking? By bike? Other? ________

This activity works well with classes of 20 to 40 students. If the class is much larger than 40, you might want to analyze the data from a random sample of the students. If the class is quite small, you might want to combine the data with those from another class.

Note that there can be issues of interpretation with some of the questions. For example, what constitutes a "date"? What does it mean to say that you smoke? You might ask the class how they would have worded the smoking question. A good alternative to "Do you smoke?" is "Have you smoked a cigarette within the past 24 hours?" (Someone once answered the haircut question by saying ".33". This confused everyone until the respondent explained that he had spent 20 minutes, or .33 of an hour, on his last haircut. Ambiguous questions can be confusing; having an ambiguous question on the list allows you to raise this issue after the data have been collected.)

Collecting data in class provides a good opportunity to discuss the concepts of *validity* and *reliability*. "What time did you go to bed last night?" might yield reliable answers, but one can question whether the data provide valid estimates of such things as sleep patterns, study patterns, etc. That is, you could make the claim "I know that some of you study more than others because some of you went to bed later than others" and then wait for the class to argue that such a claim is not reasonable. At this point you can discuss why the bed time item does not provide a valid measure of time spent studying.

The item that asks "What is your pulse?" might yield data of dubious reliability. Indeed, this could be checked by having the students measure their pulses twice, noting the difference between the first reading and the second reading. You might also want to discuss the extent to which pulse is a valid measure of one's general state of health. The issues of reliability and validity can be raised with each of the items on the questionnaire.

Extensions

This activity provides a good lead-in to exploratory data analysis. Choose a question from the list and analyze the responses immediately. This will demonstrate how one can apply exploratory data analysis techniques and learn something quickly about a dataset. For example, if the question "What is your pulse?" is on the questionnaire, you can create a stem and leaf diagram of the data as the students announce their pulses. Of course, if this is the first time the class has seen a stem and leaf diagram, then you must take some time to explain how you are creating the plot.

If you construct a stem and leaf diagram of pulses, you might then want to construct a back-to-back stem and leaf display. Ask the class to suggest a nominal variable with which to divide the pulse data into two groups, for example, coffee drinkers and non-coffee drinkers.

The question "How much money do you have with you right now, in change only?" often leads to a good male/female comparison. One way to do this is to construct a dot plot of the data, using different plotting symbols or different colors of chalk for men and women—without telling the class that you are doing this. They will catch on that the symbols or colors provide for a comparison of men and women. (If no one notices the presence of two symbols or colors, then point it out and ask them to speculate why there are two; then explain it.) This can lead to side-by-side dot plots or side-by-side box plots, if you wish.

Likewise, responses to the haircut question often differ dramatically between men and women. You might want to construct a scatter plot of haircut spending versus pocket change, after the data have been loaded into a computer (see the Technology Extension section below), and examine (a) the trend between these and (b) the relative locations of men and women on the scatter plot.

You might ask students to construct a 2 × 2 table of, for example, smoking by gender or having a job by whether or not they plan to study more than 5 hours per week for the course. Other relationships can be studied early in the course and returned to later, after the students have learned about more statistical tools to help with their analyses.

The "Choose a random number in the range 1 to 20" item lets you explore the extent to which people behave like random number generators. You can look at things like the percentage of odd numbers, how often numbers below 10 were chosen, etc.

After all the questionnaires are returned, you can make the data available to the class. Data analysis tasks can then be assigned as homework or as laboratory work.

Technology Extension

You can put the data from the entire class on a computer file that is available to the students and ask them to analyze the data with the aid of a statistics package. If your software requires a numerical code for categorical data (e.g., 0 for men and 1 for women), be aware of the difficulty some students will have with this coding scheme. Explain how the coding works, and warn them not to compute things like the average major of a student if you have an item that asks for their major and the responses are coded as Business = 1, Science = 2, etc.

You can ask for parallel box plots of the heights of men and of women, of the haircut expenditures of men and of women, or of the pulses of smokers and nonsmokers. You can ask for assessments of the normality of any of the distributions, using normal probability plots, and have the students investigate the effects of transforming the variables.

You could ask for a scatter plot to investigate whether there is a relationship between height and haircut expenditure and point out that there may be an effect due to gender. If you are using software that allows you to link graphs, you can generate the scatter plot and then generate a bar chart of gender. Link these two graphs, and use the bar chart to select first the women and then the men, noting in each instance which points in the scatter plot are highlighted. If your software does not allow you to link graphs, you could use different plotting symbols for men and women.

More generally, you could simply ask the students to investigate relationships between the variables. How much guidance they will need will depend on what topics have been covered in class and on how easy it is for them to use the software available.

Sample Results from Technology Extension

Here are box plots, by gender, for responses to the question, "How much did you spend on your last haircut (including the tip)?"

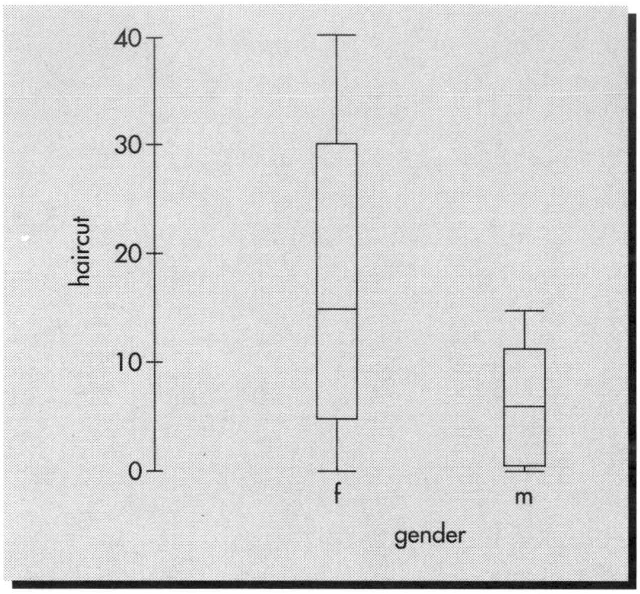

Sample Assessment Questions

1. Suppose someone claims, "Serious students buy lots of books. We can tell how good students are by asking them how much they spent on textbooks this semester." Discuss this claim. That is, discuss the use of textbook expenditure as a measure of the quality of a student.

2. Which method of measuring a person's pulse is likely to be more reliable: (a) count for 6 seconds and multiply by 10, or (b) count for 60 seconds? Why?

Getting to Know the Class

SCENARIO

No two people are alike. Some people are taller than others. Some people love classical music, and others do not. Indeed, people vary from one another in many interesting ways. A good way to begin thinking about data is to collect and then analyze data from those around you. In this activity we will collect data from members of the class. This will lead to some graphical methods of data analysis and a discussion of topics like the wording of questions. Once the data are compiled you can explore them and see where you fit within the class distribution of the variables measured.

Question

How do you compare to the other students in the class?

Objectives

The goals of this activity are to learn some of the tools of exploratory data analysis and to address some concepts in measurement.

Activity

Your instructor will give you a questionnaire to answer. Please answer the questions anonymously; feel free to skip any question you feel uncomfortable answering. These data are being collected for educational purposes and will form a database for the class to analyze.

Wrap-Up

1. Write a brief summary of what you learned in this activity about exploratory data analysis.

2. Write a brief summary of what you learned in this activity about measurement.

3. Suppose you wanted to know what proportion of the students at your college are vegetarians. How would you construct a question that would give you the information you need? What kinds of things could go wrong in writing the question? Would your question provide a valid and reliable measure of the health of the respondent?

Measurement Bias

NOTES FOR THE INSTRUCTOR

Statistical Setting

This lesson should come early in the term but after students have had experience with variability in measurements and ways to describe distributions of data, both graphically and numerically. Emphasize that bias can be seen even in highly variable data sets and that bias is a function of the measurement system rather than a property of a single measurement. Demonstrate that comparisons of centers of distributions are often valid even when the measurements are biased, and this is an important reason for designing comparative experiments.

Prerequisites for Students

Students should have some knowledge of basic data displays, such as dot plots and stem plots, and basic numerical summaries of center and spread, such as the median and interquartile range. They should have experience in describing distributions of data.

Materials

Students will need two strings of different lengths. It is recommended that each be a little over 36 inches long.

Procedure

1. Making measurements and collecting data
   Hold up a string so that the students can see its full length, but so they cannot get close enough to actually measure it against their arms or hands or books. This activity works well if the string is a little longer than 36 inches, which is the standard against which students seem to be forming their guesses. A string of around 45 inches in length should show the effect of bias; student guesses should center much closer to 36. You might want to collect the student data at the end of one

period and prepare it for presentation to the students, individually or collectively, during the next period. This is especially important if you have a large class; large amounts of class time should not be used for collecting and organizing data.

2. Describing the data
Experience suggests that the distribution of guesses will display a reasonable amount of symmetry and mound shapedness, if there are enough observations. This may require the pooling of data from two or more classes. In fact, you can keep the data archived from one term to the next. Students are sometimes surprised to find that their biased guesses are about the same as those from other classes and even other terms.

   Outliers are often present. These should be discussed, and some may be removed from the data set, if the class agrees. A common reason for outliers is that a student did not get the rules correct and was guessing on the wrong scale or was guessing something other than the length of the string.

3. Collecting a second set of data
The second string should be of a different length from the first one and should be clearly identified as a different string (perhaps by color). However, it may only be a couple of inches longer or shorter than the other one, as a 2-inch difference seems to be enough to be picked up by eye.

4. Drawing conclusions about bias
The "correct" length of the string should be explained in terms of how you determined it. Usually, the correct length is well above the center of the distribution of guesses, and this leads to a discussion of bias. Students are confused by the idea that the measurement system we are using is considered to be biased even though some individual students guessed quite well. The bias is an attribute of the class rather than the individual.

Sample Results from Activity

The plots below show typical distributions of guessed string lengths from students. String A was 45 inches and B was 47 inches in length.

```
Stem-and-leaf of A      N = 47
Leaf Unit = 1.0
    1  2 8
    5  3 0000
    6  3 3
   10  3 4555
 (17)  3 66666666666666667
   20  3 88888888
   12  4 000
    9  4 23
    7  4 5
    6  4
    6  4 88888
    1  5 1
```

```
Stem-and-leaf of B      N = 47
Leaf Unit = 1.0
    1  2 8
    3  3 00
    3  3
    3  3
   12  3 666666667
   17  3 88888
   22  4 00000
  (9)  4 222222333
   16  4 555
   13  4 7
   12  4 8888888
    5  5 00
    3  5
    3  5 5
    2  5
    2  5 8
    1  6 0
```

Sample Assessment Questions

1. Individual students or small groups can be assigned a measurement task, such as measuring the area of a table top or a wall of the classroom. Of the many ways to do the task, students should agree on two methods. Then, have them make a series of measurements of the area in question by each method. There will be variability and bias in the results for each data set. The analysis should include a description of the data sets along with a statement about variability and bias in each. Which measurement method would they use if they were to do the project again? Are the potential biases large enough to have a serious effect on the result, say, if they were to buy paint to paint the wall or a glass top for the table?

2. Ask students to find an article in the media, or assign them one that you have found, to critique for possible bias in the measurements used. The critique should include a discussion of possible factors causing the bias and the possible effect of the bias on the conclusions of the article. (Such articles are attached for your use).

Where is the bias?

Lead tests deemed erroneous

By DONYA CURRIE
Sun staff writer

> The results add weight to fears that faulty testing has cost milllions of dollars to clean up housing projects.

Lead-based paint is less of a threat in Gainesville public housing projects than earlier tests reported.

The results received Tuesday from an independent laboratory add weight to officials' fears that faulty testing has cost millions of dollars to clean up housing projects where lead was not a health threat.

Karen Godley, director of maintenance for the Gainesville Housing Authority, said testing done this year by two different companies showed erroneously high levels of lead in several apartment complexes across the city. The problem—a hand-held testing device now under investigation by the federal government for misleading officials in two ways: some tests have shown high levels of lead where the substance posed no health threat; other tests may have failed to detect illegal levels that endangered residents.

Congress banned lead-based paint in 1978 because, if ingested, toxic levels can cause brain damage and learning disabilities in children. Lead testing has become a multimillion dollar industry nationwide, with the most money going to companies that garner cleanup contracts of up to $2 million for even a midsized housing project.

Gainesville officials contracted with the Miami-based Accutest in 1993 to conduct random tests of city housing projects. When a local company followed up by testing 12 different apartments, almost every result was different. A third test, done this month using paint chip samples instead of a less-expensive, hand-held device, showed lower levels of lead than first detected.

"Before we spent millions or hundreds of thousands of dollars abating lead, we wanted to make sure the results were accurate," said Godley, who expects to replace 100 apartment doors containing high levels of lead.

Officials with the Federal Department of Housing and Urban Development are in the process of rewriting rules for lead testing at public housing projects. In the meantime, Florida housing officials have called for a statewide investigation of test results that may either be putting residents at risk or funneling public money to unnecessary cleanups.

Jim Walker, special assistant in HUD's Jacksonville office, expects Florida officials to take action soon by advising the state's 81 public housing authorities on how to ensure accurate testing. Stories published in The Orlando Sentinel in early April showed Accutest had problems at several housing authorities, reporting high levels of lead where the substance was not a threat and possibly giving cleanup contracts to a related company.

Accutest owner David Mingus denies the charges and came to Gainesville two weeks ago to help with the paint-chip testing. He was not available for comment Tuesday but told The Sun in an earlier interview that he stands by the accuracy of his company's tests.

Godley said she hopes the new HUD guidelines will require only paint-chip testing instead of tests done using a hand-held X-ray device. Housing authorities will pay more for lab analysis of paint chips scraped from apartment walls and doors, but the test results will be accurate, she said.

"I think most of what we're proving right here is that the machine is not reliable," Godley said. "I think the company (Accutest) followed the guidelines. The problem is with the machine."

HUD officials are considering changing requirements for lead testers. Florida is one of many states that does not require certification for testers. To enter the business, a person need only purchase a $10,000–$20,000 testing machine and complete a week-long course from the manufacturer.

Source: Gainesville Sun, 4/27/94

Instrument flaw explains mystery in color vision

NEW YORK (AP)—For 45 years, scientists thought that people's color vision varied according to the seasons. And it left them baffled.

Now science has come up with an answer to one of nature's enduring mysteries: The instrument used for the test in 1948 was flawed.

A study in today's issue of the journal Nature found that the instrument's readings are affected by room temperature. So if a testing room got warmer and cooler through the year, the device would seem to indicate that people's color vision varied with the seasons, researchers said.

Source: THE NEWS-TIMES, (DANBURY, CT) THURSDAY, JUNE 10, 1993

3. Optical illusions are related to bias. Here is an activity on optical illusion that you might want to try. Have students collect data from the device explained here and then write a report on the results.

For most of us, our eyes are deceiving. We do not really see what we think we are seeing. There are many demonstrations of optical illusions that help prove this point, and some of them are well suited to studies of bias in a measurement process or device. This activity will consider one such demonstration.

One simple device that works well to show bias in our visual judgments is illustrated on the attached pages. The goal is to make the line with open arrows at the ends equal in length to the line with closed arrows at the ends. The arrow with closed ends is of fixed length, drawn on the cardboard in advance. The line with open arrows slides out from behind the front cardboard so that the person making the "measurement" is free to stop it at any point. The length of the fixed line with closed arrows is the "truth." It is the job of the participant to make the length of the line with open arrows as close to truth as possible.

The size of the device can change depending on how it is to be used. For small groups an 8-inch square background is large enough, but class demonstration may require a larger one. Rules for sliding out the back arrow can be formulated by the class. Two that work well are

a. slide the arrow out until you think the lengths match and then stop, with no "backsliding" allowed.
b. slide the arrow back and forth at will until you think the lengths match.

Students may work in groups to collect data, but each student should repeat the procedure three or four times so that individual biases and variability can be assessed. Plot the measurements of the length of the open arrowed line on a dot plot for each student; then combine the data to make a plot for each group. Discuss the notion of bias in the measurement process.

Front side masters for the Jacket and the Slide.

Duplicate this page to produce enough devices for your class to conduct the visiual bias experiment.

Cut on the solid lines to produce the front sides of the sleeve and the slide.

Staples in the marked locations hold the front and back panels together and allow for the slide to move in the desired path.

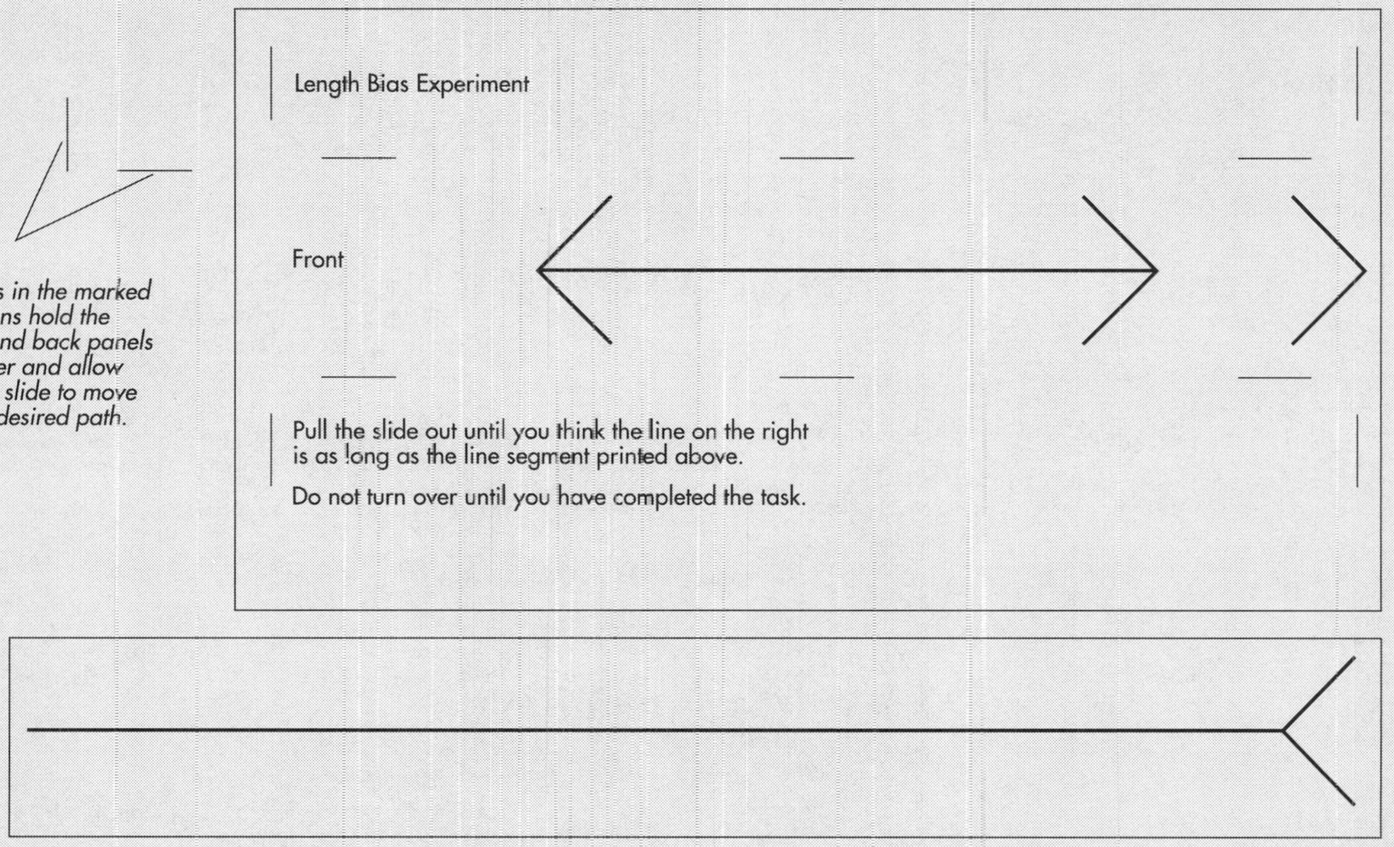

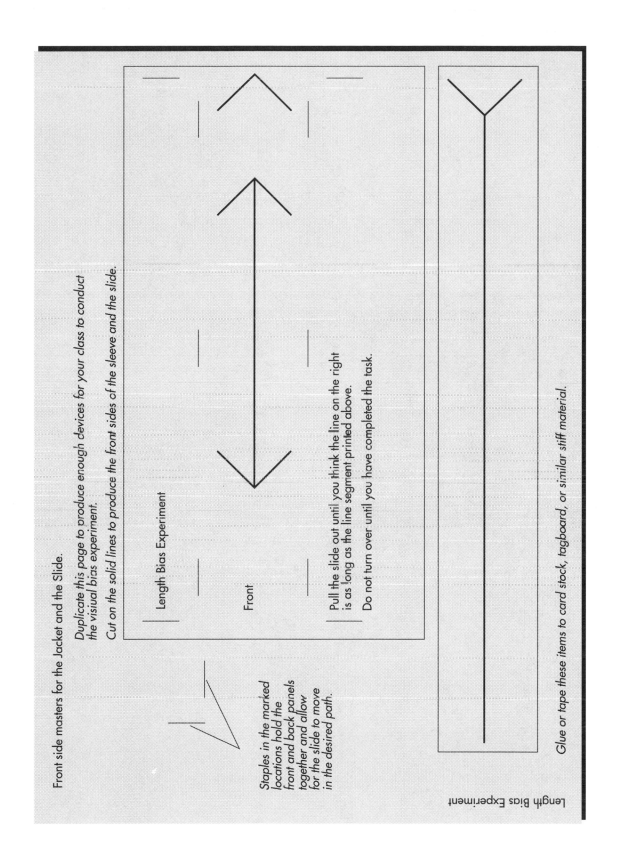

Length Bias Experiment
Front
Pull the slide out until you think the line on the right is as long as the line segment printed above.
Do not turn over until you have completed the task.

Length Bias Experiment

Glue or tape these items to card stock, tagboard, or similar stiff material.

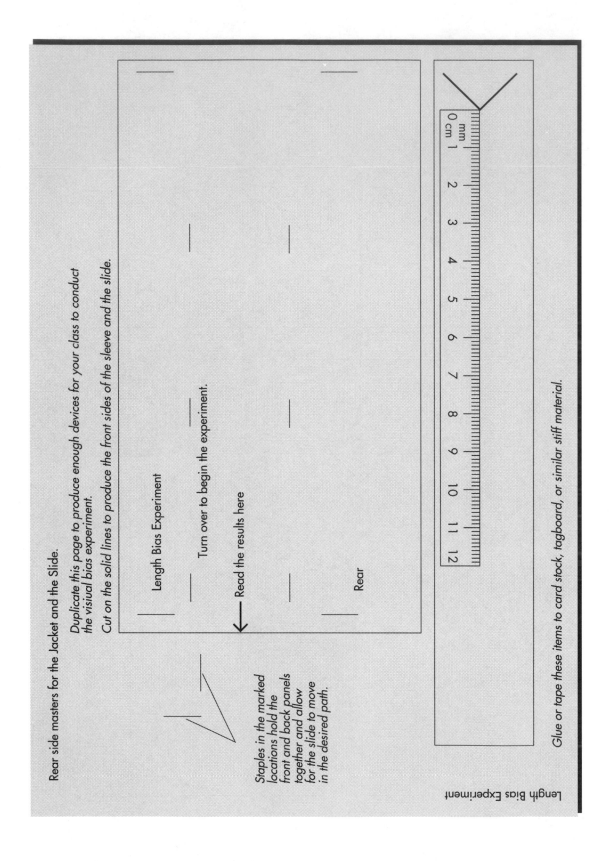

Rear side masters for the Jacket and the Slide.

Duplicate this page to produce enough devices for your class to conduct the visiual bias experiment.

Cut on the solid lines to produce the front sides of the sleeve and the slide.

Staples in the marked locations hold the front and back panels together and allow for the slide to move in the desired path.

Length Bias Experiment

Turn over to begin the experiment.

← Read the results here

Rear

Length Bias Experiment

Glue or tape these items to card stock, tagboard, or similar stiff material.

Measurement Bias

SCENARIO

You open your program before the start of a basketball game and see the rosters for each team along with the height of each player. What thought runs through your mind? Maybe it's, "Hmmm, I wonder if these players are really that tall." Since there is an advantage to having tall basketball players on your team, sometimes these figures are "stretched" a bit toward the larger values. This is an example of **bias.** Interestingly enough, if the opposing team has even taller players, you might be concerned about the outcome of the game. Even though both sets of heights might be biased, the comparison between the sizes of the teams might be fair.

Question

What is measurement bias, and how can it be detected?

Objectives

In this lesson you will see that measurement systems are often subject to measurement bias. Your awareness for the potential of bias in data that you analyze should be heightened as you think about possible sources of bias. Bias is a property of the measurement system and cannot be reduced simply by taking more measurements. The only way to measure the bias is to compare the measurements with an independent source of "truth" outside the measurement system you are using. Comparisons between two sets of measurements are often little affected by the bias in each.

Activity

1. Collecting the data
   Look carefully at the string (marked A) the instructor is holding out straight. Without using any measuring instruments (except your eyes), write down the length of the string to the nearest whole inch.

   Length of string A = __________

   The instructor will collect the guessed string lengths and provide you with the data for your class. The data will be used in the next part of the activity.

2. Describing the data graphically
   a. Make at least two different plots of the data on string length.
   b. Describe the plots of the data in terms of symmetry versus skewness of the distribution; clusters and gaps that might be present; and outliers that might be present, including a possible reason for the outliers.

3. Describing the data numerically
   a. Compute the following numerical summaries of the data: mean and median; standard deviation; interquartile range.
   b. Which of these measures seems to provide the best description of center? Why?
   c. Which of these measures seems to provide the best description of variability? Why?

4. Collecting and summarizing another set of data
   a. The instructor is now holding another string, string B. As before, write down the length of this string to the nearest whole inch.

   Length of string B = __________

   The instructor will collect the guessed string lengths and provide you with the data for your class.
   b. Describe the data for string B using the graphical and numerical techniques you found most useful in the analysis of data from string A.

5. Making comparisons
   From your analysis of the two sets of data, decide which is the longer string. On what basis did you make the decision? Are you sure that your decision is correct?

6. Determining the bias
   a. The instructor will provide you with the "correct" lengths for each string. Plot the correct values on the plots of the data made previously. What do you see?

b. It is quite likely that the true value is not at the center of the data display. This discrepancy between the center of the measurements and the true value is called **bias.** Bias is a property of the measurement system, not of an individual person making a guess. Does the "system" of guessing string lengths appear to be biased? What factors might be causing the bias?

c. Does knowledge of the bias change your answer or reasoning on part 5?

Wrap-Up

1. Select another activity in which the measurements may be affected by bias. Collect data within the class or outside the class, as directed by the instructor.

> Possible choices for measurements that are likely to be biased are
>
> - the guessed height of a doorknob as viewed from across the room,
> - the guessed length of the hall outside the classroom,
> - the guessed distance to some object across a parking lot or lawn,
> - the perceived minute ("Tell me when one minute passes"),
> - the guessed circumference of your head.

For the set of data you collect,
a. analyze it by constructing appropriate plots and numerical summaries,
b. describe the key features of the data in words,
c. describe what factors might contribute to the bias,
d. plan and carry out a method for assessing the amount of bias.

2. Write a brief summary of what you learned in this activity about measurement bias.

Extensions

1. Find a printed article using data that are subject to a bias that could have a dramatic effect on the conclusions reached in the article. Summarize the conclusions in the article that are based on data, discuss possible biasing factors in the way the data were collected or analyzed, and explain how the bias in the data might affect the conclusion. (Be sure to separate bias in the data from biased reporting of conclusions for other reasons.)

2. Testing laboratories, such as those that test blood samples or drinking water, are quite concerned about keeping their measuring equipment accurate. Contact a testing laboratory and ask them about their procedures for reducing bias in the measurements they produce. Give a brief report to the class.

The Shape of the Data

NOTES FOR THE INSTRUCTOR

Statistical Setting

Measurement activities of this type should come early in the course, soon after the students have some basic skills at plotting and summarizing data. In fact, these activities can be used to introduce graphs and numerical summaries of data. Students should see that accurate measurements are not easy to make and that any measurement system, no matter how precise, will produce variable measurements at some level. If a major source of variation can be identified, the measurement system can sometimes be improved, or at least "standardized" so that measurements can be fairly compared.

Prerequisites for Students

Students must have some familiarity with plotting data and calculating numerical summaries of data before they can follow through on the instructions given here.

Materials

Measurements are to be made on tennis balls, in addition to objects that all students should have with them (change, their head, and a textbook). Rulers and tape measures, both in metric units, must be provided.

Procedure

Have individual students or small groups of students choose one or more of the settings in part 1 for making their measurements. (You may want to assign these to groups.) Make sure that all four settings are being used so that comparisons can be made for the Wrap-Up.

POCKET CHANGE

Typically, this data set will produce a large amount of variability. To add a little interest, you may want to keep track of measurements for males and females so that comparisons can be made. Have students speculate, first, about who tends to carry more change. The main source of variation here is the nature of the elements (people) themselves. Some carry more change than others. There is no way to reduce this variation, and, in fact, the variation is an interesting phenomenon to study.

TENNIS BALL DIAMETER

Provide some students or groups with a tennis ball and a tape measure and other groups with a tennis ball and a ruler. Let them decide upon a method for making the measurement. You might want each student (or group) to make more than one measurement so that they have an internal check on their own variability. Keep track of the different methods so that variability can be compared across methods. Discuss with the class which method of measurement seems to be best. Here, the major component of variation is the measurement system, not the actual diameter of the tennis balls (although the "actual" diameter of a fuzzy ball is hard to define and will vary from ball to ball).

HEAD CIRCUMFERENCE

As data on head circumference are accumulated for the class, you can discuss the fact that both sources of variation are present here. The measurements are somewhat difficult to make and will vary for an individual, and the head sizes actually differ from person to person. Discuss the importance of each person's making multiple measurements and summarizing them, perhaps by taking the average, before combining the data to see the variation in head sizes for the class. Try this averaging approach, and see if the shape of the plot changes.

THICKNESS OF A PAGE

This is the most difficult of the measurements to make. Students will try different schemes, but most will eventually settle on pinching together a number of pages, measuring the total thickness, and then dividing by the number of pages. The variation often changes with the number of pages chosen, since larger thicknesses are easier to measure accurately. You may want to keep track of the number of pages used to see how that affects the variation in page thickness measurements.

Sample Results from Activity

The following plots and summaries show data on page thickness from two different classes. Each student was to make 10 measurements on page thickness and report all of them. In Class I (Figure 1) three of the students divided by the wrong number (for-

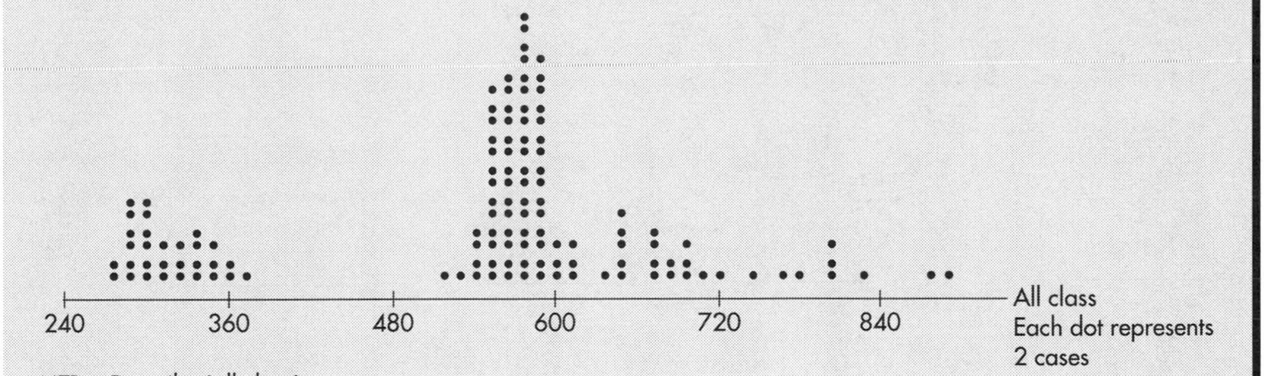

MTB > Describe 'all class'.

	N	Mean	Median		STDEV	
All class	128	541.1	570.0		143.2	480 = .0480mm

	Min	Max	Q1	Q3
All class	280.0	889.0	534.7	596.0

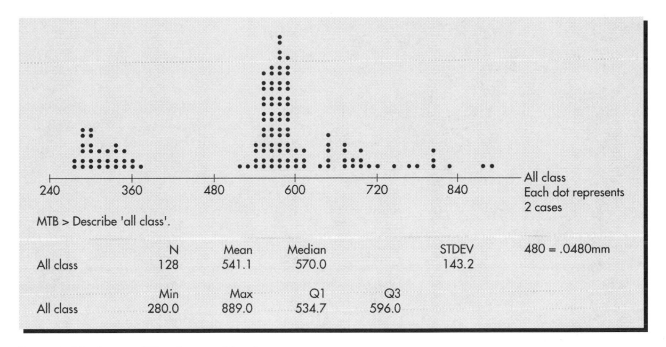

Figure 1: Thickness of Text Paper—Class I

getting that a single page has a number on each side). These measurements are eliminated for Figure 2. In Class II (Figure 3) some of the students measured in inches rather than centimeters, and the two different systems are shown separately. Students quickly see that it is difficult to accurately measure something as small as the thickness of a page.

Sample Assessment Questions

1. Measure the area of your desktop with a meter stick to the closest square centimeter. Do this five separate times.
   a. Comment on the variability among your five measurements.
   b. Comment on the sources of variability for the process of measuring the area of a desktop.
   c. How would you combine the five trials into a single measure of area to report to the rest of the class?

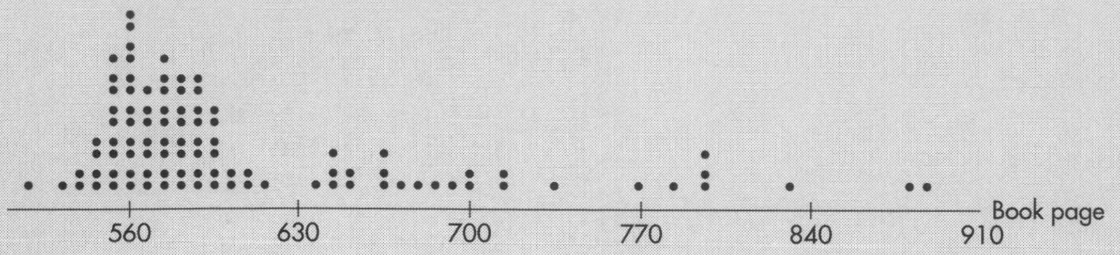


```
MTB > Describe 'book page'.

                  N      Mean    Median              STDEV
Book page        98    610.10    580.00              78.16

                Min       Max        Q1        Q3
Book page    521.00    889.00    559.00    645.25
```

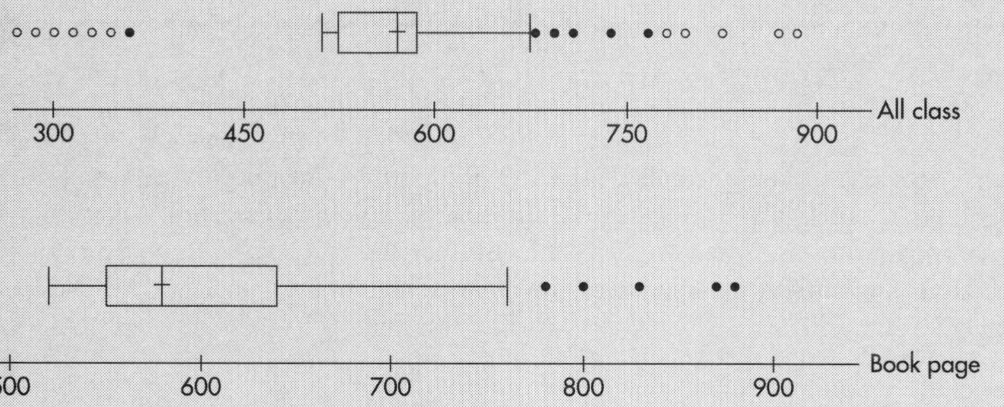

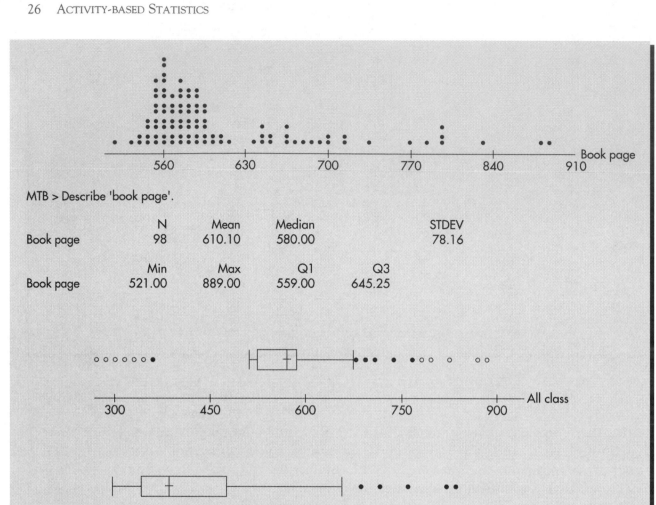

Figure 2: Thickness of Text Paper—Class I with Three Sets Removed

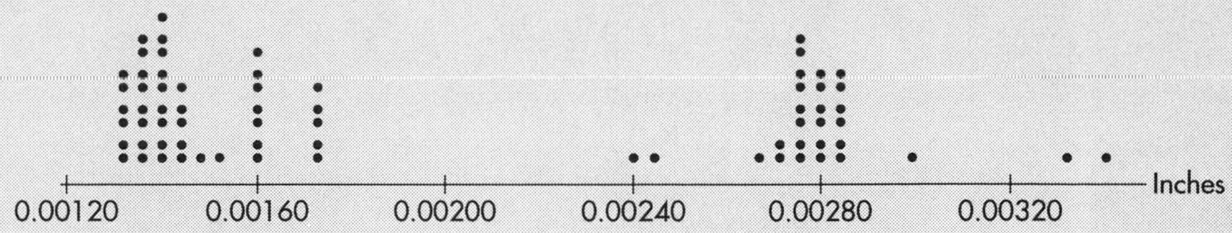

MTB > Describe 'inches'.

	N	Mean	Median	STDEV
Inches	70	0.00199	0.00160	0.00008

	Min	Max	Q1	Q3
Inches	0.00130	0.00340	0.00140	0.00276

MTB > Dot plot 'mm'.

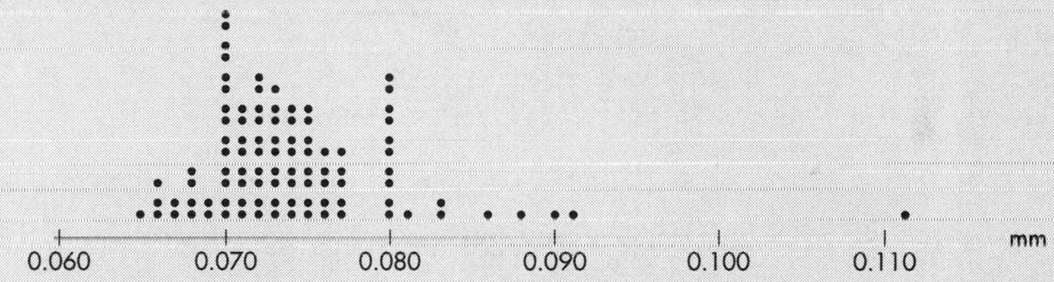

MTB > Describe 'mm'.

	N	Mean	Median	STDEV
mm	97	0.07425	0.07300	0.00629

	Min	Max	Q1	Q3
mm	0.06500	0.11100	0.07000	0.07600

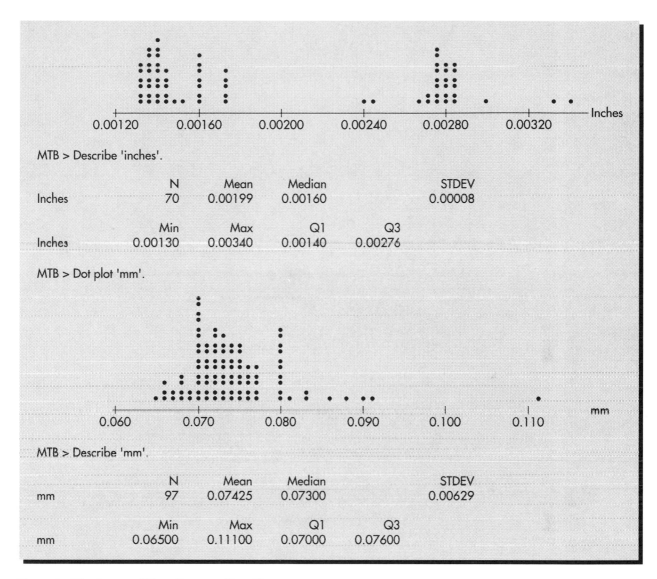

Figure 3: Thickness of Text Paper—Class II

2. To determine how much sleep students get on a typical night, an instructor asked his class to report how many hours they slept last night. The data are shown below.

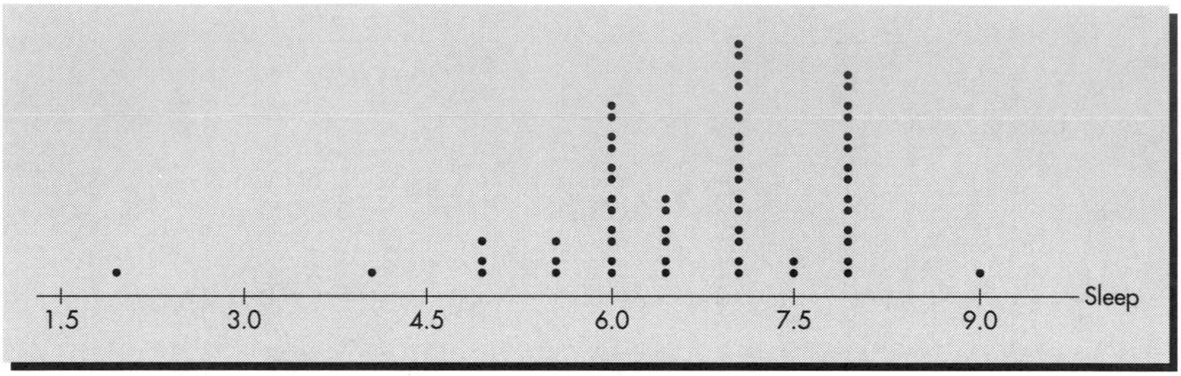

a. Describe two different sources of variation in these data.
b. How would you suggest the measurements be made if the goal were to find out how much sleep students get on a typical night?

3. Would you want variability to be high or low for each of the following variables? Explain your decision.
a. Age of trees in a national forest
b. Diameter of new tires coming off one production line
c. Scores on an aptitude test given to a large number of job applicants
d. Daily rainfall

The Shape of the Data

SCENARIO

Look out a window close to where you are right now. Wouldn't it be dull if all the people you see were the same size, or if all the cars were the same size, or if all the trees were the same size? Virtually everything around us is subject to variability. Sometimes variability is good, as in sizes of people, and sometimes it is not so good, as in the diameter of the cylinders in the engine of your car. Since variability is always with us, we must learn to describe it and work with it, rather than ignore it. How can we adequately describe the important features of variable measurements? That is a key question in statistics and the focus of this activity.

Question

How can we succinctly describe the key features of a set of measurements, such as the shape of the data distribution, the center of the data, the variability of the data, and any "unusual" data points?

Objective

The goal of this lesson is to gain experience in describing sets of measurements in the face of variability, which is present in virtually all measurement problems. Distributions of data can be described graphically by looking at symmetry versus skewness, clusters and gaps, and outliers as shown in appropriate plots of the data. Distributions of data can be described numerically in terms of center, variability or spread, and unusual observations, sometimes called outliers.

Activity

1. Getting ready
   Look out your window again. Describe three instances of variability in what you see. Does the variability strike you as good or not so good? What seems to be the cause (or causes) of the variability?

2. Collecting data
   Collect data for one or more of the following as directed by the instructor.
   a. Count the dollar value of the change you are carrying on your person (in your pocket, wallet, or pocketbook).
   b. Measure the diameter of a tennis ball, to the nearest millimeter, with the measuring equipment supplied by the instructor.
   c. Measure the circumference of your head, to the nearest millimeter, with the tape measure provided.
   d. Measure the thickness of a single page of your statistics book, to the nearest thousandth of a millimeter, using only a ruler.

3. Analyzing the data
   For each of the settings chosen from the list in step 2, follow the instructions provided below.
   a. Give the measurement you obtained to the instructor; he or she will provide you with the set of measurements for the class.
   b. Construct a plot of these measurements that shows the shape of the distribution. Describe the shape of the distribution.
   c. Construct a box plot for the set of measurements. Describe any interesting features of this box plot. Are there any outliers shown by the IQR rule used in box plots?
   d. Find the mean of the measurements, and compare it to the median found for the box plot. Explain why these differ (if, indeed, they do).
   e. Find the standard deviation of the measurements. Identify any observations that are more than two standard deviations away from the mean. Are these "unusual" observations the same ones identified by the box plot? Does the standard deviation appear to be a reasonable measure of variability for these data?
   f. Discuss the sources of variation for these measurements.

Wrap-Up

Sometimes variation in measurements is due primarily to the fact that the elements being measured vary. This is true, for example, in counting the number of chairs per classroom in your school; classrooms have differing numbers of chairs, but, presumably, they can be accurately counted within each room.

Sometimes variation is due primarily to the fact that the measurement device or system produces seemingly different results on the same element. This is true, for example, in measuring the weight of an object on a balance. Two people using the same balance can come up with slightly different weights for the same object. In fact, one person may come up with differing weights for the same object measured repeatedly over time. To see this, have two people measure your height.

Both types of variation enter most measurement problems, but often one of them predominates. It is important to recognize these differing sources of variation if a measuring system is to be improved.

1. Look back at the measurement problems in the activity for this lesson. For each one, identify the primary source of variation as being from variation among the elements or variation within the measurement system. If both are major contributors, state this as well.

2. For each of the measurement problems presented in the activity, discuss ways of reducing variation, if there are any, if these activities were to be completed again.

3. Select one of the measurement activities and run it again, making use of the variation-reduction suggestions from part 2. Did your suggestions work?

Extensions

Sources of variation increase as the system producing the measurements becomes more complicated.

How many drops of water will fit on a penny?

a. Guess how many drops you think will fit on a penny. Give your guess to the instructor.

b. Using the penny, eyedropper, and cup of water provided, place as many drops of water as possible on the penny. Remember to count while carefully placing the drops. (Teamwork may help here.)

c. Give your experimental result to the instructor, who will provide you with the data from the class.

d. Analyze these data by graphical and numerical methods.

e. Write a paragraph summarizing the distribution and any other interesting features of these data and commenting on the sources of variation in this experiment.

f. How would you refine the instructions for this problem in order to reduce variability?

Matching Graphs to Variables

NOTES FOR THE INSTRUCTOR

Statistical Setting

This activity can be used early in a course as a way to help students learn to read and understand graphs and to relate features of a distribution to the shape of a graph. It takes practice to develop skill in describing key features of data sets; these data provide an opportunity for such practice.

Prerequisites for Students

Students should have some familiarity with box plots and histograms.

Materials

The only materials needed are copies of the student activity sheet, which contains a listing of the variables and graphs of each of them.

Procedure

Distribute the student activity sheet to the students. As a group, discuss the two introductory graphs and variables, making sure that students see how the graphs are related to the distributions of the variables before proceeding to the main activity. Then give the students time to complete the main activity, preferably working in small groups.

Other variables (and accompanying graphs) could be used in place of those given here. Data collected on the students themselves are interesting to them.

Sample Results from Activity

The variables and graphs match as follows: a—5, b—3, c—1, d—4, e—2.

Note that there is an unlikely bump in the histogram of the fertility data at 6 months, which is probably due to rounding; that is, when the data were collected some women in the 5-month and 7-month categories probably rounded and reported "half a year." Scores on an easy exam will be skewed to the left, with 100 serving as an upper bound. The Olympic medal distribution is strongly skewed to the right. The class height data are bimodal, with one mode for men and one for women. This leaves the SAT data matched with histogram number 2, which shows a fairly bell-shaped distribution.

Sample Results from Extensions

1. Answers will vary. Students should note that graphs 3 and 4 are skewed to the right and these distributions have outliers; graph 5 is skewed to the left; and graphs 1 and 2 are reasonably close to being symmetric.

2. For graphs 3 and 4 the mean is greater than the median, since the long right-hand tail pulls the mean up. For graph 5 the mean is somewhat less than the median, since this graph is skewed to the left. For graphs 1 and 2 the mean and the median are nearly equal, since the graphs are nearly symmetric.

Sample Assessment Questions

1. Consider the selling prices of houses sold in the United States during the past year. Draw a rough sketch of the histogram for this variable.

2. Here is a histogram for some data. Name a variable that might have led to this histogram, and explain why the histogram for your variable would have this shape.

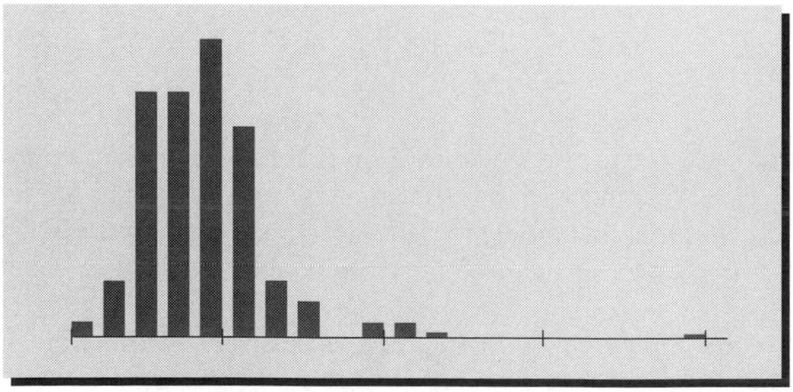

3. Repeat question 2, but this time name a variable that not only has the right shape, but also is measured in the right units. That is, name a variable for which the smallest observation is near zero, the largest observation is between 75 and 100, and the shape is as indicated. Explain your reasoning.

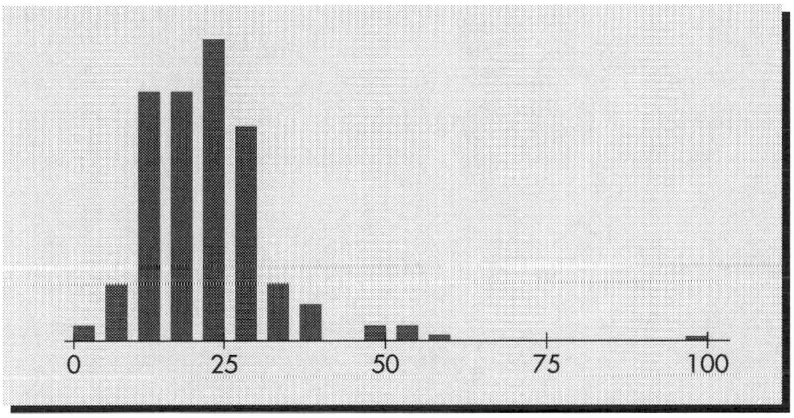

4. Repeat question 2 for the following histogram.

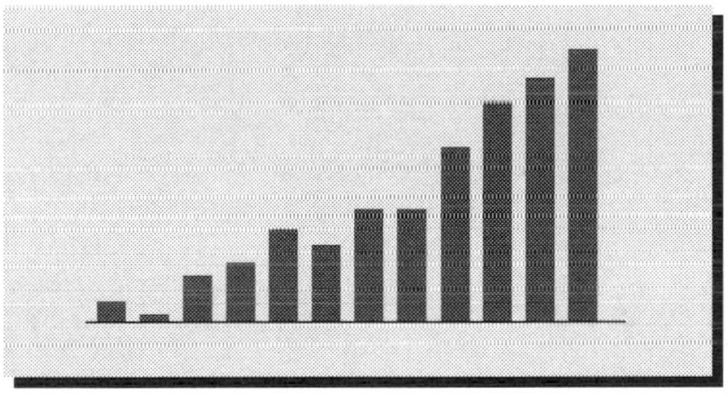

Matching Graphs to Variables

SCENARIO

Some people earn a lot of money each year, so there are some families with very large incomes relative to the rest of the population. Thus, if we were to collect family income data from a sample of Americans and then construct a histogram, we would expect the histogram to be skewed to the right. If we think about the type of data we would be likely to obtain, we can say something about the shape of the histogram without actually collecting the data. Can we do the same with other variables?

Question

Can we deduce the likely shape of the histogram of each of several variables?

Objectives

The purpose of this activity is to learn how features of distributions are related to graphs of the data. After completing this activity, you should be able to sketch the shape of the histogram for a variable by thinking about the nature of the data.

Activity

1. Warm-up
   Consider the following two variables:
   A. age at death of a sample of 34 persons;

B. the last digit in the social security number of each of 40 students.

We wish to match these variables to their graphs:

I.

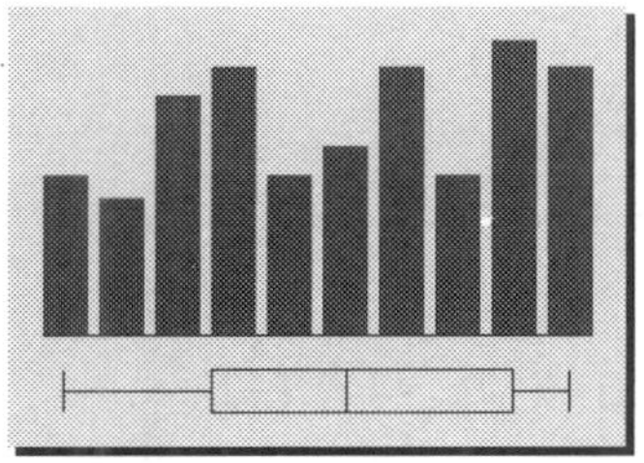

II.

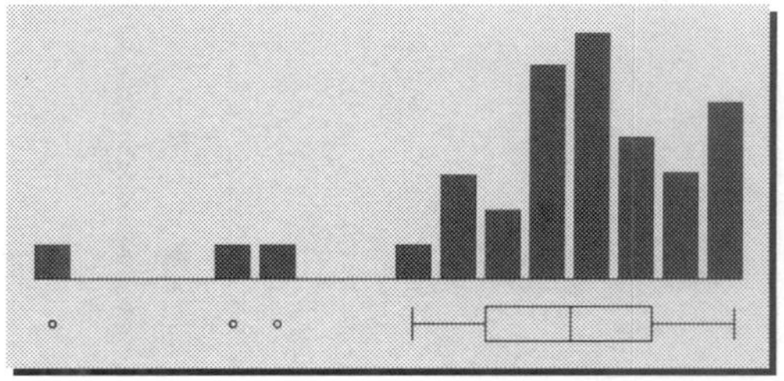

We know that there are relatively few deaths among young people; the death rate rises with age. Thus, we would expect the histogram of the age at death data to be skewed to the left, with most of the observations being in a large group toward the right of the graph. We expect the histogram to have a small left-hand tail. Hence, we match A with II. On the other hand, the social security data should have a distribution that is close to uniform on the integers 0, 1, . . . , 9. Thus, we match B with I.

2. Main activity
   Consider the following list of variables and graphs:
   a. scores on a fairly easy examination in statistics;
   b. number of menstrual cycles required to achieve pregnancy for a sample of women who attempted to get pregnant. Note that these data were self-reported

14 ACTIVITY-BASED STATISTICS GUIDE

from memory. (Data from S. Harlap and H. Baras (1984), "Conception—waits in fertile women after stopping oral contraceptives," *Int. J. Fertility*, **29:**73–80.)

c. heights of a group of college students;
d. numbers of medals won by countries in the 1992 Winter Olympics;
e. SAT scores for a group of college students.

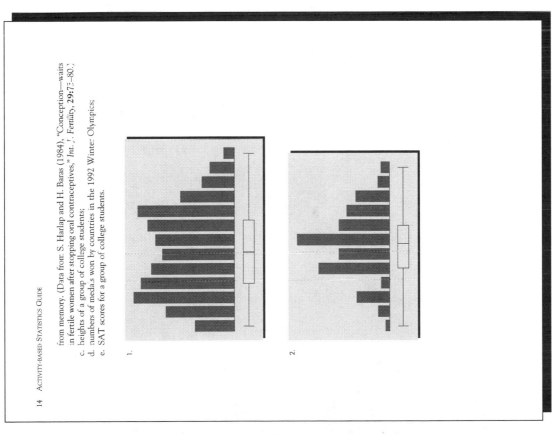

1.

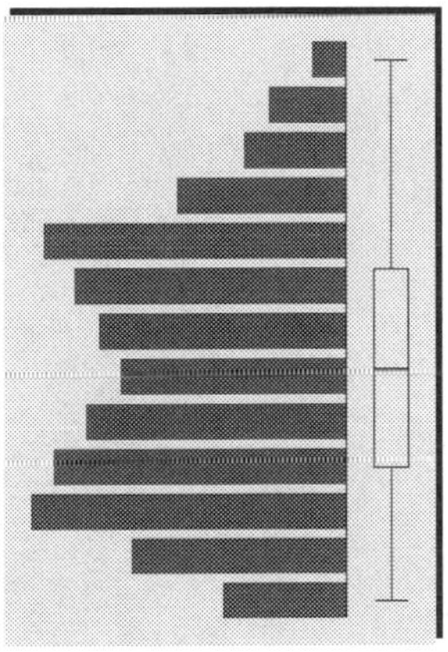

2.

15

3.

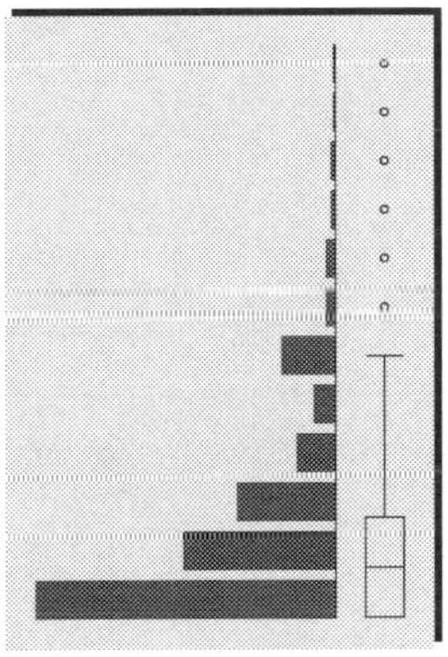

4.

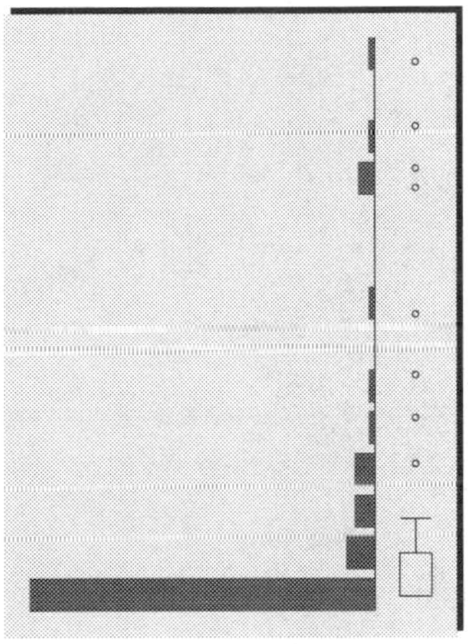

5.

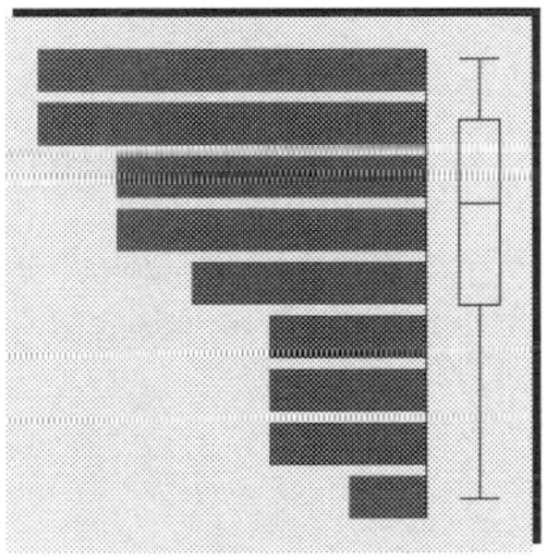

Use your knowledge of the variables (i.e., ask yourself if the distribution is likely to be symmetric or skewed, etc.) to match the variables with the graphs.

Wrap-Up

1. Write a brief summary of what you learned in this activity about how features of distributions are related to graphs of the data.

2. Name two variables that have symmetric distributions, two that have distributions skewed to the right, and two that have distributions skewed to the left.

Extensions

1. After you have matched the graphs to the variables, describe each distribution and any unusual features (e.g., is the distribution skewed to the right? Is it symmetric? Are there outliers? Why might this be so?).

2. In each case estimate whether the mean is greater than, less than, or equal to the median, and explain your reasoning.

Matching Statistics to Graphs

NOTES FOR THE INSTRUCTOR

Statistical Setting

This activity can be used early in a course, shortly after the concepts of mean, median, and standard deviation have been introduced. The activity helps students estimate the mean, median, and standard deviation by looking at histograms. The activity also helps students see how box plots are related to histograms. You might want to use this activity in conjunction with the "Getting to Know the Class" activity, in which you will have collected data on the students. Histograms of those variables can be used to augment the artificial data here.

Prerequisites for Students

Students should be familiar with box plots and histograms, along with the concepts of mean, median, and standard deviation. It is not necessary, however, that they know how to compute these values.

Materials

The only materials needed are copies of the student activity sheet, which contains a listing of the variables and graphs of each of them.

Procedure

Distribute the student activity sheet to the students, and have them work in groups. After each group has had a chance to complete the activity, ask the class what the answers are, and ask them to justify their answers.

Sample Results from Activity

In part 1 the variables and graphs match as follows: a—1, b—4, c—5, d—2, e—3, f—6. In part 2 the histograms and box plots match as follows: a—2, b—3, c—4, d—1.

Sample Assessment Questions

1. Estimate the mean, median, and standard deviation of each of the distributions graphed below.

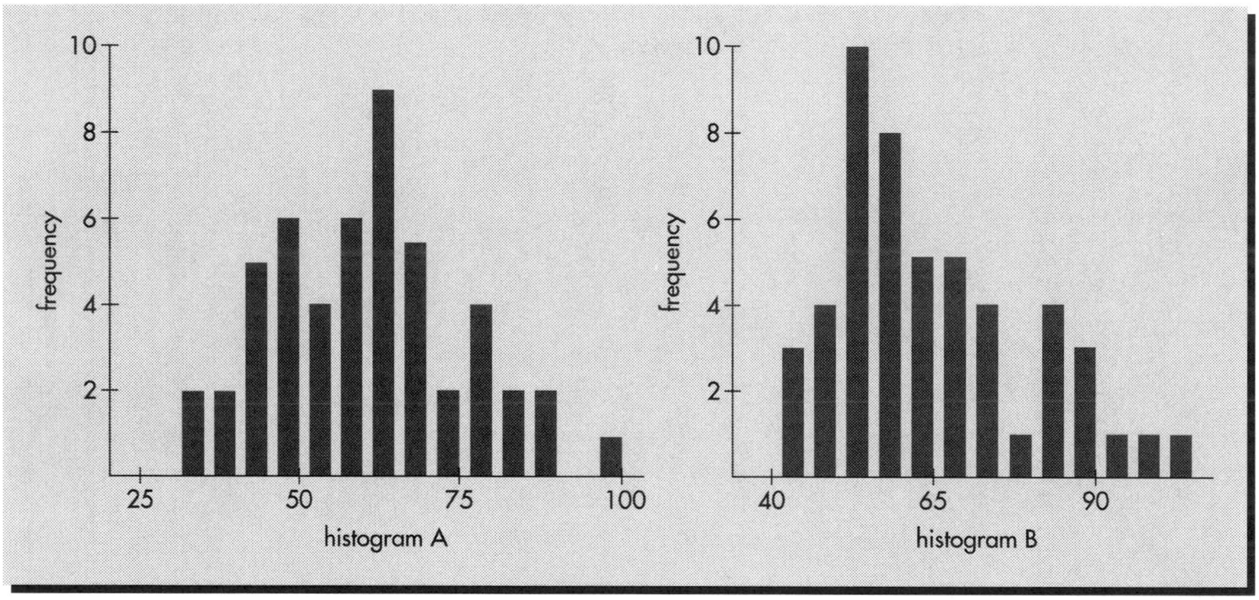

2. Sketch the histogram of a variable for which the mean is greater than the median.

Matching Statistics to Graphs

SCENARIO

Sometimes it is reported that the average lawyer earns a lot of money; at the same time, the median income of lawyers is reported to be more modest. One year the average team in the National Football League made a sizable profit while the median team barely broke even. People often confuse the mean and the median of a distribution. How are these statistics related to the shape of the distribution?

Question

Can we estimate the mean, median, and standard deviation of a distribution by looking at the histogram?

Objective

The goal of this activity is to learn how summary statistics are related to graphs of data and how box plots are related to histograms. After completing this activity, you should be able to recognize when and how the mean of a distribution differs from the median. You should also be able to sketch an accurate box plot of a distribution after seeing the histogram, and vice versa.

Activity

1. Consider the following group of histograms and summary statistics. Each of the variables (1–6) corresponds to one of the histograms.

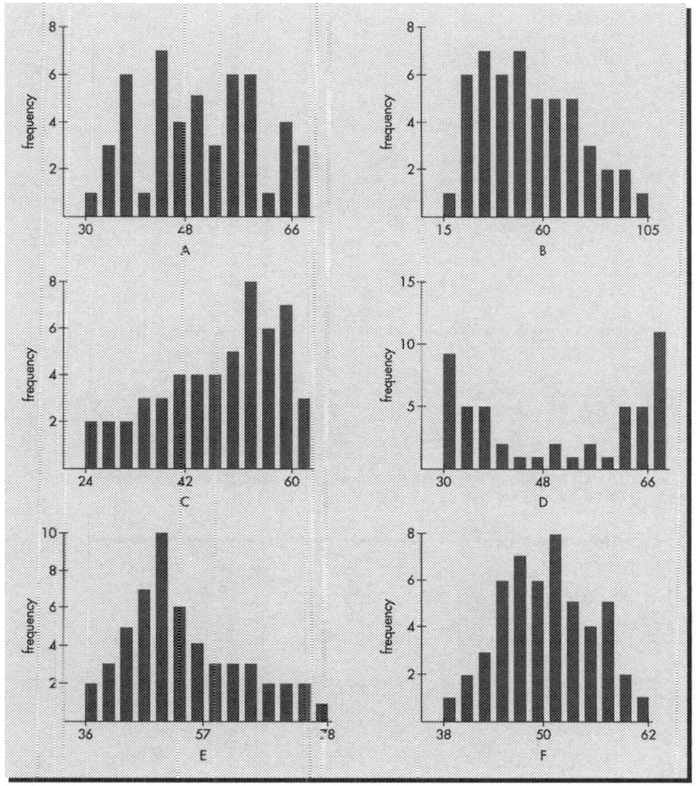

Variable	Mean	Median	Standard Deviation
1	50	50	10
2	50	50	15
3	53	50	10
4	53	50	20
5	47	50	10
6	50	50	5

Write the letter of the histogram next to the appropriate variable number in the above table, and explain how you made your choices.

2. Consider the following group of histograms and box plots.

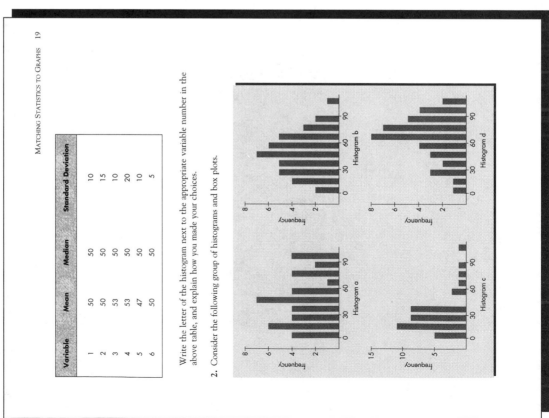

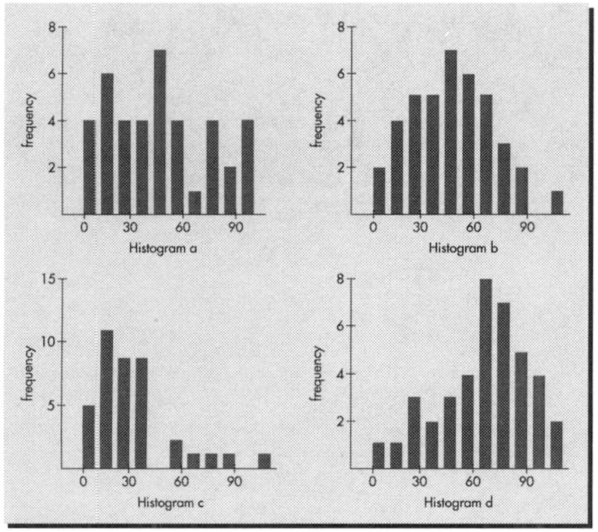

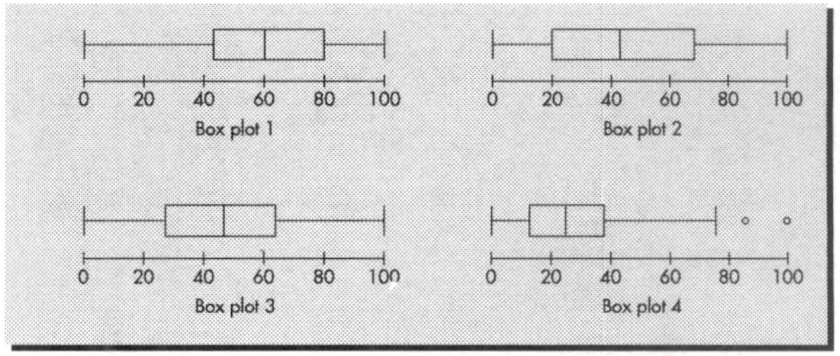

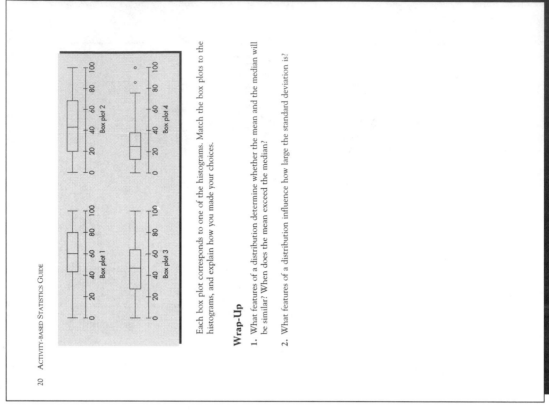

Each box plot corresponds to one of the histograms. Match the box plots to the histograms, and explain how you made your choices.

Wrap-Up

1. What features of a distribution determine whether the mean and the median will be similar? When does the mean exceed the median?

2. What features of a distribution influence how large the standard deviation is?

Let Us Count

NOTES FOR THE INSTRUCTOR

Statistical Setting

This activity can be used early on in the course to illustrate variability, or in conjunction with a discussion of sampling distribution of sample means.

Prerequisites for Students

Students should know how to construct and interpret stem and leaf plots, and they should be familiar with means and standard deviations as measures of location and dispersion.

Materials

Bags of small pasta shells, a half-cup measure, and a one-cup transparent plastic measure will be needed.

Procedure

This activity is most effective for classes/groups of 20 to 40 students, where students can watch others measure and not have to stand around waiting too long. Having several teams measure the shells simultaneously can cut down on the time required, although this approach increases the costs of the materials. The shells recommended measure approximately 1″ from tip to tip and are irregular in shape, which makes it difficult to decide when one has reached a half-cup. This results in the variability in the counts. Substituting more regularly shaped objects, such as macaroni, may not work as well.

The differences in the means and standard deviations will vary with each class. Common cause variation for the process is reflected in the variation due to the measurement device used. The source of special cause variation is the different measurers. Although not included in the activity, you may ask the students to decide which method gives more consistent answers by computing the coefficient of variation for each of the two measurement devices and comparing their values.

In choosing between the two methods, you could consider the trade-off between using a process with a lower mean or using a process with less variability.

Sample Assessment Questions

1. In attempting to control a process, why is it important to systematically check both the center and the variability of key measurements of the process?

2. It was suggested in the activity that five measurements should be averaged and the averages checked against the boundaries. What is the advantage of working with the small samples of five as opposed to simply checking individual measurements against a boundary? Which method would tend to show "out of control" cases more often?

3. A costly process might be shut down if a sample mean lies beyond two standard deviations from the overall mean, according to the rule used in this activity. Some quality control plans replace the two standard deviation limits with three standard deviation limits. Discuss the pros and cons of using three standard deviation boundaries instead of two.

Let Us Count

SCENARIO

We all know that when we repeat an action we may not get the same result. An athlete may not run a mile in the exact same time twice, you may not get the same number of french fries in two orders at McDonald's, or two scoops of ice cream will not be exactly the same. Therefore, "variability" is present in the outcomes of all repeated actions. **Process variability,** where measurement plays a role, is a major concern in the real world. The main objective in process control is to identify the sources of variation and look for ways to control the variation.

Questions

Which process of measurement is better? How do differences in measurement processes affect the variability in the outcomes?

Objectives

This activity illustrates the presence of variability in a process and analyzes this variability. You will see how to compare the results of two measurement processes, as well as the patterns and sources of variation in each process.

Activity

Suppose you are running an ice-cream parlor. You notice that there is variability in the amount of ice cream scooped out in a half-cup serving depending on how the half-cup is measured out. As a manager, you need to know which way of measure-

ment varies as little as possible, so that you can plan and run your business better. Ice cream, however, cannot be used in a classroom experiment. You will use pasta shells instead.

1. Collecting the data
   The class will be divided into an even number of teams of at least two students each, and the teams will be separated into two groups. Within the first group, each team measures out a half-cup of shells and counts the number of shells. The process here is that one person pours shells into the half-cup measure until the person thinks that it is full. The second person then counts the number of shells and records it without informing the first person of this number. Each team repeats the experiment five times and computes the mean and the standard deviation for its measurements.
   In the second group, students use a transparent plastic measure. One person in the team pours shells into the one-cup measure until the person thinks it has reached the half-cup mark by holding the cup in the air and checking the level. The second person counts the number of shells and records it. Again, each team repeats the experiment five times and computes the mean and the standard deviation.

2. Plots and boundaries
   a. Construct stem and leaf plots of the mean counts of the teams for each group, and combine these two plots in a back-to-back stem and leaf plot.
   b. Construct stem and leaf plots for the standard deviations for each group, and combine these in a back-to-back stem and leaf plot.
   c. Compute the two boundaries for these means, $\bar{\bar{x}} - 2\bar{s}$, and $\bar{\bar{x}} + 2\bar{s}$ for each group, where $\bar{\bar{x}}$ = overall mean for the group, and $\bar{s}$ = average of all the standard deviations for that group.
   d. How many of the sample means for each group fall outside these boundaries?

3. Analyzing the results
   a. What do the stem and leaf plots of the sample means and standard deviations indicate about the variation within each group and between the groups? Do both processes vary about the same? Are there differences in the patterns of variation?

   b. Variability in processes is often divided into two types:

   - common cause—that which is part of the system or process and affects everyone in the process;
   - special cause—that which is not part of the system or process all the time or does not affect everyone in the system.

Identify the different sources of variation. Which would you consider to be common cause variation and which special cause variation?

c. Compare the overall means and the boundaries for the two groups. Which of these two methods would you recommend and why?

Wrap-Up

This type of analysis is often used by industry to understand the processes that manufacture products or provide various services. This activity looked at a simple process, investigated the variability, and checked for unusual observations. There is a theoretical basis for using this approach. Answering the following questions should help you understand this basis.

1. Why did we use the boundaries of $\bar{\bar{x}} - 2\bar{s}$ and $\bar{\bar{x}} + 2\bar{s}$ to check for unusual team means? Should we expect many team means to fall outside these limits? What is the approximate probability of a sample mean's falling outside the boundaries that you have used?

2. From knowledge of the behavior of sample means, can you explain why the distance between the boundaries will get smaller for larger sample sizes? What does that imply about the variability of the sample means?

Extension

Instead of measuring the amount of shells, would weighing the shells result in a process with less variation? Depending on the scales and weights available to you, take repeated samples of the shells of a specified weight, count the number of shells, and construct a stem and leaf plot and a control chart. Note that here you have only one measurement process. Assume that you are the quality manager. Write a report to the production manager outlining the results of your experiment and the extension activity and recommending the measurement process.

Getting Rid of the Jitters: Finding the Trend

NOTES FOR THE INSTRUCTOR

Statistical Setting

This activity can be used to introduce students to the analysis of data collected over time. While the standard graphs such as stem plots and box plots provide some information, they often are not useful for studying the behavior of a time series that does not display a well-defined trend. The activity can be included in a discussion of the use of graphs and data summaries. Although there are several methods of smoothing data, the activity uses only 3-point and 5-point moving averages since moving averages are simple to understand and compute. Students could be encouraged to explore other ways of "smoothing" the data.

Prerequisites for Students

As the main activity involves studying graphs and doing a short simulation, it can be used early in the course since students do not need to know a lot of statistics to complete the activity. Students should be familiar with graphing data and interpreting graphs. They should also have some experience in summarizing data using measures such as the mean and the median, so that they can appreciate the usefulness of moving averages.

Materials

A random number table or a random number generator is needed for this activity.

Procedure

The activity can be assigned individually or as a group activity.

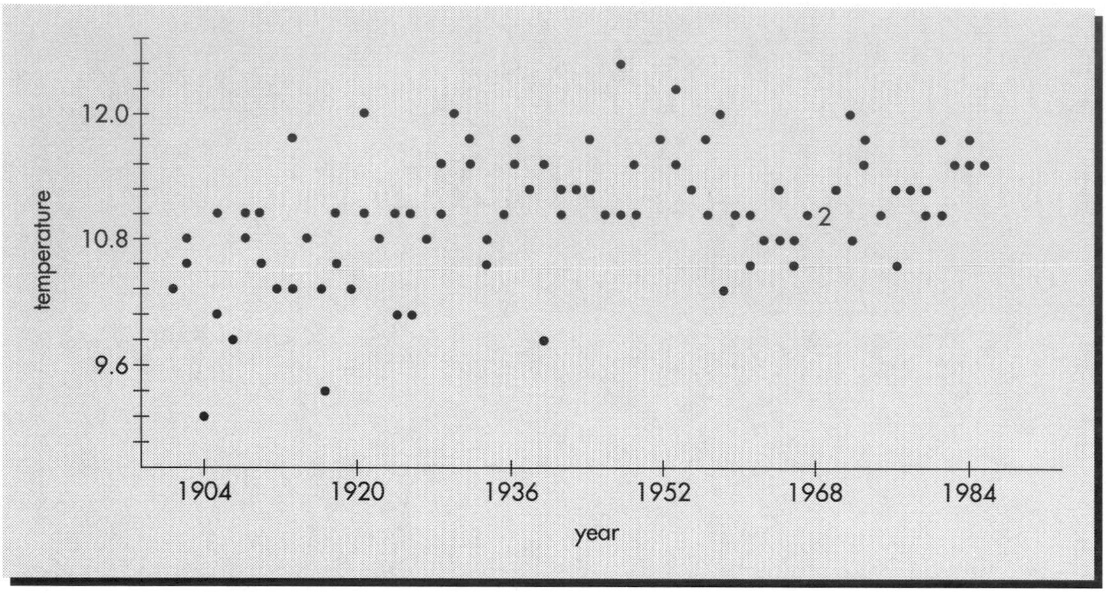

Figure 1: Average annual temperature (degrees C) versus year.

Sample Results from Activity

The smoothed data plots, Figures 2 and 3, completed to 1987 are shown below. Figure 3 shows the trends most clearly. The plot indicates an upward drift in the average annual temperatures until about 1950, followed by a downward movement that is be-

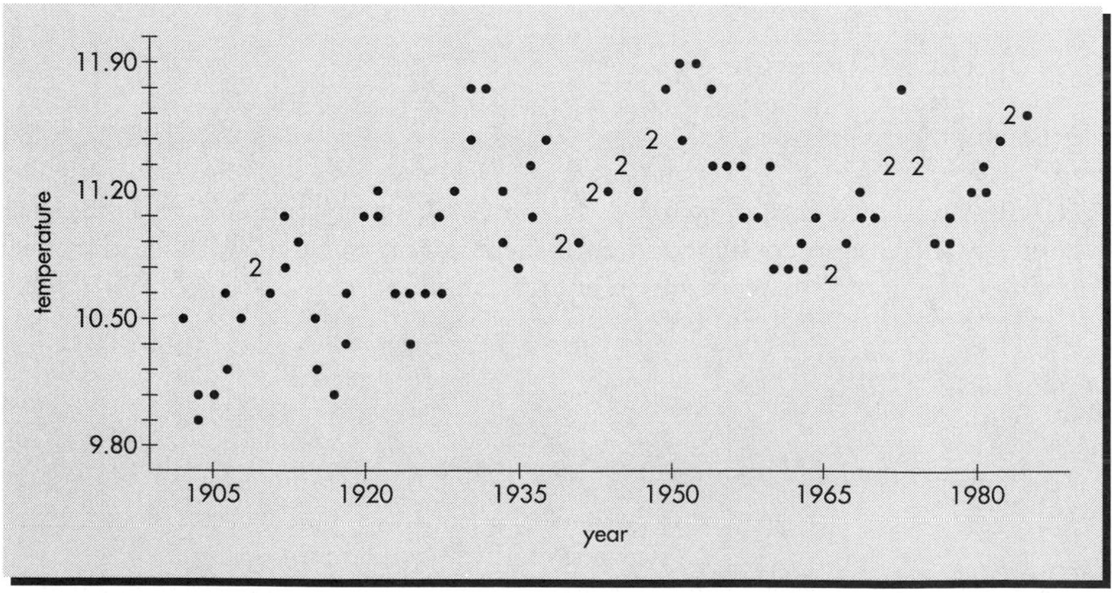

Figure 2: Three-point moving averages of average annual temperature versus year.

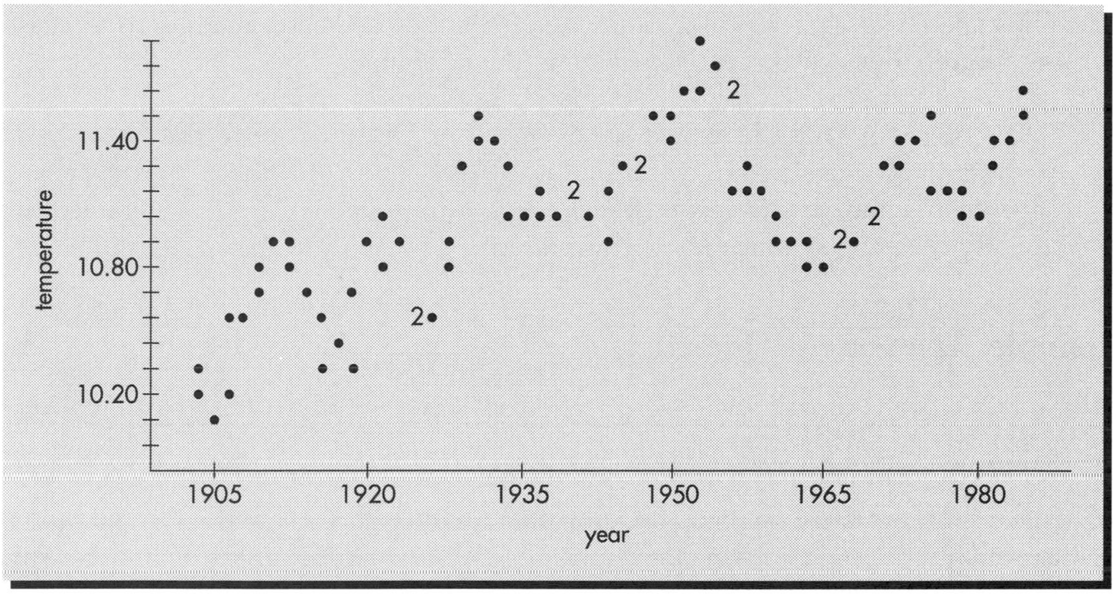

Figure 3: Five-point moving averages of average annual temperature versus year.

ginning to rise again in recent years. During the last 20 years, the annual average temperature has been higher than the mean temperature for the century 14 times.

When comparing the seasonal data to the annual, it should be observed that a very cold winter or a very warm summer will affect the annual average far more than unusual temperatures in the spring or fall.

Wrap-Up

The Wrap-Up provides practice in detecting the trend in a time series. This series is also included in the activity "Levels of CO_2," where students will fit a straight line to the series. Plots of the series together with the 3-point and 5-point moving average series are shown here in Figures 4, 5, and 6, respectively. Interesting patterns are the steady increase in consumption until the Arab oil embargo and then the effects of conservation and the recession on energy consumption.

Extension

1. A 5-point moving average series can be expected to fluctuate less than a 3-point moving average series and therefore provide a better indication of the trend. However since we are averaging over 5 adjacent points we will not have the smoothed values for the first two and the last two observations in the original series as compared to the 3-point moving averages where we miss only one observation at each end.

2. The time series that was used does not increase or decrease steadily, so forecasting future values with this series may not be advisable.

3. The probability that one would see average temperatures above the mean in 15 of the last 20 years purely by chance is 0.0148. Therefore there does appear to have been some warming over the century, although there is considerable controversy over the significance of this pattern.

Sample Assessment Questions

1. What are some of the advantages of looking at smoothed data from a time series? What are some disadvantages?

2. In recent years there has been flooding on a number of rivers in the United States. In studying historic data on the heights of rivers, do you think it would be good to smooth the time series? Explain.

3. Suggest ways of smoothing time series data other than by using running averages, as described in this activity

References

1. Karl T.R., Baldwin R.G. and Burgin M.G. (1988) Time series of regional seasonal averages of maximum, minimum and average temperature, and diurnal temperature range across the United States: 1901–1984. Historical Climatology Series 4–5. National Climatic Data Center. National Oceanic and Atmospheric Administration. National Environmental Satellite, Data and Information Service. Asheville, North Carolina.
2. Basic Petroleum Data Book. Petroleum Industry Statistics, Volume XIV, Number 1, January 1994, American Petroleum Institute, Washington D.C.

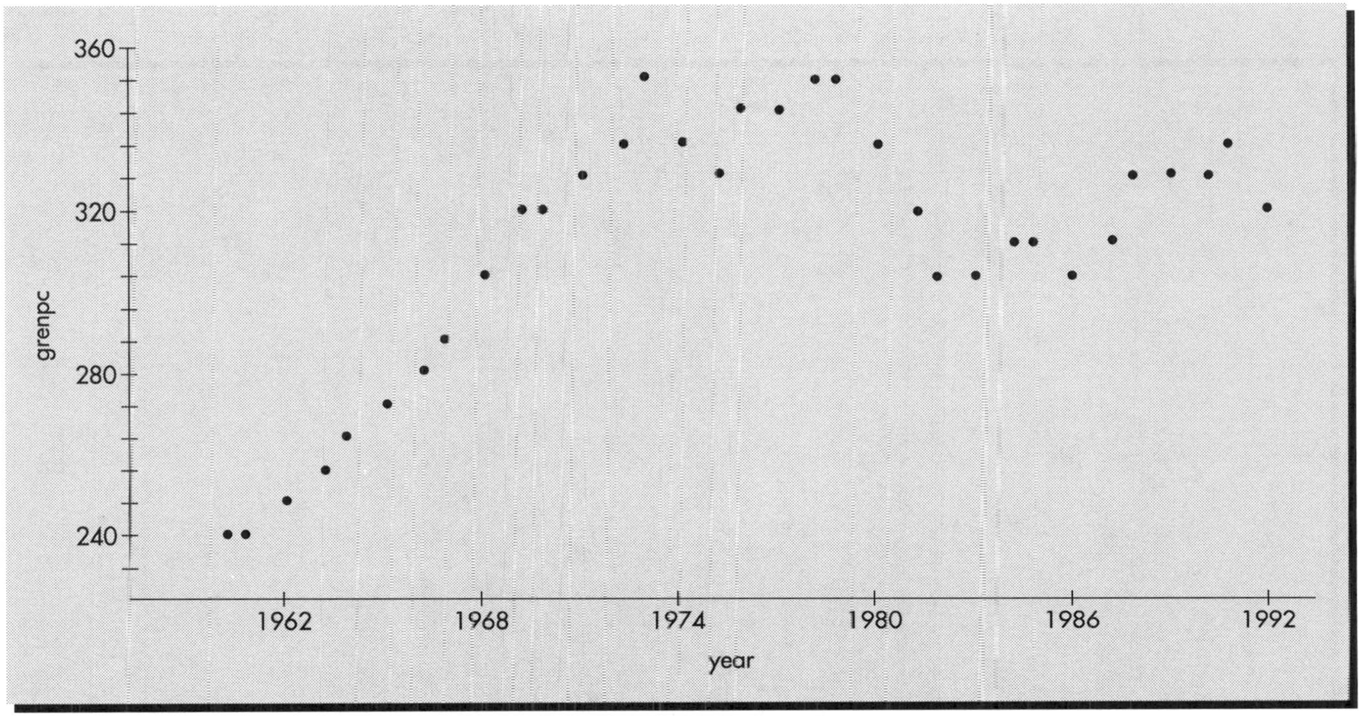

Figure 4: Per capita gross energy consumption in the United States.

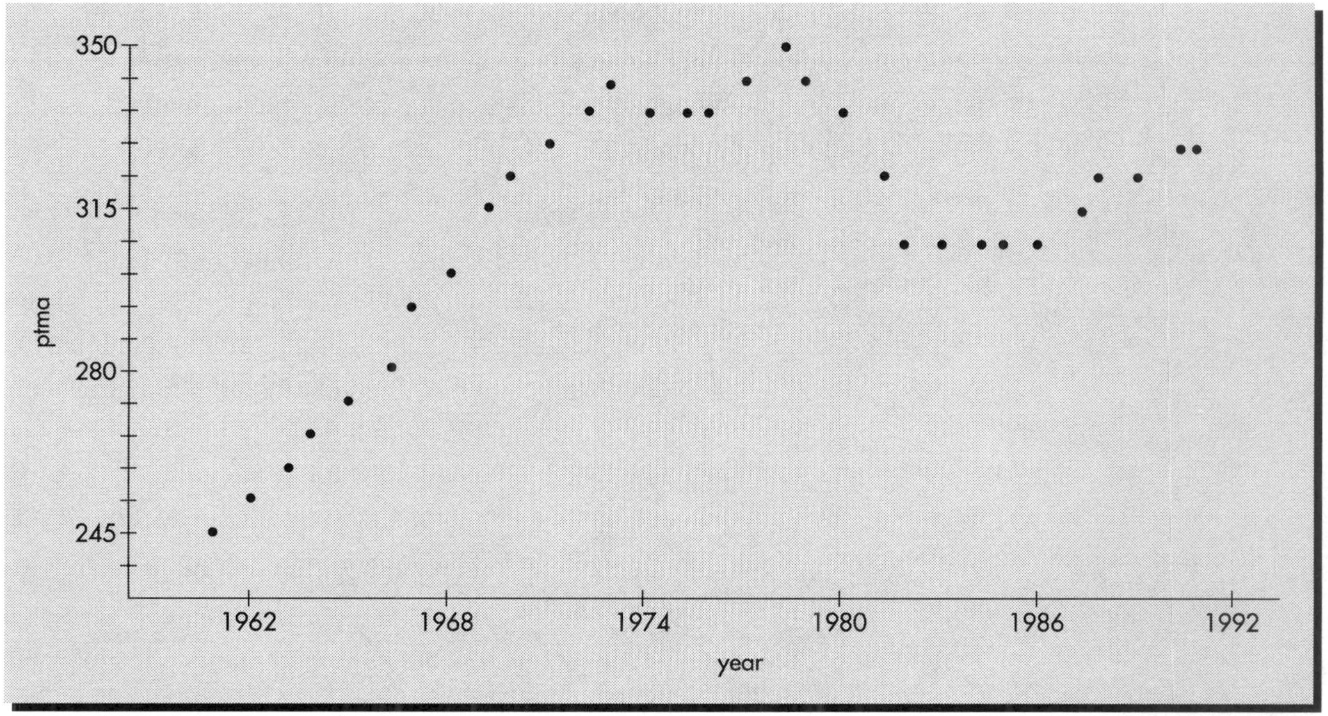

Figure 5: Three-point moving average for data in Figure 1.

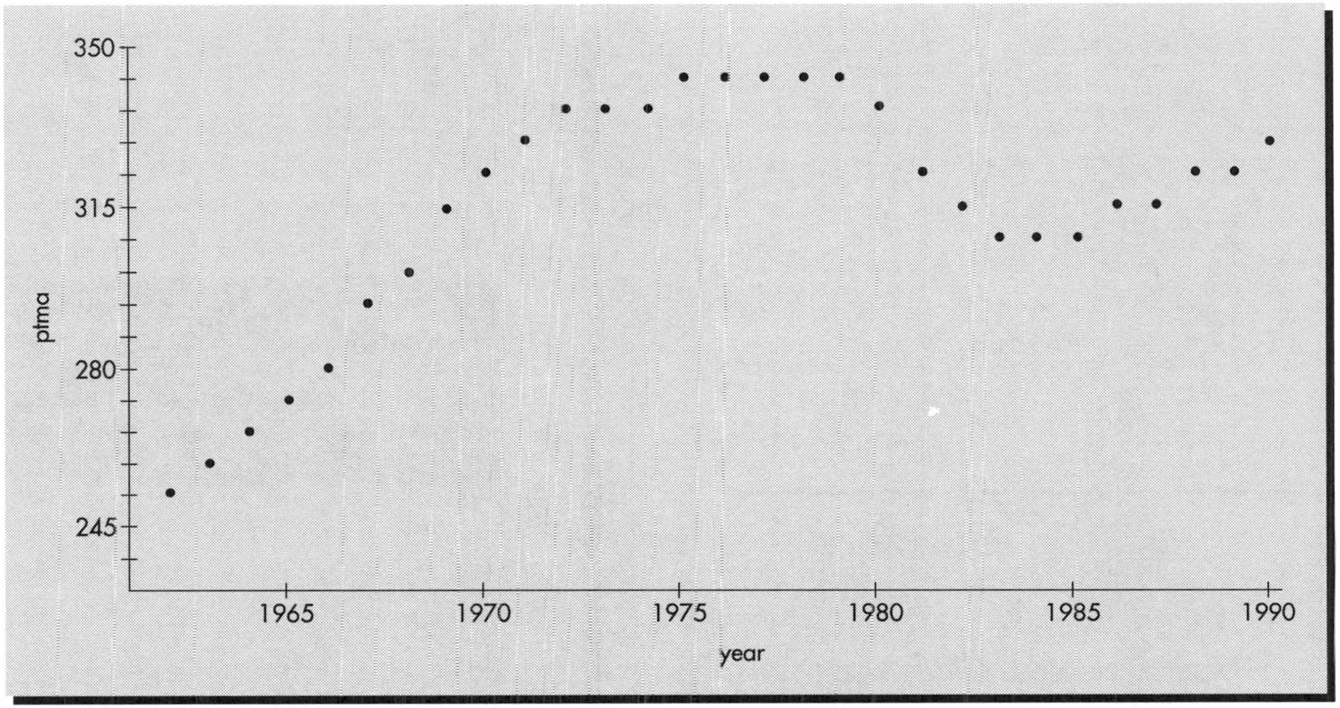

Figure 6: Five-point moving average for data in Figure 1.

Getting Rid of the Jitters: Finding the Trend

SCENARIO

Is there a trend toward global warming, or is the relatively warm weather of recent years just part of the variability expected in a process like the weather? This question is much debated partly because it is difficult to see long-term trends in data that is subject to many short-term fluctuations (sometimes called "jitters"). Some of the fluctuations can be removed, however, by "smoothing" the data. In this lesson, moving averages will be used to smooth temperature data from the coastal northeastern United States so that trends can be spotted more easily.

Question

How can a long-term trend be detected in data that is subject to short-term "jitters"?

Objective

Moving averages will be used to smooth time series data so that trends can be more easily observed.

Activity

Data on temperatures for the coastal northeast of the United States are shown in the accompanying table on pp. 29-31. The data are average temperatures by year and by season for the years 1901 through 1987.

1. Study the table of data for a few minutes. Can you spot any trends just by looking at these columns of numbers?

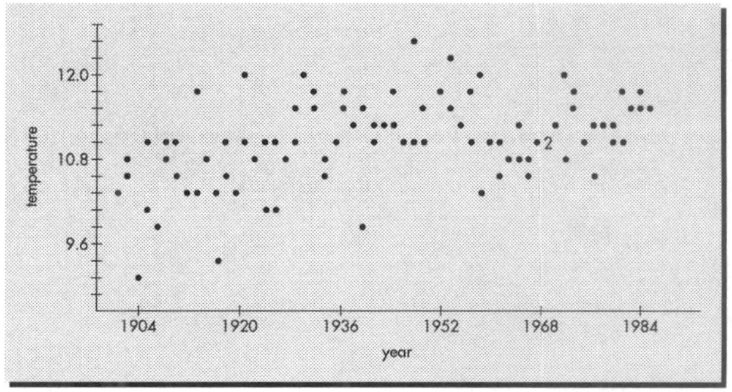

Figure 1: Average annual temperature (degrees C) versus year. Note the software has replaced two points that are very close together with a 2.

2. A plot usually helps clarify patterns in data. Figure 1 shows the average annual temperatures plotted against years. Discuss any patterns or trends you see in this plot.

3. One way to smooth data that come ordered in time is to average each observation with its nearest neighbors. Thus, a 3-point moving average replaces each data point by the average of that point and its two nearest neighbors, one on each side. Using the 1902 annual temperature as an example, the 3-point moving average would replace the value 10.69 with the average of 10.30, 10.69 and 10.46, which is 10.48. (Note that the value at the beginning of a series cannot be replaced by such an average.) Figure 2 shows the 3-point moving average for the annual temperature data up to 1968. What patterns or trends are beginning to emerge?

4. A 5-point moving average replaces a data point by the average of that point and its two nearest neighbors on each side. For example, the 1903 value would be replaced by the average of 10.30, 10.69, 10.46, 9.20 and 10.11, which is 10.15. Figure 3 shows the 5-point moving average for the temperature data up to 1968. Are the patterns or trends showing up more strongly? Describe these trends.

5. Using the data from the table, complete the 3-point (Figure 2) and 5-point (Figure 3) moving averages for the period 1968 to 1987. What patterns or trends do you see?

6. The mean temperature for the century is 11.043. How have the temperatures during the period 1968–1987 behaved relative to this mean?

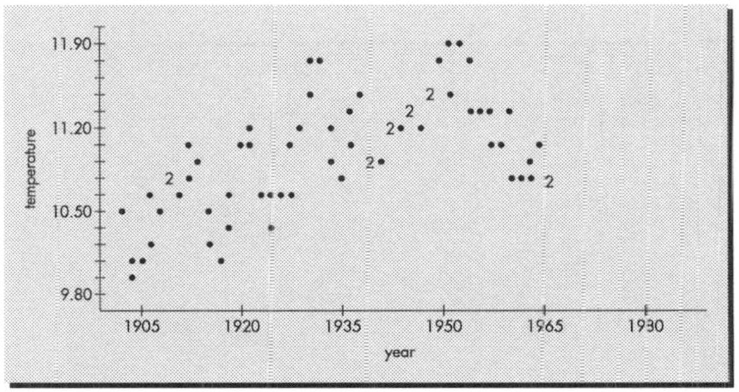

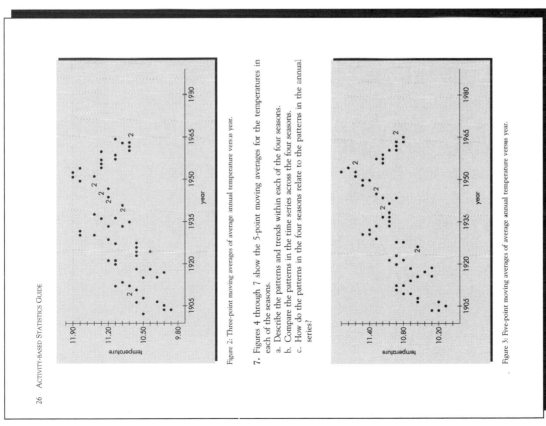

Figure 2: Three-point moving averages of average annual temperature versus year.

7. Figures 4 through 7 show the 5-point moving averages for the temperatures in each of the seasons.
   a. Describe the patterns and trends within each of the four seasons.
   b. Compare the patterns in the time series across the four seasons.
   c. How do the patterns in the four seasons relate to the patterns in the annual series?

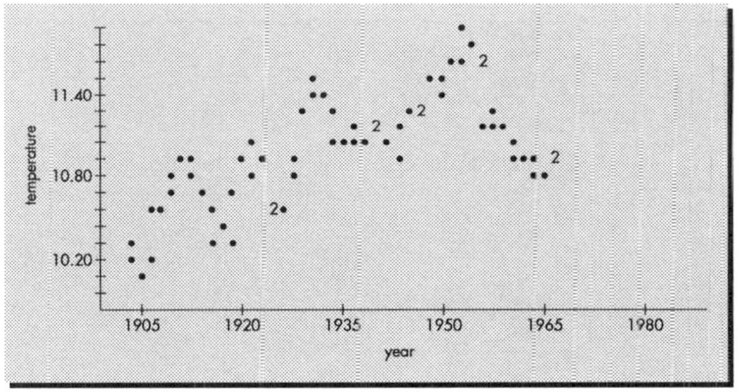

Figure 3: Five-point moving averages of average annual temperature versus year.

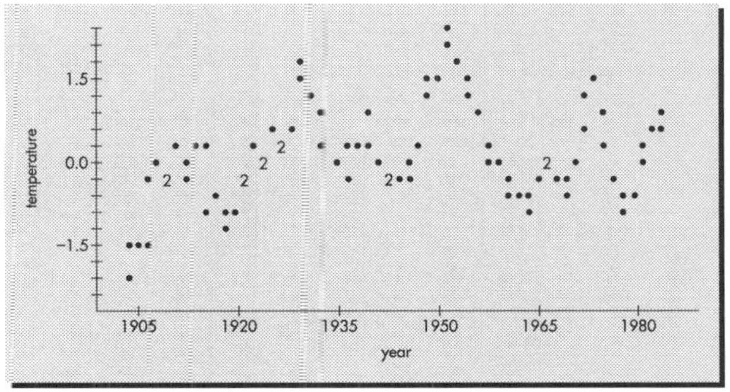

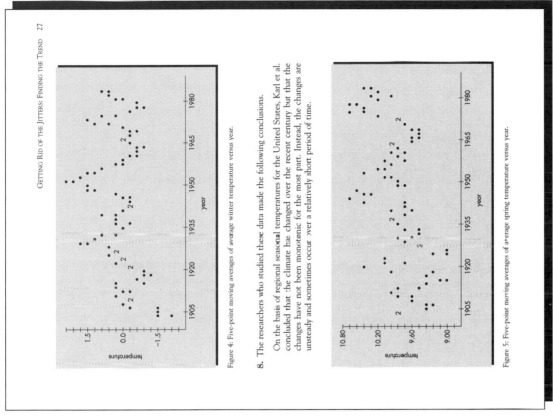

Figure 4: Five-point moving averages of average winter temperature versus year.

8. The researchers who studied these data made the following conclusions.

On the basis of regional seasonal temperatures for the United States, Karl et al. concluded that the climate has changed over the recent century but that the changes have not been monotonic for the most part. Instead, the changes are unsteady and sometimes occur over a relatively short period of time.

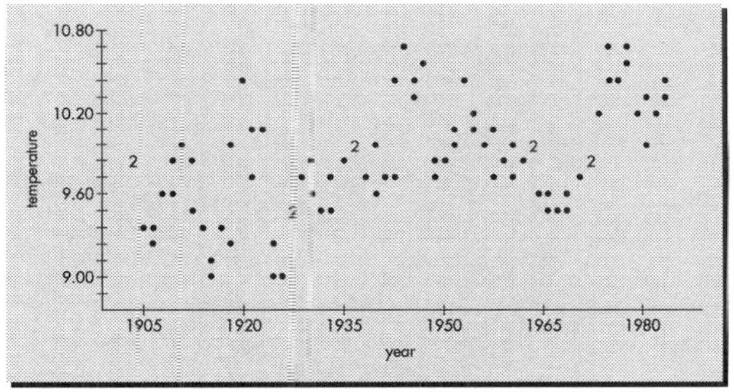

Figure 5: Five-point moving averages of average spring temperature versus year.

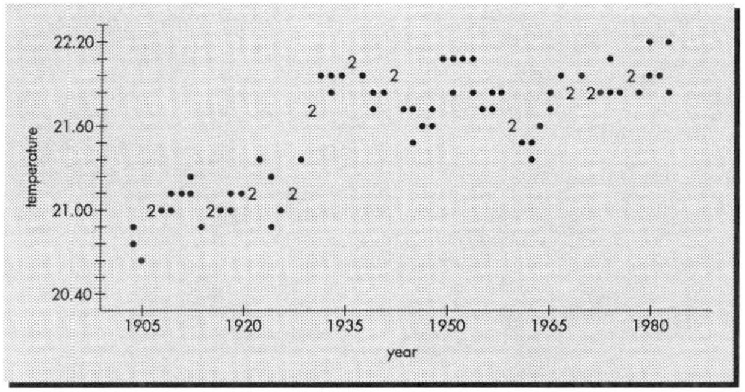

Figure 6: Five-point moving averages of average summer temperature versus year.

For the aggregated east region, which also includes this subregion, Karl et al. reported that the average annual time series can be divided into three epochs: a warm epoch from the early 1920s to the mid 1950s, preceded and followed by periods during which temperatures were generally at or below the mean for the century.

Do the data support these conclusions? Explain.

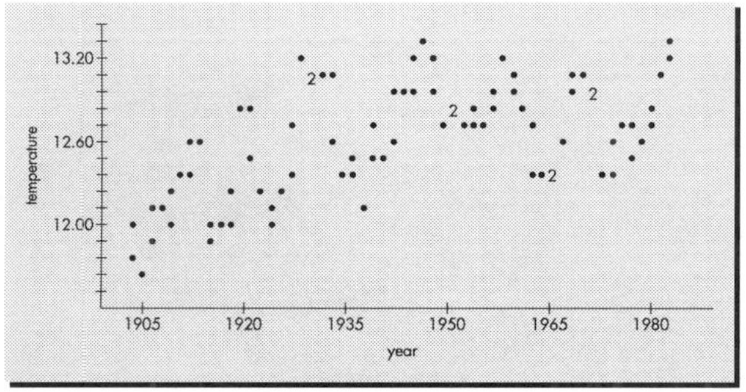

Figure 7: Five-point moving averages of average fall temperature versus year.

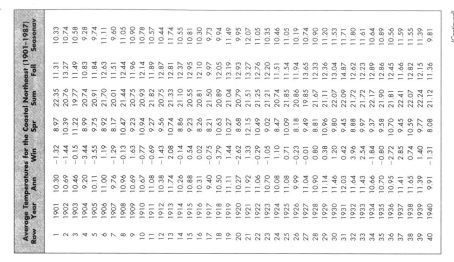

Average Temperatures for the Coastal Northeast (1901–1987)

Row	Year	Ann	Win	Spr	Sum	Fall	Seasonav
1	1901	10.30	−1.32	8.97	22.35	11.31	10.33
2	1902	10.69	−1.44	10.39	20.76	13.27	10.74
3	1903	10.46	−0.15	11.22	19.77	11.49	10.58
4	1904	9.20	−3.44	8.99	20.74	10.83	9.28
5	1905	10.11	−3.55	9.75	20.91	11.84	9.74
6	1906	11.00	1.19	8.92	21.70	12.63	11.11
7	1907	9.76	−1.29	8.17	20.01	11.51	9.60
8	1908	10.96	−0.13	10.47	21.44	12.44	11.05
9	1909	10.69	1.63	9.23	20.75	11.96	10.90
10	1910	10.67	−0.77	10.94	20.93	12.14	10.78
11	1911	11.08	−0.69	9.27	21.82	11.89	10.57
12	1912	10.38	−1.43	9.56	20.75	12.87	10.44
13	1913	11.74	2.08	10.74	21.33	12.81	11.74
14	1914	10.26	−0.14	8.86	21.10	12.37	10.55
15	1915	10.88	0.54	9.23	20.55	12.95	10.81
16	1916	10.31	0.02	8.26	20.81	12.10	10.30
17	1917	9.40	−0.75	8.21	21.50	9.97	9.73
18	1918	10.50	−3.79	10.63	20.89	12.05	9.94
19	1919	11.11	1.44	10.27	21.04	13.19	11.49
20	1920	10.27	−2.62	8.68	20.79	12.93	9.95
21	1921	11.92	1.33	12.15	21.51	13.27	12.07
22	1922	11.06	−0.29	10.49	21.25	12.76	11.05
23	1923	10.70	−1.05	9.02	21.21	12.20	10.35
24	1924	10.08	1.10	8.47	20.74	11.51	10.46
25	1925	11.08	0.71	10.09	21.85	11.54	11.05
26	1926	9.99	−0.23	8.18	20.86	11.94	10.19
27	1927	11.04	−0.01	9.49	19.85	13.65	10.74
28	1928	10.90	0.80	8.81	21.67	12.33	10.90
29	1929	11.14	0.38	10.96	21.11	12.36	11.20
30	1930	11.46	1.20	9.80	22.07	13.04	11.53
31	1931	12.03	0.42	9.45	22.09	14.87	11.71
32	1932	11.64	3.96	8.88	21.72	12.62	11.80
33	1933	11.43	2.54	9.97	21.72	12.23	11.61
34	1934	10.66	−1.84	9.37	22.17	12.89	10.64
35	1935	10.70	−0.80	9.58	21.90	12.86	10.89
36	1936	10.95	−2.72	10.70	21.81	12.45	10.56
37	1937	11.41	2.85	9.45	22.41	11.66	11.59
38	1938	11.65	0.74	10.59	22.07	12.82	11.55
39	1939	11.39	1.40	9.77	22.24	12.15	11.39
40	1940	9.91	−1.31	8.08	21.12	11.36	9.81

(Continued)

Average Temperatures for the Coastal Northeast (1901–1987)

Row	Year	Ann	Win	Spr	Sum	Fall	Seasonav
41	1941	11.40	0.12	9.90	21.47	14.13	11.41
42	1942	11.26	0.13	11.31	21.79	13.11	11.58
43	1943	10.99	−0.30	9.38	22.81	11.89	10.94
44	1944	11.19	0.09	9.98	22.31	12.49	11.21
45	1945	11.26	−1.07	11.87	21.37	13.28	11.36
46	1946	11.71	0.34	11.12	20.57	14.01	11.34
47	1947	11.06	1.17	9.17	21.52	13.29	11.28
48	1948	11.02	−1.57	10.06	21.60	13.21	10.82
49	1949	12.46	3.12	10.52	22.99	13.03	12.42
50	1950	11.07	2.74	8.50	21.14	12.67	11.26
51	1951	11.61	1.32	10.24	21.63	12.59	11.44
52	1952	11.80	2.14	10.03	22.32	12.33	11.83
53	1953	12.35	2.63	10.95	21.99	13.34	12.23
54	1954	11.55	2.30	10.13	21.48	13.12	11.76
55	1955	11.30	0.27	10.99	22.50	12.49	11.56
56	1956	11.06	−0.15	8.43	21.34	12.45	10.51
57	1957	11.73	1.56	10.74	22.02	12.93	11.81
58	1958	10.36	0.12	9.50	21.29	12.59	10.87
59	1959	11.94	−1.20	10.90	22.27	13.64	11.40
60	1960	10.95	1.72	8.98	21.70	12.94	11.34
61	1961	11.12	−1.52	9.14	21.77	14.20	10.90
62	1962	10.46	−0.37	10.18	20.98	11.66	10.61
63	1963	10.69	−1.81	10.38	21.52	13.07	10.79
64	1964	11.22	−0.75	10.31	21.36	12.45	10.84
65	1965	10.86	0.12	9.65	21.30	12.33	10.85
66	1966	10.91	0.22	9.18	22.24	12.33	10.99
67	1967	10.49	0.63	8.29	21.43	11.38	10.43
68	1968	11.04	−0.94	10.27	22.14	13.35	11.20
69	1969	11.02	−0.45	9.98	22.13	12.39	11.01
70	1970	11.05	−1.31	9.53	21.98	13.83	11.01
71	1971	11.36	−0.45	9.03	21.81	13.89	11.07
72	1972	10.79	1.59	9.15	21.04	11.66	10.86
73	1973	11.99	1.29	10.68	22.70	13.44	12.01
74	1974	11.44	1.62	10.58	21.55	12.07	11.45
75	1975	11.68	2.06	9.48	22.04	13.59	11.79
76	1976	10.97	1.21	11.06	21.84	10.78	11.22
77	1977	11.37	−2.30	11.85	22.05	13.16	11.19
78	1978	10.51	−1.84	9.12	21.83	12.42	10.38
79	1979	11.25	−0.94	10.96	21.32	13.40	11.19
80	1980	11.15	0.43	10.44	22.51	12.42	11.43
81	1981	11.26	−0.29	10.58	22.35	11.94	11.17

Average Temperatures for the Coastal Northeast (1901–1987)

Row	Year	Ann	Win	Spr	Sum	Fall	Seasonav
82	1982	11.06	−0.79	9.82	21.03	13.04	10.76
83	1983	11.64	1.96	10.07	22.65	13.33	11.99
84	1984	11.62	0.49	8.95	22.31	13.09	11.21
85	1985	11.68	1.16	11.57	21.51	14.22	12.11
86	1986	11.59	−0.07	11.02	21.96	12.68	11.40
87	1987	11.53	0.33	10.64	22.61	12.43	11.50

Reference: Karl, T.R., C.N. Williams, Jr. and F. T. Quinlan. 1990. United States Historical Climatology (HCN) serial temperature and precipitation data. NDP - 019/R1. Carbon Dioxide Information Analysis Center. Oak Ridge National Laboratory. Oak Ridge, Tennessee.

Wrap-Up

Time series data on productivity, unemployment and other economic statistics play an important role in assessing the state of the economy and formulating public policy. Graphs of such data can be particularly helpful in understanding how the data has changed over time. However, as you saw in Figure 1, the short term fluctuations made it difficult to observe the long term drift in the data. By averaging out some of these "jitters" we were able to observe the underlying trend in the temperatures. The data in the following table gives the gross energy consumption per capita in the U.S. measured in million Btu. Gross energy includes the energy generated by primary fuels such as petroleum, natural gas and coal, imports of their derivatives plus hydro and nuclear power. The changes in temperatures have been attributed by some to be in part caused by increasing fossil fuel consumption. Use moving averages to smooth the time series and comment on the trend that you observe. How does it compare to the trend in the temperature data? Does it indicate some positive trends in our energy consumption habits?

Extension

1. You have used three-point and five-point moving averages to smooth the time-series. State the advantages and disadvantages of the five-point average over the three-point average.
2. Time series are often used to forecast future values. Can these series give very exact forecasts? Why not?
3. Could the higher than average temperatures in the last 20 years have been due entirely to chance? You can check that out by doing a simulation where the chance of the average temperature being above or below the century mean is 1/2. Toss a coin 20 times, once for each year and note the number of times you get a head. We are assuming that a head in a coin toss represents the event that the temperature is above the century mean. Combine your results with the rest of the

Year	Gross Energy per Capita	Year	Gross Energy per Capita	Year	Gross Energy per Capita	Year	Gross Energy per Capita
1960	242.4	1969	316.7	1978	350.9	1987	311.8
1961	242.1	1970	323.7	1979	349.7	1988	324.2
1962	249.3	1971	326.9	1980	333.8	1989	325.3
1963	255.3	1972	339.7	1981	322.7	1990	326.9
1964	263.3	1973	350.5	1982	305.8	1991	334.7
1965	271.2	1974	338.9	1983	301.3	1992	321.5
1966	283.3	1975	326.4	1984	313.7		
1967	289.9	1976	340.7	1985	309.6		
1968	303.9	1977	346.5	1986	307.8		

class and get an estimate of the chance of observing 15 or more heads. Can the fact that in the last 20 years there were 15 years with the annual average temperatures above the mean for the century (11.043) reasonably be attributed to chance?

References

1. Karl T.R., Baldwin R.G. and Burgin M.G. (1988) Time series of regional seasonal averages of maximum, minimum and average temperature, and diurnal temperature range across the United States: 1901–1984. Historical Climatology Series 4–5. National Climatic Data Center. National Oceanic and Atmospheric Administration. National Environmental Satellite, Data and Information Service. Asheville, North Carolina.
2. Basic Petroleum Data Book. Petroleum Industry Statistics, Volume XIV, Number 1, January 1994, American Petroleum Institute, Washington D.C.

Flick the Nick: Observing Processes Over Time

NOTES FOR THE INSTRUCTOR

Statistical Setting

This activity can be used as an introduction to the problem of measuring the performance of a process over time. Time series plots should be treated separately from scatter plots, as we are not necessarily looking for linear trends and correlation in time series data. Also, time series data can be turned into a line graph since there is usually only one data point per time unit.

The important ideas involve looking for key features, such as trends, center, variability, and unusual observations, in the time series data that might indicate improvement or deterioration in the process being studied. The plots here are forerunners of control charts, and the language of quality improvement can be used in conjunction with this activity. It is not designed, however, to introduce all the terminology associated with classical control charts.

Prerequisites for Students

Students should have experience with the basic exploratory data tools (dot plots, stem plots, box plots, scatter plots) as well as knowledge of and experience with summarizing data by use of the mean and the standard deviation.

Materials

You will need a nickel coin for each student or group, a table top at least five feet long, and a ruler or meter stick for measuring in centimeters. The floor can be used if a table top is not available. You should make a line on each table using chalk or tape about two feet from the far end.

Procedure

The activity works well in small group settings, with two or three students per group. It can be assigned to be completed outside of class, as the set-up is quite simple and students can have fun completing it on their own. Another possibility is to have one group demonstrate the activity for the class and then have other groups work on their own to see how their results compare. For a small class, this could be a whole-class activity.

At least 25 observations are needed in order to see anything of interest in the time series plots. Emphasize the fact that time order is important and that data must be recorded and plotted in order. If time allows, you might let students practice a little more and compare other strategies for flicking the nickel with accuracy. In fact, this could develop into a design-of-experiment activity.

Each single player should show some improvement over 25 trials. If that is the case, there should be less variability toward the end of the time series. This may not happen, of course, and you should be prepared to discuss plots that show no improvement or deterioration over time.

Trends should not be so significant as to preclude the use of the mean as a center line and the standard deviation as a measure of variability around the center line. It is quite likely that some points will be "out of control." Box plots and their associated outliers should also be used in the analysis and discussion.

Sample Results from Activity

The 25 results for one student (in centimeters) read as follows:

$$-36, -22, -28, -14, 46, 38, 32, 30, 0, -4, -2, -14, -15,$$
$$-13, 6, -17, 13, -18, 4, -21, 22, -17, 10, 5, -6$$

The plot of these data over time is shown below. Notice that this person was conservative for a while, perhaps not wanting to shoot off the table, and then overadjusted to produce some large values. The last half of the data shows some control of the process, as the player alternates from positive to negative in a regular fashion. This is caused, no doubt, by the player's adjustment after each trial.

The mean of these data is -1.24, indicating that, overall, the player is still conservative, and the standard deviation is 21.28. Only the 46 is more than two standard deviations from the mean, and this occurs right at the point at which the player seemed to change strategies to overcome the negative trend. From observation 12 on, the mean is about -5 and the standard deviation is about 13. These values may summarize this player's performance, after warm-up, quite accurately.

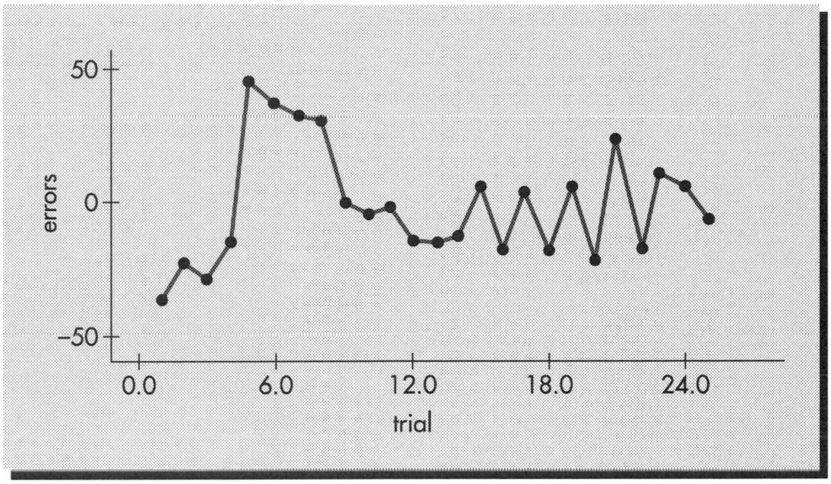

Sample Assessment Questions

1. Have each student collect data on some interesting daily activity, such as time spent studying, sleeping, or exercising, calories consumed, or the price of a stock he or she is watching. After two or three weeks have passed, have each student analyze the data and write a report on what he or she has learned about this aspect of his or her life. A data sheet like the one attached here may be of help.

2. To monitor the "quality" of the statistics class itself, have students record data on the class and plot the points over time. Have the students analyze the data periodically and write a short report on class quality. Some suggested variables to measure are

   starting time and ending time,
   number of students arriving late,
   number of students leaving early,
   daily attendance,
   number of questions students ask,
   number of different students asking questions,
   number of times the instructor does not know the answer to a question.

Flick the Nick: Observing Processes Over Time

SCENARIO

"The number of traffic accidents on that highway seems to be going up every month. Things are getting out of control!"

How often have you heard reference to trends in behavior over time? How often have you heard reference to the fact that something seems to be "out of control"? In many quantitative problems, the data are collected over time. In such cases, the time order of the data and the patterns generated across time are important considerations. If accidents really are on the increase, then a community might want to implement new traffic control strategies. If, however, last month's accident total is exceedingly large for some explainable reason, like road construction or bad weather, then changes in traffic control might not be warranted. For such a solution, you need to be able to explain meaningfully patterns in data collected over time and recognize data points that depart from the norm in an unusual way.

Question

Can I improve my play of a game over time?

Objectives

In this lesson you will gain experience with plotting data over time and describing patterns in such time series data. Identifying unusual data points, or points "out of control," is an important aspect of studying processes over time.

Activity

1. Collecting the data

   You will be playing a game that involves flicking a nickel across a table, using your thumb and forefinger. The goal is to come as close as possible to a line marked on the table, about three feet from the edge, at which the nickel will be flicked. After each flick, the distance from the edge of the nickel to the goal line is to be measured to the closest centimeter. If the nickel has not crossed the line, the distance from the line is recorded as a negative number. If the nickel has crossed the line, the distance is recorded as a positive number. (The class should agree on a rule for handling shots that go off the table.)

   Work in groups of at least two students each to facilitate the measurement process. Each person flicks the nickel at least 25 times in succession. Record the distance between the nickel and the goal line each time, making sure to keep the data in correct time order.

2. Describing and interpreting the data
   a. Make a plot of your distances from the goal line (vertical axis) against the number of the trial (horizontal axis).
   b. Do you see any pattern in the plot? If so, explain its nature.
   c. Compare your plot with those of other students. What differences do you see among the patterns?
   d. For your entire data set, construct a box plot and identify any outliers that might have occurred. If you have outliers, where in the ordered sequence did they occur? What is a possible cause of the outliers?
   e. Looking at your data, is it reasonable to describe the center of the measurements by the mean and the variability in the measurements by the standard deviation? Why or why not? If it is reasonable, identify any data points that are more than two standard deviations from the mean. Where in the ordered sequence did they occur?

3. Drawing conclusions

   The goal is to learn something about the process at hand by studying patterns in the time series. Use the time series plot constructed in step 2 to answer the following questions.
   a. What pattern in the data would suggest improvement over time? Did you improve over time? Among the members of the class, find one that showed marked improvement over time.
   b. What pattern in the data would show a stable process over time (one with no obvious improvement or deterioration)? Find at least one plot that shows a relatively stable process.
   c. What pattern in the data would show deterioration of the process over time (the process is actually getting worse with time)? Can you find a plot that shows deterioration?

d. Taking into account the various patterns of the data, discuss possible rules for selecting a winner of the "Flick the Nick" competition. The class should reach a consensus and "award" an appropriate prize to the winner.

Wrap-Up

Consider the following situations for generating data over time. You are to think about what the data might look like if you were to carry out these activities; you need not actually carry them out.

A. You are to perform a task that takes some skill (like threading a needle or programming a VCR). The measurement of interest is the time it takes to perform the task. You are to repeat the task many times and record the time to completion for each trial.

B. You are to close your eyes and guess when one minute has passed. A partner with a stopwatch measures your guessed time and tells you if your guess was too long (more than 60 seconds) or too short (less than 60 seconds). Then, you repeat the procedure, keeping this up for a number of trials. The measurement recorded for each trial is $(x - 60)$, where x is your guessed time (in seconds) for a minute.

For each situation, A and B, the time data are to be graphed against the trial number.

1. What will the graphs look like if each system is stable, meaning that no improvement or deterioration takes place over the trials?

2. What will the graphs look like if there is improvement over time?

3. What will the graphs look like if there is deterioration (you actually get worse) over time?

4. Write a statement that will guide other students on how to describe and interpret data from a process that is measured repeatedly over time.

Extensions

1. The rules of the game have been changed. The objective of the game now is to flick the nickel as close as possible to the goal line *without going over the line*. Any flick that causes the nickel to go past the line automatically disqualifies the player. Still, the data should be recorded as earlier, with positive and negative values, so that all the flicks can be analyzed.

a. Plot your distances from the goal line against the number of the trial. What pattern do you see?
b. Compare your plot with those of other students in the class. In what ways do the plots differ?
c. What pattern in these plots shows improvement over time?

2. Describe another game or task in which one might improve over time. Collect data on various people who are just learning the game, and describe the patterns of their play over time.

Ratings and Ranks

NOTES FOR THE INSTRUCTOR

Statistical Setting

The purpose of this lesson is to show the difference between a rating and a rank and how the two can be related. It stands alone in its content and is not tied to the usual topics on introductory statistics, but it should be considered a part of data exploration. It is important that students see the arbitrary nature of composite ratings, which are often used in ranking places or individuals. The lesson can serve as an introduction to the concept of weighted averages.

Prerequisites for the Students

None, if weighted average is explained in the context of this lesson.

Materials

A copy of the *Places Rated Alamanc.*

Procedure

Discuss the definitions of the five variables used in this lesson so that students have a clear understanding of them. In particular, students should understand which components of a given rating are actual data and which are arbitrary weights.

Discuss the idea of students' choosing their own weights for the five variables of this lesson so that they see the idea clearly. You might work through an example using your own weights for the five variables.

Weighted averages is likely to be a new idea for the students. Make sure they understand this concept. Their grade point average serves as a good example.

A graphing calculator or a spreadsheet on a computer helps simplify the calculations here. If you want to introduce spreadsheets into your course, here is a good way to do it.

Sample Results from Activity

The rankings of the cities, after calculating a weighted average of ranks across the five categories, are shown below for three different students. One simply weighted all five categories equally (weights of .2 for each) to produce R1. Another thought housing, jobs, and crime were more important than the others and chose weights of .3, .4, and .3, respectively, for these categories. This weighting produced R2. A third student was interested only in pursuing an education, along with some recreation, and weighted these two categories at .5 each. This weighting produced R3. Notice that R1 and R3 tend to favor the large cities and R2 tends to favor the smaller ones. Students should compare results like these and discuss the subjectivity involved in rating places or objects.

	R1	R2	R3
Boston	2	8	1
Washington	1	4	3
Atlanta	3.5	5	4
San Diego	3.5	7	2
Terre Haute	6	3	7.5
Lincoln	7.5	6	5
Greenville	7.5	2	7.5
Salem	5	1	6

The actual ratings of the NFL quarterbacks are given on the following page.

ALL-TIME LEADING PASSERS (THROUGH 1989–1990 SEASON)

Player	Att	Comp	Yards	TD	Int	Rating
Joe Montana	4059	2593	31054	216	107	94.0
Dan Marino	3650	2174	27853	220	125	89.3
Boomer Esiason	2285	1296	18350	126	76	87.3
Dave Krieg	2842	1644	20858	169	116	83.7
Robert Staubach	2958	1685	22700	153	109	83.416
Bernie Kosar	1940	1134	13888	75	47	83.415
Ken O'Brien	2467	1471	17589	96	68	83.0
Jim Kelly	1742	1032	12901	81	63	82.74
Neil Lomax	3153	1817	22771	136	90	82.68
Sonny Jurgensen	4262	2433	32224	255	189	82.6
Len Dawson	3741	2136	28711	239	183	82.6
Ken Anderson	4475	2654	32838	197	160	81.9
Danny White	2950	1761	21959	155	132	81.7
Bart Starr	3149	1808	24718	152	138	80.5
Fran Tarkenton	6467	3686	47003	342	266	80.4
Tony Eason	1536	898	10987	61	50	80.3
Dan Fouts	5604	3297	43040	254	242	80.2
Jim McMahon	1831	1050	13335	77	66	79.2
Bert Jones	2551	1430	18190	124	101	78.2
Johnny Unitas	5186	2830	40239	290	253	78.2

Sample Assessment Questions

1. The NCAA uses the following formula to establish a rating for college quarterbacks:

$$\text{Rating} = (\%\text{COMP}) + 3.3(\%\text{TD}) - 2.0(\%\text{INT}) + 8.4(\text{YDS/ATT}).$$

a. Compare this formula to the one the NFL uses. Comment on the similarities and differences between the two. Does an interception carry more weight in the NFL or in the NCAA?

b. Find the statistics for two college quarterbacks, and produce their ratings. How do they rank compared to each other?

2. Select five brands of automobiles that are of interest to you.
   a. Obtain data for each on price, resale value, gas mileage, interior room, and trunk space.
   b. Weight these five variables according to their importance to you.
   c. Calculate a composite rank for each automobile, and rank them on this composite measure.
   d. Find another student who has a different ranking, and comment on why the two of you differ.

Ratings and Ranks

SCENARIO

"Did you like the movie?"

"Yes, I'd rate it at about four stars."

"How did it compare to the two movies you saw last week?"

"I think this one ranks second among the three."

Ratings and ranks; we love them and regularly use them to gauge our thoughts and feelings about places and events in our lives. Cars are rated on various performance criteria and then ranked against one another. Athletes are similarly rated on their performance and then ranked in comparison to others. In olympic figure skating, for example, a high rating produces a rank close to the top of the scale, with the highest rating producing the winner (the person ranked first). The Nielsen Co. reports that a certain TV show has 23% of the market for its time slot. Is this information in the form of a rating or a rank? The Nielsen Co. reports that the TV show has the second highest percentage of the viewing audience for that time slot. Is this information in the form of a rating or a rank? How did your home town fare in the most recent rating of places to live?

Questions

What are ratings and what are ranks? How are the two related?

Objectives

In this lesson you will see how various factors may be combined to form a numerical rating of a person, place, or object. The method of combining these factors is often quite arbitrary, and this should be taken into account when ratings are interpreted. Thus, you will discover that changes in the relative weights of factors making up a rating can have a dramatic effect on the conclusions. Finally, you will see that it is convenient to turn ratings into ranks when information is to be summarized for potential users.

Activity

1. Understanding ratings

   A widely referenced source of ratings and ranks is the *Places Rated Almanac* (see References). In that book, all metropolitan areas of the United States are rated in 10 categories, including housing cost, jobs, education, crime, and recreational facilities. A brief description of how the ratings are established for these five categories follows. (Data on these ratings, for selected cities, will be presented later.)

   Housing (H): the annual payments on a 15-year, 8% mortgage on an average-priced home after making a 20% down payment.

   Jobs (J): the product of the estimated percentage increase in new jobs from 1993 to 1998 and the number of new jobs created between 1993 and 1998, added to a base score of 2000.

   Education (E): the total of enrollment in two-year colleges divided by 100 plus enrollment in private colleges divided by 75 plus enrollment in public colleges divided by 50.

   Crime (C): the violent crime rate plus one-tenth of the property crime rate. (Crime rates are reported as the number of crimes per 100,000 residents.)

   Recreation (R): a score directly proportional to the number of public golf courses, good restaurants, zoos, aquariums, professional sports teams, miles of coastline on oceans or the Great Lakes, national forests, parks, and wildlife refuges, and state parks.

   a. For which categories are high ratings good and low ratings bad?
   b. Which categories might tend to produce higher ratings in larger cities?
   c. Which would have the greater effect on the education rating: an increase of 1,000 students in two-year colleges or an increase of 1,000 students in public colleges? Explain.
   d. Does it seem reasonable to put the multiplier of 1/10 in the formula for the crime rating? Explain the purpose of this multiplier.
   e. Name a city in the United States that might have a relatively high rating for recreation; name a city that might have a relatively low rating for recreation.

Table 1

City	Housing	Jobs	Education	Crime	Recreation
Boston, MA	18,903	3,456	4,176	1,051	2,278
Washington, DC	15,466	16,288	3,764	1,028	1,857
Atlanta, GA	8,676	16,777	1,692	1,474	1,822
San Diego, CA	20,322	14,772	2,335	1,266	3,800
Terre Haute, IN	4,116	2,028	290	823	1,100
Lincoln, NE	6,362	2,457	554	993	1,486
Greenville, NC	6,911	3,477	377	882	900
Salem, OR	6,226	2,787	237	869	1,784

2. From ratings to ranks
Ratings in the five categories defined in part 1 for a selection of eight cities are provided in Table 1.
 a. Review your answer to part 1b. Do the data suggest that you were correct?
 b. For which category are the ratings most variable? For which are the ratings least variable? Does this seem reasonable? Explain.
 c. With 1 denoting most desirable and 8 denoting least desirable, rank the cities with regard to housing cost (H). What features of the data are lost when you go from the actual rating to the rank?
 d. Rank the cities within each of the other four categories, with 1 denoting most desirable. Write the ranks on the table next to the ratings.
 e. Discuss the differences and similarities in rankings you made in part 2d. What are some advantages to reporting the ranks rather than the ratings?

3. Combining ranks
You will now be making extensive computations with the data from Table 1. The most efficient way to accomplish this is to enter the data into a graphing calculator as lists or into a computer spreadsheet. The goal of this investigation is for you to come up with your personal overall ranking of the eight cities as desirable places for you to live. The first step is to establish the relative importance of the five categories.
 a. Gather together 10 pennies. Divide the pennies among the five categories on Table 1 according to the relative importance you assign to the categories. (Place pennies on the data table.) For example, if you think that all five categories are equally important, then lay two pennies on each column. If you think that housing and jobs are equally important, and nothing else matters, then place

five pennies on housing and five on jobs. The number of pennies assigned to a category divided by ten (which makes a convenient decimal) is your personal weighting for that category.
 b. Using your weights as established in part 3a, calculate the **weighted average** of the ranks for each city.
 c. Rank the cities according to your weighted average ranks from part 3b.
 d. Compare your ranking of the cities with those of other class members. Is there any one city that always ranks first?

Wrap-Up

1. Discuss the difference between a rating and a rank.

2. What questions might you ask when someone reports a new study ranking desirable spots for a vacation or desirable occupations?

3. Table 2 shows the all-time leading passers in the NFL as of 1990, ranked according to their ratings as passers. The leader is Joe Montana, with a rating of 94.0. The formula used to produce these ratings is

$$R = (1/24)\{50 + 20C + 80T - 100I + 100(Y/A)\},$$

where

R = the rating,
C = percentage of completions,
T = percentage of touchdowns,
I = percentage of interceptions,
Y = yards gained by passing,
A = attempts.

 a. Using a spreadsheet on a computer or a graphing calculator, add columns to the data table for C, T, I, and Y/A.
 b. Using the constructed data columns, find the ratings (R) for the other quarterbacks on the list. Is the listing in the correct order?
 c. Note that the list goes through the 1989–1990 season only. Find the passing statistics for your favorite NFL quarterback from last season. How does he compare to the all-time leaders?
 d. Discuss the relative weights in the formula for quarterback ratings. You might begin by answering the following questions:
  i. Why do some factors have a minus sign in front of them?
  ii. How much more important is a touchdown than a simple completion?
  iii. Does a touchdown carry more weight than an interception?
  iv. Why use yards per attempt rather than just total yards passing?

Table 2 ALL-TIME LEADING PASSERS (THROUGH 1989–1990 SEASON)

Player	Attempts	Completions	Yards	TD	Interceptions	Rating
Joe Montana	4,059	2,593	31,054	216	107	94.0
Dan Marino	3,650	2,174	27,853	220	125	
Boomer Esiason	2,285	1,296	18,350	126	76	
Dave Krieg	2,842	1,644	20,858	169	116	
Roger Staubach	2,958	1,685	22,700	153	109	
Bernie Kosar	1,940	1,134	13,888	75	47	
Ken O'Brien	2,467	1,471	17,589	96	68	
Jim Kelly	1,742	1,032	12,901	81	63	
Neil Lomax	3,153	1,817	22,771	136	90	
Sonny Jurgensen	4,262	2,433	32,224	255	189	
Len Dawson	3,741	2,136	28,711	239	183	
Ken Anderson	4,475	2,654	32,838	197	160	
Danny White	2,950	1,761	21,959	155	132	
Bart Starr	3,149	1,808	24,718	152	138	
Fran Tarkenton	6,467	3,686	47,003	342	266	
Tony Eason	1,536	898	10,987	61	50	
Dan Fouts	5,604	3,297	43,040	254	242	
Jim McMahon	1,831	1,050	13,335	77	66	
Bert Jones	2,551	1,430	18,190	124	101	
Johnny Unitas	5,186	2,830	40,239	290	253	

Roger W. Johnson, How Does the NFL Rate the Passing Ability of Quarterbacks? *The College Math Journal*, vol. 24, no. 5, Nov. 1993, p. 451.

Extensions

1. Discuss other ways that a composite ranking of the cities could be obtained. Discuss their advantages and disadvantages.

2. Find a copy of the *Places Rated Almanac*.
   a. What method is used to produce a composite ranking of the cities in the United States?
   b. Select cities of interest to you and repeat the second investigation ("From Ratings to Ranks") as described in Activity 2. You may want to use different categories, according to your interests.

References

1. Roger W. Johnson (1993), "How does the NFL rate the passing ability of quarterbacks?" *College Mathematics Journal*, **24**:451–453.
2. David Savageau and Richard Boyer (1993), *Places Rated Almanac*, New York: Prentice Hall Travel.

Predictable Pairs: Association in Two-Way Tables

NOTES FOR THE INSTRUCTOR

Statistical Setting

This activity should follow an activity or discussion on exploring frequencies and relative frequencies for a single categorical variable. The spirit of the activity is exploratory data analysis, not inference. Students are to begin thinking about summarizing frequency data on two-way tables and interpreting relative frequencies of three types: joint, marginal, and conditional.

As students are encouraged to view the data in the world around them, they will realize that they are exposed to much more frequency and relative frequency data than continuous measurement data. Thus, it is important that they be able to make sense out of such categorical data.

Prerequisites for Students

Students should have experience with categorical data and summaries in terms of frequencies and relative frequencies for single variables, and they should know the difference between a proportion and a percent.

Materials

You will need a set of questions with simple categorical responses which pertain to issues of interest to the students in the class. (These questions can be formulated by the class.)

Procedure

The selection of questions and the collection of data should be done as a whole-class activity. Choose questions that have fairly unambiguous yes/no answers, but ones in which the class might have some interest. Direct the discussion to the extent that some pairs of questions that appear to be associated are, in fact, selected.

Collect data separately on each individual question first. The data on frequencies can be quickly collected by a show of hands; all students get the same data for their analyses. When the students, individually or in groups, begin to look at two-way tables for summarizing information on a pair of questions, they will realize that something is missing. Let them make this discovery and decide what to do about the missing data.

The idea of calculating conditional relative frequencies and relating them to association between the variables is difficult for students to grasp. Discuss a number of examples from the class, making sure some tables show strong association (at least in an exploratory sense) and some show little association. Have students express, both orally and in writing, what they are seeing in the tables and how they are relating this to association.

Sample Results from Activity

Students were asked for a show of hands on the question, "How many of you have jobs (outside of being a student)?" Then, they were asked for a show of hands on the question, "How many of you own a motor vehicle?" Setting up the appropriate two-way table and asking them to fill it in caused them to realize that the joint information was missing. One more question had to be asked of the class.

The data from a group of 40 participants turned out as follows:

		Job		
		Yes	No	Totals
Vehicle	Yes	8	2	10
	No	18	12	30
	Totals	26	14	40

The marginal total of 10 tells us how many own vehicles, and the marginal relative frequency (or proportion) of 10/40 = .25 tells us that 25% of the group own vehicles. Similarly, 26/40 = .65 tells us that 65% of the class have jobs.

The joint frequency of 8 tells us that 8 people have a job *and* own a vehicle. The joint relative frequency of 8/40 = .20 tells us that 20% of the group said "yes" to both of these questions.

What about vehicle ownership among those who have a job, as compared to those who do not have a job? This information is obtained by looking at the conditional relative frequencies within each column. If we look only at people with a job in our group, the data show 8/26 = .31 as the proportion who own a vehicle. (This is often said to be the proportion who own vehicles *conditional* on the fact that they have jobs.) If we look at people who do not have jobs, then 2/14 = .14, or 14% own vehicles.

The conditional proportions can be calculated across rows rather than down columns. Here, the proportion of people who have jobs, conditional on the fact that they own a vehicle, is 8/10 = .8. The proportion who have jobs conditional on the fact that they do not own a vehicle is 18/30 = .60.

So, after all of this data collection and summarization, can we answer the question, "Does having a job appear to be associated with owning a vehicle?" The marginal information sheds no light on the issue, because it describes the answers to the questions one at a time. The joint information tells us a little; 8 out of 40 people saying "yes" to both questions may indicate that many who own vehicles may also have a job. But, what is *really* interesting is that these 8 come from only 10 vehicle owners in the entire group. Thus, the conditional proportion of 80% of vehicle owners having a job, as opposed to only 60% of non-vehicle owners having a job, shows that there appears to be an association between vehicle ownership and having a job.

The conditional proportions can be analyzed the other way. The fact that 31% of those who have a job also own a vehicle as opposed to only 14% of those who do not have a job is, likewise, grounds for suggesting association between the two variables.

What if the data turned out to look like the following?

		Job	
		Yes	**No**
Vehicle	Yes	6	4
	No	20	10

Would this be a situation showing strong association between the two variables? Have students use arguments similar to those introduced above to make the case for or against association.

Sample Assessment Questions

1. Have students find an article from a newspaper or magazine that contains responses to categorical variables. (Most opinion polls are set up this way.) Then have them discuss possible associations among the variables in the article. Can

these associations be measured from the data presented in the article? (Often they cannot because the articles tend not to give the joint information.)

2. Another question on the survey of teenagers was, "Do your peers care about staying away from marijuana?" The population projected frequencies (in thousands) are as follows.

	NS	EX	FS	CS
A Lot	7,213	2,693	75	857
Somewhat	2,482	1,861	109	1,102
A Little	744	542	27	298
Don't Care	1,878	1,550	119	1,312

a. What proportion of those who smoked think their peers care a lot about staying away from marijuana?
b. What proportion of the current smokers think their peers care a lot about staying away from marijuana?
c. Among those who think their peers care a lot about staying away from marijuana, what proportion have never smoked?
d. Among those who think their peers don't care about staying away from marijuana, what proportion have never smoked?
e. Do you think perceived peer attitudes toward staying away from marijuana are associated with the smoking status of the teenager? What proportions help to justify your answers?

3. Additional questions on the Teenage Attitudes and Practices Survey were asked directly of those being interviewed. Some of the data are reported in the form of percentages, rather than frequencies. Two examples are as follows:

"DO YOU BELIEVE CIGARETTE SMOKING HELPS REDUCE STRESS?"

	NS	EX	FS	CS
Yes	12.0	18.7	29.8	46.5
No	84.9	78.5	68.9	51.7
Don't know	3.0	2.5	1.6	1.6

"DO YOU BELIEVE ALMOST ALL DOCTORS ARE STRONGLY AGAINST SMOKING?"

	NS	EX	FS	CS
Yes	80.1	78.8	80.1	80.5
No	17.3	18.8	17.3	16.7
Don't know	2.5	2.3	2.6	2.6

a. How were these percentages calculated; what do they mean? Are these percentages joint, marginal, or conditional?
b. Do the opinions on whether or not cigarette smoking helps reduce stress appear to be associated with the smoking status of the person responding? Write a paragraph justifying your answer.
c. Do the opinions on doctors being against smoking appear to be associated with the smoking status of the person responding? Write a paragraph justifying your answer.

4. Heart attacks aren't the only cause for concern in the Physician's Health Study. Another is that too much aspirin can cause an increase in strokes. Among the aspirin users on the study, 119 had strokes during the observation period. Within the placebo group, only 98 had strokes. Place this data on an appropriate two-way table, and comment on the association between aspirin use and strokes, as compared to the association between aspirin use and heart attacks.

5. What about smoking as it relates to heart attacks and the use of aspirin? The following two tables show the number of heart attacks for each treatment group separated out according to whether the participant was a current smoker or had never smoked.

CURRENT SMOKERS

		Aspirin	Placebo
Heart Attack	Yes	21	37
	No	1,192	1,188

NEVER SMOKED

		Aspirin	Placebo
Heart Attack	Yes	55	96
	No	5,376	5,392

Is aspirin as effective a preventative among current smokers as it is among those who never smoked? What can we say about the rate of heart attacks among the current smokers as compared to those who never smoked?

Predictable Pairs: Association in Two-Way Tables

SCENARIO

The results of a survey might provide lots of information on characteristics of students at your school, such as what proportion eat breakfast regularly, like first-period classes, have a job, or own a car. Does a "yes" response to one of these characteristics help us predict what the response of that same person will be to another characteristic? Is a person who eats breakfast likely to have an early class? Is a person who has a job likely to own a car? Looking for associations between categorical variables is an important part of the analysis of frequency data. How to begin the process of measuring association is the subject of this lesson.

Question

What features of the data help us to establish association between two categorical variables?

Objectives

In this lesson you will begin to look at associations between two categorical variables by constructing two-way frequency tables. From such tables, you will calculate and interpret marginal frequencies and relative frequencies, joint frequencies and relative frequencies, and conditional frequencies and relative frequencies.

Activity

This is a class activity. All the data can be collected quickly from a few simple questions that will be asked of all class members.

1. Collecting the data
   As a class, formulate simple questions that can be answered either "yes" or "no." Some suggestions are as follows:
   Do you have a job outside of school?
   Do you own a motor vehicle?
   Do you participate regularly in a sport?
   Do you play a musical instrument?
   Did you see the recent movie (A)__________?
   Did you see the recent movie (B)__________?

   Collect the total "yes" responses and the total "no" responses to each question from all students in the class. (This can be done quickly by a show of hands.)

2. Describing and interpreting the data
   a. Two questions with categorical responses are said to be **associated** if knowledge of the response to one question helps in predicting the response to the other question. For example, knowing that a student owns an automobile may increase the likelihood that the student has a job. If that is the case, then automobile ownership and job status are associated. From the list established by the class, choose a pair of questions whose answers you think might be associated.
   b. Arrange the data for the chosen pair of questions in a two-way frequency table, as illustrated here:

		Question A Yes	Question A No	Total
Question B	Yes	a	b	$a + b$
	No	c	d	$c + d$
	Total	$a + c$	$b + d$	$a + b + c + d = n$

   From the class data already collected, which of the values in the two-way table can be filled in with observed frequencies

   c. How many more pieces of information do you need to complete the table? Obtain this information and complete the table.

d. Does it appear that the two questions you chose are associated? What features of the data in the two-way table give evidence for or against association?

3. Drawing conclusions

a. What does a/n tell you? (Fill in your numbers for a and n.) What does c/n tell you? This is **joint information** of the two questions.
b. What does $(a + c)/n$ tell you? This is **marginal information** on Question A.
c. What does $(a + b)/n$ tell you? This is marginal information on Question B.
d. What does $a/(a + c)$ tell you? This is **conditional information** on Question B for those who are known to respond "yes" to Question A.
e. What does $b/(b + d)$ tell you?
f. Which of the three types of information supplied by the two-way table (joint, marginal, or conditional) gives the greatest evidence for or against association between two questions in a pair? Explain your reasoning.

4. Summarizing results

Write a summary of your analysis for the data collected on two different pairs of questions used in your class. Address the issue of association in your summary.

Wrap-Up

1. Define a variable with ordered categorical responses, such as "How frequently do you drink alcoholic beverages?" The responses could be categorized as none, seldom (less than once per week), or often. Then, define a second variable with similar response categories, such as "How often do you drink soft drinks?" Have students analyze the results of data collected from the class (or from another convenient group) by looking at a two-way table. A good way to set up such a table for ordered categories is as follows:

		Drink Alcohol Never	Seldom	Often
	Often	*a*	*b*	*c*
Drink Soft Drink	Seldom	*d*	*e*	*f*
	Never	*g*	*h*	*i*

a. What type of association would be suggested by large values of (c, e, g)?
b. What type of association would be suggested by large values of (a, e, i)?
c. Do you see any relationship between the concepts of **association** between categorical variables and **correlation** between measurement variables? Explain.

2. The 1989 Teenage Attitudes and Practices Survey (see Ref. 2) obtained completed questionnaires, either by telephone or mail, from a randomly selected group of 9,965 12–18-year-olds living in households across the country. One type of question asked the teenagers being interviewed about the perceived behavior of their peers. The teenagers answering the questions are classified as

NS = never smoked,
EX = experimenter with smoking,
FS = former smoker,
CS = current smoker.

One question was, "Do your peers care about keeping their weight down?" The data shown are the population projections (in thousands) calculated from the sample responses.

	Care about keeping weight down NS	EX	FS	CS
A Lot	6,297	3,613	197	2,114
Somewhat	2,882	1,677	90	793
A Little	1,441	625	16	354
Don't Care	1,709	822	33	377

a. Why would a question be formed in terms of peer behavior rather than as a direct question to the person being interviewed?
b. Approximately how many teenagers in the United States never smoked (as of 1989)?
c. What proportion of teenagers thought their peers cared a lot about keeping their weight down?
d. Approximately how many teenage current smokers were there in the United States in 1989? What proportion of them thought their peers cared a lot about keeping their weight down?
e. Among those that cared a lot about keeping their weight down, what proportion never smoked? What proportion were current smokers?
f. Do you think perceived peer attitudes toward keeping weight down are associated with the smoking status of the teenager? Calculate appropriate proportions to justify your answer.

3. Let's now apply our skills at analyzing frequency data to an important medical experiment. Joseph Lister, a British physician of the late 19th century, decided that something had to be done about the high death rate from postoperative complications, which were mostly due to infection. Based on the work of Pasteur, he thought that the infections had an organic cause, and he decided to experiment with carbolic acid as a disinfectant for the operating room. Lister performed 75 amputations over a period of years. Forty of the amputations were done with carbolic acid, and 35 were done without carbolic acid. For those done with carbolic acid, 34 of the patients lived; for those done without carbolic acid, 19 of the patients lived. Arrange these data on an appropriate table, and discuss the association between mortality rate and the use of carbolic acid.

4. The ELISA test for the presence of HIV antibodies in blood samples gives the correct result about 99.7% of the time for blood samples that are known to be HIV-positive. The test is correct 98.5% of the time for blood samples that are known to be HIV-negative.
   a. Suppose 100,000 blood samples contain 10% that are HIV-positive. All are tested with ELISA. Set up a two-way table to show what the outcomes of the tests are expected to look like. What proportion of the samples that test positive are really expected to be positive?
   b. Repeat the instructions for part a for a set of 100,000 blood samples that are assumed to contain 20% that are HIV-positive.
   c. How do the results for parts a and b compare? Comment on the possible social implications of mass testing of blood samples.

Extensions

1. Is hand dominance associated with eye dominance? Have each person in the group check eye dominance by focusing on an object across the room through a small opening, as made by outstretched hands, with both eyes open. Close one eye and see if the object appears to move. If it does not move, the open eye is dominant. Now, collect data on the following two-way table:

		Eye Dominance Left	Right
Hand Dominance	Left	a	b
	Right	c	d

   a. Does there appear to be strong association between hand dominance and eye dominance?
   b. What would large values in positions b and c tell you?
   c. What would large values in positions a and d tell you?

2. Does aspirin really help prevent heart attacks? During the 1980s, approximately 22,000 physicians over the age of 40 agreed to participate in a long-term health study for which one important question was to determine whether or not aspirin helps to lower the rate of heart attacks (myocardial infarctions). The treatments of this part of the study were aspirin or placebo, and the physicians were randomly assigned to one treatment or the other as they entered the study. The method of assignment was equivalent to tossing a coin and sending the physician to the aspirin arm of the study if a head appeared on the coin. After the assignment, neither the participating physicians nor the medical personnel who treated them knew who was taking aspirin and who was taking placebo. This is called a *double-blind experiment*. (Why is the double blinding important in a study such as this?) The physicians were observed carefully for an extended period of time, and all heart attacks, as well as other problems, that might occur were recorded. All of the data can be summarized on two-way tables.

   Other than aspirin, there are many variables that could affect the rate of heart attacks for the two groups of physicians. For example, the amount of exercise they get and whether or not they smoke are two prime examples of variables that should be controlled in the study so that the true effect of aspirin can be measured. The following table shows how the subjects (from Ref. 1) eventually divided according to exercise and to cigarette smoking.

		Aspirin	Placebo
Exercise Vigorcusly			
	yes	7,910	7,861
	no	2,997	3,060
Cigarette smoking			
	never	5,431	5,488
	past	4,373	4,301
	current	1,213	1,225

   a. Do you think the randomization scheme did a good job in controlling these variables? Explain in terms of association or lack of association.
   b. Would you be concerned about the results for aspirin having been unduly influenced by the fact that most of the aspirin-takers were also nonsmokers? Explain.
   c. Would you be concerned about the placebo group possibly having too many nonexercisers? Explain.

3. The results of this study report that 139 heart attacks developed among the aspirin users and 239 heart attacks developed in the placebo group. This was said

to be a significant result in favor of aspirin as a possible preventive for heart attacks. To demonstrate this difference, place the data on heart attacks on an appropriate two-way table. (Remember, the 22,000 participants were about evenly split between aspirin and placebo.) What are the appropriate conditional proportions to study if we want to compare the rates of heart attacks for the two treatment groups? Do these proportions turn out to be quite different?

References

1. "The final report on the aspirin component of the ongoing physicians' health study," *The New England J. Medicine* (1989), **231**(3):129–135.
2. Karen Allen and Abigail Moss (1993), "Teenage tobacco use," *Advance Data*, no. 224 (Feb. 1), National Center for Health Statistics.

Matching Descriptions to Scatter Plots

NOTES FOR THE INSTRUCTOR

Statistical Setting

In this activity students learn that the correlation coefficient and regression line can change dramatically when an influential observation is added to the set of data. This important concept is often given short shrift in the introductory statistics class.

Prerequisites for Students

This activity can be used as soon as students have learned about regression and that the correlation measures spread about the least-squares regression line.

Materials

None. This activity is self-contained.

Procedure

Each student should have a copy of the student activity sheet. The activity is best done in small groups.

Sample Results from Activity

1. The descriptions and scatter plots match as follows: a—1, b—5, c—2, d—4, e—3.

The set of points graphed in Scatter plot #1 are as follows:

2.95826	1.66802
3.18038	1.26812
3.89815	3.42542
3.92966	5.91132
4.11099	3.00204
4.26061	5.42325
4.57334	3.42135
5.08053	2.66203
5.36271	6.71185
5.48393	5.84860
5.71988	5.35038
5.73496	6.75695
6.20179	5.96230
6.25993	6.29150
7.37081	8.62256

The additional point (15.5, 19.725) was added for Scatter plot #2.
The point (5, 19.725) was added to Scatter plot #1 to make Scatter plot #3.
To make Scatter plot #4, the y-values of the points in Scatter plot #1 were reversed (so that $y = 1.66802$ is paired with $x = 7.37081$, etc.).
To make Scatter plot #5, the point (8, 10) was added to Scatter plot #4.

2. a. (3,7) 0.73
   b. (2,6) 0.71
   c. (10,0) −0.70
   d. (10,6) 0.22
   e. (10,14) 0.96
   f. (100,0) −0.84
   g. (100,6) 0.02

Sample Results from Wrap-Up

1. Without the United States, the correlation is $r = 0.055$ and the equation of the regression line is $y = 2.41 - 0.056x$. That is, for the European countries there is almost no association between marriage rate and divorce rate. The much higher correlation with all 14 countries is the result of an influential observation, the United States, which is unlike the other countries in many ways.

2. To make the correlation close to 1, place the point far away from the other points but on the regression line. To make the correlation close to 0, place the point far

away from the other points and with the y-value of $\bar{y}$. Alternatively, place the point far above or below the other points and with the x-value of $\bar{x}$.

Sample Results from Extension

Each of the four data sets has the same correlation coefficient, $r = 0.8164965811$ ($r^2 = 2/3$) and the same regression equation, $y = 3.0 + 0.5x$. However, examining the scatter plots makes it clear that this model can be appropriate only for the first data set. The second data set looks quadratic, concave down. The third data set has an influential observation, which should be examined and possibly discarded as the remaining 10 points are collinear. The fourth data set is different from the remaining three as there are only two different values of x. There isn't enough information to speculate on a model. The single point (19, 12.5) determines the slope of the regression line.

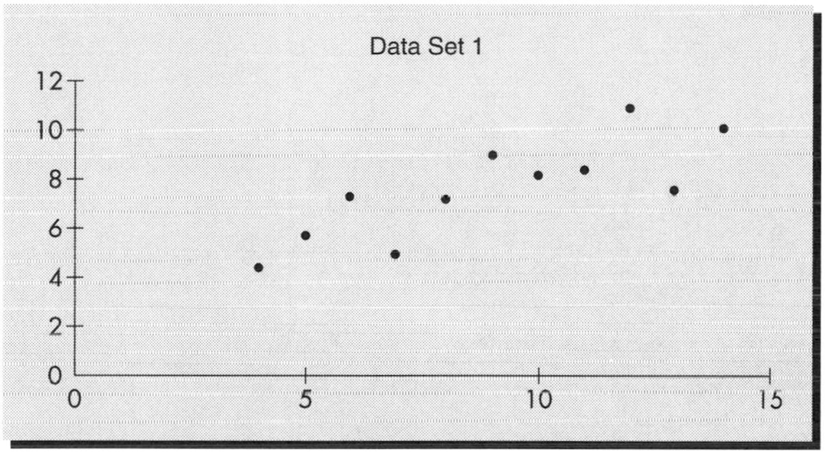

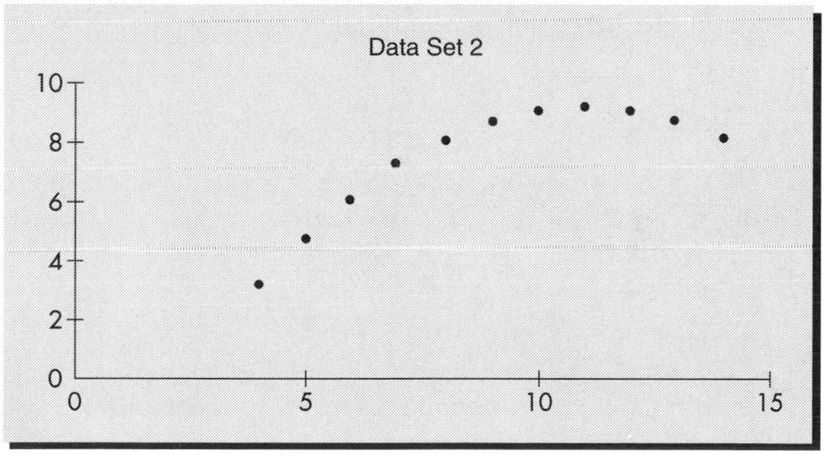

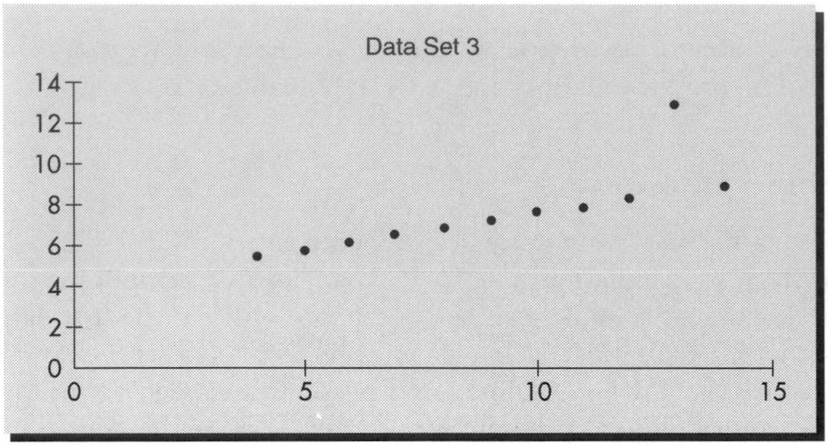

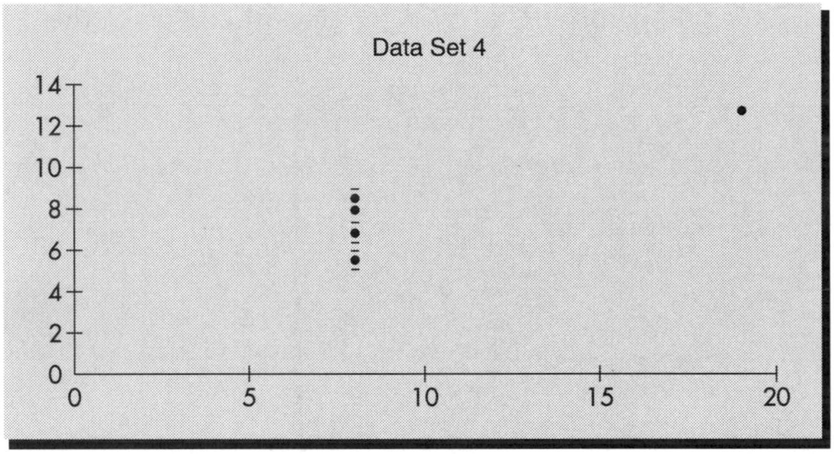

Sample Assessment Question

Draw on a scatter plot an example of a set of points that have negative correlation but so that by removing one point, the correlation becomes positive. Circle the influential point.

Reference

The Anscombe data sets are discussed in Sanford Weisberg (1985), *Applied Linear Regression*, second ed., New York: Wiley.

Matching Descriptions to Scatter Plots

SCENARIO

Table 1 and the scatter plot in Figure 1 show marriage and divorce rates (per 1,000 populations per year) for 14 countries. The correlation coefficient for these data is $r = 0.597$, and the equation of the least-squares regression line is $y = -0.7 + 0.5x$.

Table 1 MARRIAGE AND DIVORCE RATES

Marriage Rate per 1,000 People per Year	Divorce Rate per 1,000 People per Year
5.6	2
6	3
5.1	2.9
5	1.9
6.7	2
6.3	2.4
5.4	0.4
6.1	1.9
4.9	2.2
6.8	1.3
5.2	2.2
6.8	2
6.1	2.9
9.7	4.8

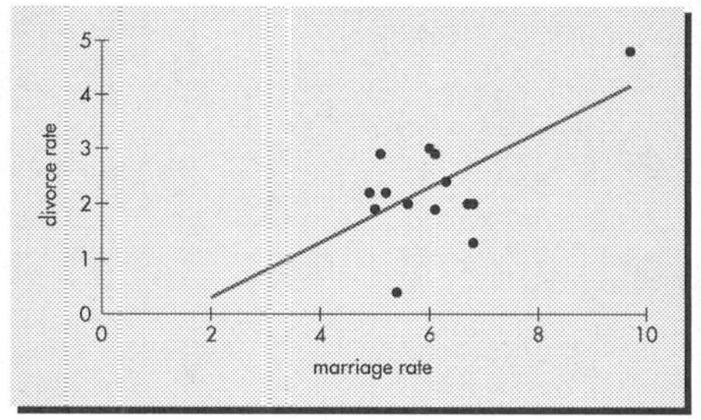

Figure 1: Scatter plot of marriage and divorce rates.

Question

On the basis of the regression line, predict the divorce rate for a country with a marriage rate of 8 per 1,000. How much conviction do you have in this prediction?

Objective

In this activity, you will learn how one point can influence the correlation coefficient and regression line.

Activity

1. Match each of the five scatter plots to the description of its regression line and correlation coefficient. The scales on the axes of the scatter plots are the same.
   a. $r = 0.83$, $y = -2.1 + 1.4x$
   b. $r = -0.31$, $y = 7.8 - 0.5x$
   c. $r = 0.96$, $y = -2.1 + 1.4x$
   d. $r = -0.83$, $y = 11.8 - 1.4x$
   e. $r = 0.41$, $y = -1.4 + 1.4x$

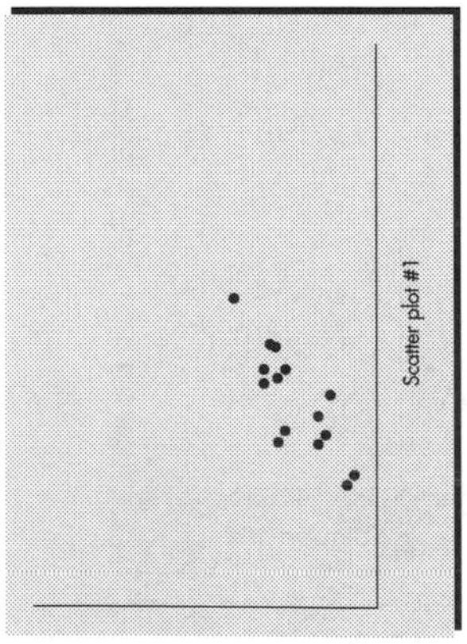

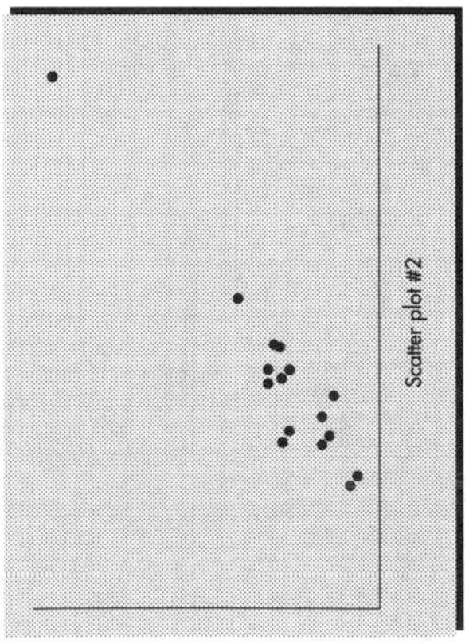

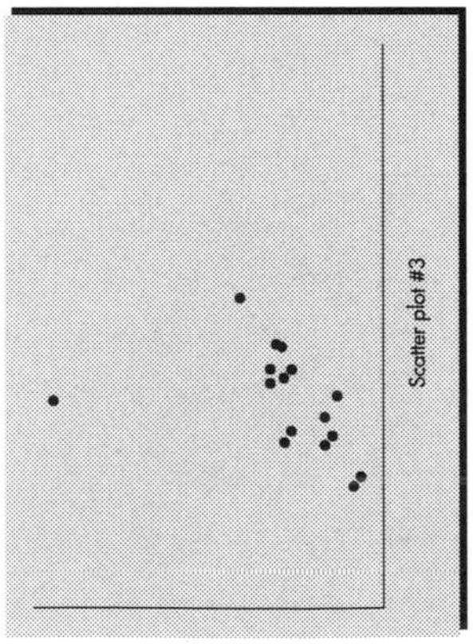

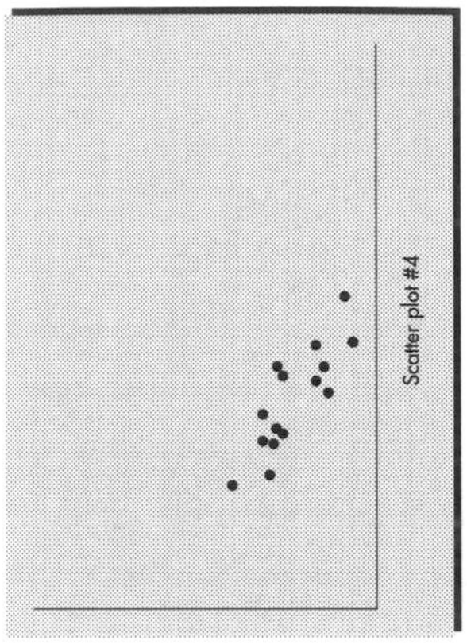

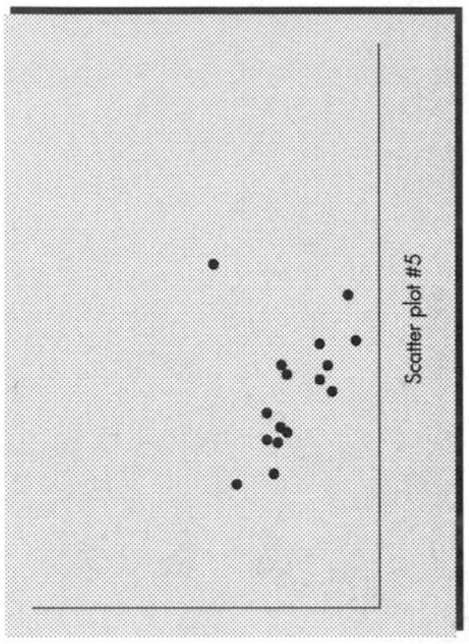

2. For the nine points on the scatter plot below, $r \approx 0.71$, $r^2 = 0.5$, and the equation of the least-squares regression line is $y = 4.00 + 1.00x$.

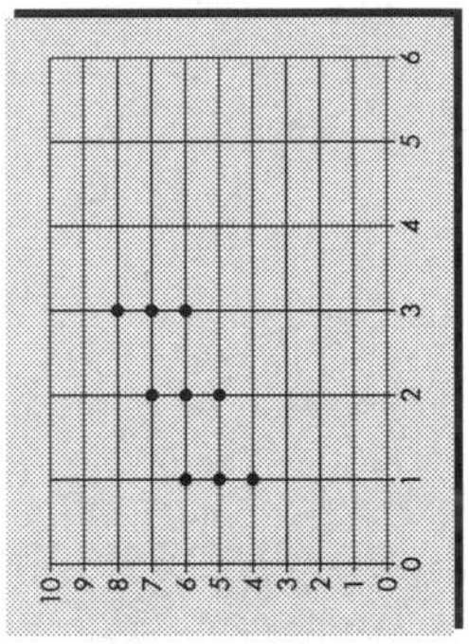

A tenth point is added to the original nine. Match each of the points below with the correlation coefficient that would result if that point were added. Try not to calculate the new correlation coefficient, but rather reason out which r must go with each point.

Correlation coefficients: −0.84   −0.70   0.02   0.22   0.71   0.73   0.96

a. (3, 7)
b. (2, 6)
c. (10, 0)
d. (10, 6)
e. (10, 14)
f. (100, 0)
g. (100, 6)

Wrap-Up

1. Look again at the scatter plot in the scenario. The country at point (9.7, 4.8) is the United States. The other 13 countries are all in Europe. Compute the correlation coefficient and the regression equation when the United States is removed from the data. What is your conclusion?

2. Where would you place a new point on a scatter plot to make the correlation coefficient as close to 1 as possible? As close to 0 as possible?

Extension

Analyze the four data sets in the following table. What do they have in common? Why are they of interest? What do they illustrate?

Data Set 1		Data Set 2		Data Set 3		Data Set 4	
x	y	x	y	x	y	x	y
10	8.04	10	9.14	10	7.46	8	6.58
8	6.95	8	8.14	8	6.77	8	5.76
13	7.58	13	8.74	13	12.74	8	7.71
9	8.81	9	8.77	9	7.11	8	8.84
11	8.33	11	9.26	11	7.81	8	8.47
14	9.96	14	8.10	14	8.84	8	7.04
6	7.24	6	6.13	6	6.08	8	5.25
4	4.26	4	3.10	4	5.39	19	12.50
12	10.84	12	9.13	12	8.15	8	5.56
7	4.82	7	7.26	7	6.42	8	7.91
5	5.68	5	4.74	5	5.73	8	6.89

References

1. Frank Anscombe (1973), "Graphs in statistical analysis," *American Statistician*, **27**:17–21. (This is the source of the data sets in the Extension.)
2. The marriage rate and divorce rate data come from James M. Landwehr and Ann E. Watkins (1995), *Exploring Data*, revised ed. Palo Alto, CA: Dale Seymour Publications. Their sources were the United Nations publications *Demographic Yearbook 1990* and *Monthly Bulletin of Statistics*, June 1991.

RANDOM BEHAVIOR

What's the Chance? Dependent and Independent Trials

NOTES FOR THE INSTRUCTOR

Statistical Setting

This activity can introduce students to the relative frequency definition of probability, as well as demonstrate the difference between dependent and independent trials.

Prerequisites

The activity can be used before or after students have learned the basic concepts in probability.

Materials

Thumbtacks and small dixie cups or small jars are needed for this activity.

Procedure

This is an easy activity that can be completed in the class. It is suitable for both large and small classes. Before you start, you may wish to ask the students to guess the probability that a tack lands point down—have a prize for those who are closest to the answer from the activity. For small classes each student can toss the tack one at a time, while a few may do the second experiment. You need at least 150 to 200 tosses to see the convergence of the relative frequencies.

Sample Results from Activity

The following results were obtained using a short thumbtack and a small jar, about 1.5″ in diameter.

TACKS TOSSED ONE AT A TIME

# of tosses	10	20	30	40	50	60	70	80	90	100	110	120	130	140	...	200
# point down	3	7	13	16	20	27	31	37	44	47	54	59	63	69	...	102
Relative frequency	.3	.35	.43	.4	.4	.45	.44	.46	.49	.47	.49	.49	.49	.49	...	.51

TACKS TOSSED 10 AT A TIME

# of tosses	10	20	30	40	50	60	70	80	90	100	110	120	130	140	...	200
# point down	4	5	11	16	21	24	26	31	34	39	47	52	53	59	...	77
Relative frequency	.4	.25	.36	.40	.42	.40	.37	.38	.37	.39	.42	.43	.41	.42	...	.39

With a small dixie cup, the two sequences converged to .53 and .33. In both cases, tossing tacks 10 at a time reduced the probability that a tack landed point down from the value when a tack was tossed one at a time. This reduction is because the tacks get tangled up when they are tossed simultaneously, and so the tosses are no longer independent. Thus, the second method for estimating probability does not satisfy the assumption of independent trials, and the subsequent probability estimate is therefore not accurate.

Sample Assessment Questions

1. A good basketball player makes about 70% of her free throws. During the course of the season, her cumulative percentage of successful free throws wanders around a bit, but ends up converging to some value close to 70% for the season as a whole. In a typical season, this player gets 100 free throw opportunities. Suppose she stood at the free throw line and tried all 100 of these in succession. Do you think her percentage of successes would still be around 70%? Explain your reasoning.

2. A sample of 1000 students at a university are to be surveyed in an opinion poll. One way to select the sample is to randomly choose 1000 names from a student directory and interview each person so chosen. Another way is to randomly select 10 names from the student directory and then let each of these 10 find 9 others to bring along to the interview. Do you think the sample percentage of students favoring the current policy on parking would be the same for each method? Explain.

What's the Chance? Dependent and Independent Trials

SCENARIO

In many situations, people need the probability that a certain event will occur. For example, a public health inspector may need the probability that a well is polluted, a pollster may wish to find the probability that voters will favor limited terms for elected officials, or a quality-control manager may be interested in the probability that a product is defective in a continuous production process. As in most real-world situations, it is impossible to test all the wells, poll all the people, or check all the products. How can we estimate these probabilities, and how accurate can our estimates be? We will use a thumb tack as our prototype and see how we can estimate the probability that a tack lands point down.

Question

What is the probability that a tack lands point down?

Objective

This activity is designed to show how the probability of an event can be estimated by the relative frequency of the event and how the accuracy of the estimate depends on the method of estimation.

Activity

1. Collecting the data

a. Put a tack in a small jar or cup and toss the tack 10 times. Note the number of times it lands point down, and find the relative frequency of a tack landing point down.

Relative frequency = # of times tack lands point down/# of tosses

This is your estimate of the probability that a tack lands point down.

b. Collect the results of the tosses from the rest of the class. You want to combine the results of the tosses one student at a time and get a sequence of relative frequencies. You will now be able to complete the following table:

# of tosses	10	20	30	40	50 . . .
# landing point down					
Relative frequency					

c. Instead of tossing one tack 10 times, put 10 tacks in the cup or jar and toss the tacks *once*. Again note the number of tacks landing point down and calculate the relative frequency that a tack lands point down in this case. You now have another way to estimate the probability using a different method.

d. Collect the results of these tosses from the rest of the class and complete a table similar to that in part b.

2. Analyzing the data and drawing conclusions

Our main objective is to find the probability that a tack lands point down. *We can define the probability of a tack landing point down as the relative frequency when a tack is tossed independently a large number of times.* So the larger the number of tosses, the better our estimate of the probability.

a. We use the table in 1b to get a sequence of relative frequencies for increasing numbers of tosses. Do the relative frequencies in the sequence in 1b converge toward a single value? Can we use this value as our probability that a tack lands point down? Verify that it satisfies the definition of probability. Complete the following statement: The probability that a tack lands point down is approximately ________.

b. Let us study the sequence in 1d, when we tossed 10 tacks at a time. Complete the following statement using the results in 1d: The probability that a tack lands point down is approximately ________.

Do the two sequences in part 1 converge to the same value and give us the same probability? Are the experiments in 1a and 1c the same? Explain why

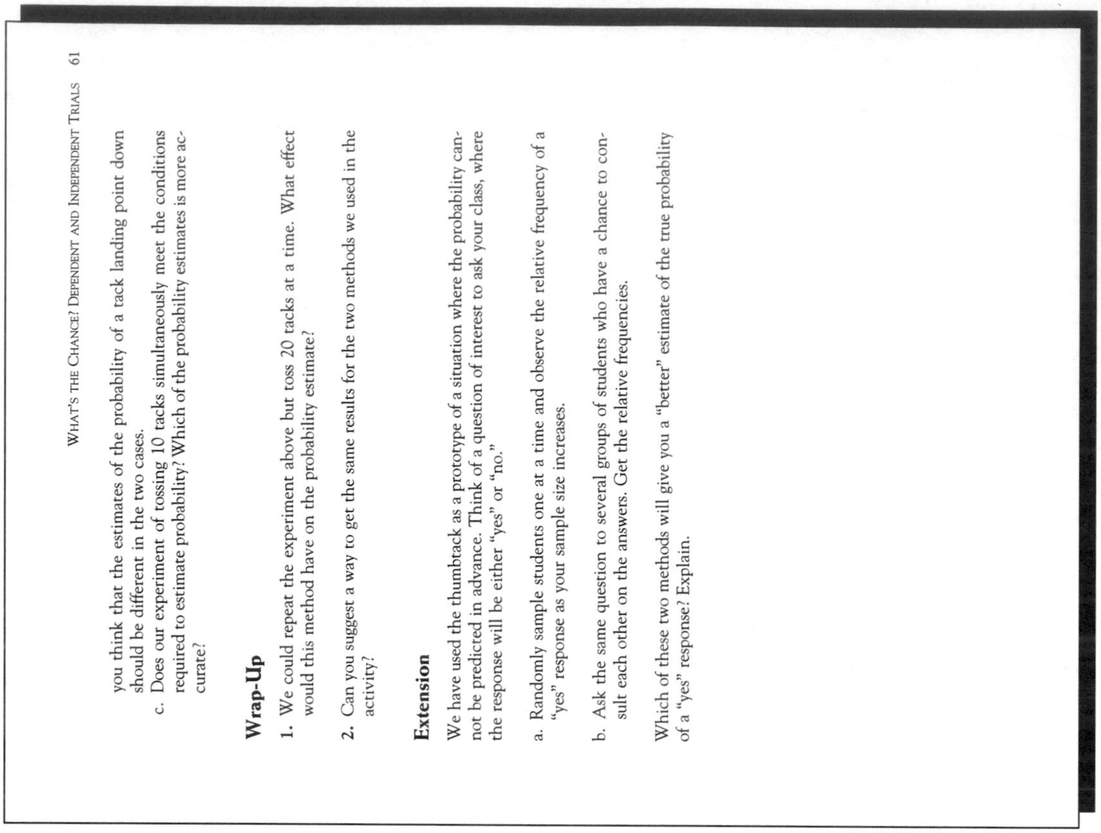

you think that the estimates of the probability of a tack landing point down should be different in the two cases.

c. Does our experiment of tossing 10 tacks simultaneously meet the conditions required to estimate probability? Which of the probability estimates is more accurate?

Wrap-Up

1. We could repeat the experiment above but toss 20 tacks at a time. What effect would this method have on the probability estimate?

2. Can you suggest a way to get the same results for the two methods we used in the activity?

Extension

We have used the thumbtack as a prototype of a situation where the probability cannot be predicted in advance. Think of a question of interest to ask your class, where the response will be either "yes" or "no."

a. Randomly sample students one at a time and observe the relative frequency of a "yes" response as your sample size increases.

b. Ask the same question to several groups of students who have a chance to consult each other on the answers. Get the relative frequencies.

Which of these two methods will give you a "better" estimate of the true probability of a "yes" response? Explain.

What Is Random Behavior?

NOTES FOR THE INSTRUCTOR

Statistical Setting

An activity like this should be used early in the discussion of probability. Two important ideas about independent random events are that a pattern does emerge in the long run (as will be illustrated by the law of large numbers and central limit theorem activities) but that the long-run pattern does not help in predicting outcomes over the short run. This activity is directed toward the second point, which is a difficult point for students to grasp.

Prerequisites for Students

Students should have some knowledge of the relative frequency interpretation of probability.

Materials

A device for generating an event with prescribed probability, such as a bead box or bag of colored candies, is needed.

Procedure

A device, such as a bead box, that generates "yellow" beads with a probability of .6 works well for the random sequence of outcomes. Events with probabilities too close to 1 can be predicted well by many rules; events with probabilities too close to .5 do not show enough variability in the performance of the rules.

The rule "choose yellow every time" should be among the rules suggested. If it is not, you should suggest it, as it will turn out to be as good as any other rule (if yellow has a probability of .5 or more). In the discussion, make clear that the fact that yellow seems to be occurring more than half of the time is important. ("Choose yellow every time" then becomes an optimal rule.) However, the information in particular patterns of occurrence does not help predict the future of independent, random events.

Sometimes, the behavior of the various rules cannot be sorted out with only 25 or 30 trials in the sequence. Longer sequences should then be generated; the analysis of the data from longer sequences could be assigned as homework.

Sample Results from the Activity

The example data here shows 100 selections from a device generating 1's 60% of the time. After the first five selections, the rule

"predict 1 for every selection"

is correct 57% of the time. The rule

"predict the outcome that has maximum frequency over all past trials, breaking ties randomly"

is correct about 42% of the time. Challenge students to try to do better than 60% correct on these data.

1 0 0 1 1 0 1 0 0 0 0 1 0 1 1 0 1 1 0 1 1 1 0 0 1 0 1 1 1 0
0 1 1 1 0 1 0 1 1 1 1 1 1 0 0 1 0 1 0 0 0 1 1 0 0 0 0 0 1 1
1 0 0 1 1 0 1 0 1 1 0 1 1 1 1 1 0 1 0 1 1 1 0 0 1 0 0 0 0 1
1 1 1 1 1 1 0 1 0 1

Sample Results from Extension

Will two people sitting at a square table with four seats most often sit next to each other or across from each other? Sample data from observations on 197 pairs in a university cafeteria produced the following results:

Sitting next to each other	134
Sitting opposite	63

About twice as many pairs sit next to each other as sit opposite each other. What probability model could produce this result? Suppose the members of each pair choose seats at random. After the first person sits down, the second person has three choices, two of which are next to the first person. Thus, the data seen above might simply support the idea that people are choosing their seats at random!

Reference

Joel E. Cohen (1973) "Turning the tables," *Statistics by Example: Exploring Data*, Reading, MA: Addison-Wesley, 87–90.

Sample Assessment Questions

1. Suppose we know that 70% of the students on a large campus are male. Random drawings are to be made to give away five tickets to a concert. Have students discuss if they can predict whether or not the first ticket given will go to a male. Can they do any better in predicting whether or not the fifth ticket will go to a male, given that the first four have already gone to males?

2. Have students discuss how the activity on predicting outcomes of random sequences relates to the process of conducting an opinion poll. Does the long-run relative frequency of responses help us to predict how the next person interviewed will respond? Does the short-run relative frequency of responses help us predict how the next person will respond?

3. Have students find articles from the media in which the "short-term" prediction scheme is used as an argument. In some of these, such as games of chance, the past does not help predict the future. In others, such as rain after a dry spell, the past might provide some information on the future. Students should be able to form clear arguments about which cases look like the random sequence activity and which do not.

4. An entrepreneur is selling advice on how to choose numbers to improve your chances of winning a lottery. If he approaches you as a potential customer, what would you say?

What Is Random Behavior?

SCENARIO

A basketball player who makes 70% of her free throws has missed four in a row. Before her next shot, someone in the stands is sure to say that she is "due" to make this one. You have lost card games to your friends every night this week; you think to yourself, "Well, next week I'm due to win a few." A couple has three children, all boys. "The next one is due to be a girl," a friend remarks. Each of these events, the free throw, the card game, the birth of a child, involves chance outcomes that cannot be controlled. In the long run, the basketball player makes 70% of her free throws, you may win 60% of the card games, and approximately 50% of the babies born are girls. But do these long-run facts help us to predict behavior of random events over the short run?

Question

Can knowledge of past events help us predict the next event in a random sequence?

Objective

In this lesson you will attempt to predict the next event in a random sequence of events, given knowledge of the past events in the sequence. The goal is to see if a pattern can be detected in a short run of random events.

Activity

1. The large box of beads in the classroom contains some that are yellow. Beads are to be selected from the box one at a time and laid on the desk, in the order of selection. Your goal is to come up with a rule for predicting the occurrence of yellow beads. To begin, take a sample of five beads from the box to serve as data for making the first prediction.

2. Construct a rule for predicting whether or not the next bead drawn will be yellow. You may use the information on the first five selections, but your rule must work for any possible sample selection of beads.

3. Record the prediction your rule generates for whether or not the next bead will be yellow. Now, randomly select the next bead. Record whether or not your prediction was correct.

4. Continue making predictions for the next selection and then checking them against the actual selection for a total of at least 25 selections (not including the first 5). Calculate the percentage of correct predictions your rule generated.

5. Compare your rule with others produced in the class. Arrive at a class consensus on the "best" rule for predicting yellow beads.

6. Do the data from the past selections help in making a prediction for the next selection? Does the particular pattern of the sequential selections help in making a prediction for the next selection? What is the best you can hope to do, in terms of correct predictions, with any decision rule? Discuss what was learned about predicting future events in a random sequence.

Wrap-Up

1. Discuss the three scenarios posed at the beginning of this activity. In light of your experience with random sequences, are the statements made about the free throws, card games, and children appropriate?

2. Try steps 1 through 6 of the Activity section over again, using the *same* random device but generating a new and different sequence of outcomes. Compare all of the prediction rules once more. Did the percentages of correct predictions remain the same as before for all rules? Did the rule originally chosen as best remain the best decision rule?

3. Try steps 1 through 6 of the Activity section with another random device generating an event with probability closer to .5. Will your ability to predict get better? Why or why not?

Extensions

1. Picture a square table for four people, with one chair on each side of the square. Two people come to the table to eat together. Do they sit beside each other or across from each other?
   a. Construct a rule for predicting whether or not two people sitting down at a square table will sit across from each other.
   b. Go to a cafeteria that has a number of square tables for four to check out your rule. What fraction of the time were you correct?
   c. Under the assumption that people choose seats at random, what fraction of the time should two people sit across from each other? How does this fraction compare with your answer to part b?

2. You are playing a card game in which some of the cards are dealt face up. Using a standard 52-card bridge deck the chance of a face card on a single draw is 12/52, or about .23. After the game has progressed a while, can you predict whether or not the next card will be a face card with a rule that is accurate more than 23% of the time? Explain.

Streaky Behavior: Runs in Binomial Trials

NOTES FOR THE INSTRUCTOR

Statistical Setting

This activity can be used as an introduction to a unit on probability and to the idea of a probability distribution.

People have many misconceptions about binomial trials. Often people do not realize how long the longest streak of successes (or failures) in a sequence of binomial trials typically is. Consequently, they feel the need to explain the streaks by abandoning the notion that they are actually observing independent binomial trials. They speak of lottery numbers, gambling tables, and sports players as being "hot" or "cold." This activity confronts students with this common misconception by showing them that most people do not realize how long the longest run of heads will be in a sequence of 200 flips of a fair coin.

Students tend to believe in streaks and slumps in sports and may get all worked up if told that when sports performances are analyzed by statisticians, the sequences look like binomial trials. That is, the lengths of streaks and slumps are no longer than we would expect if we were flipping a coin or spinning a spinner.

Prerequisites for Students

In this lesson, students will have to generate a sequence of 200 flips of a fair coin. If they are comfortable with simulation, they can do this quickly by using a random digit table. If not, the use of the random digit table can be taught at the same time.

Materials

One coin per student and a method of generating random digits (a table, calculator, or computer) are needed for this activity.

Procedure

This activity works best with large classes. If your class is small, you may want each student to do step 4 several times each so that you have 100 or so values in the probability distribution of step 5.

Students should do steps 1, 2, and 4 individually. From then on, they can work individually, in small groups, or as a class. In any case, results from steps 3 and 4 must be shared among the entire class.

Sample Results from Activity

1. Answers will vary.

2. Answers will vary, but most students will select the second sequence. In one typical class of 30 students, 23 students selected the second sequence and 7 the first one. The majority was able to convince the minority that they must be wrong!

3. Often students will choose one sequence over the other based on the longest run of heads or tails. The first sequence has a run of nine heads, and the second has a run of four heads. Most students select the second sequence as the real one because the run of nine heads in the top line of the first sequence looks suspicious.

4. Answers will vary. Beware of students faking their sequence of flips or the whole point will be lost.

5. Results from a simulation will vary somewhat, but the distribution should look something like this:

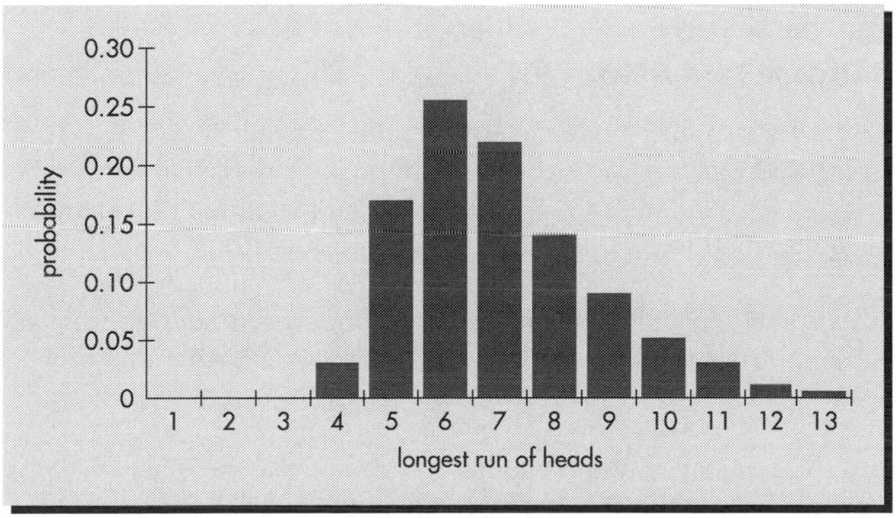

The expected longest length of a head run is about 6.6. In general, for a sequence of n binomial trials with probability p of a success and q of a failure, the mean length of the longest run of heads is approximately $\log_{1/p}(nq)$.

6. Students should now see that the first sequence is the real one. A longest head run of four, which the second sequence has, is fairly unlikely, while nine, which the first sequence has, is much more likely.

7. Most students construct sequences in which the longest run of heads and the longest run of tails are shorter than in a random sequence.
At this point construct a frequency distribution or dot plot of the longest run of heads from the students' faked sequences in step 1. Here are the results from the class of 30 students:

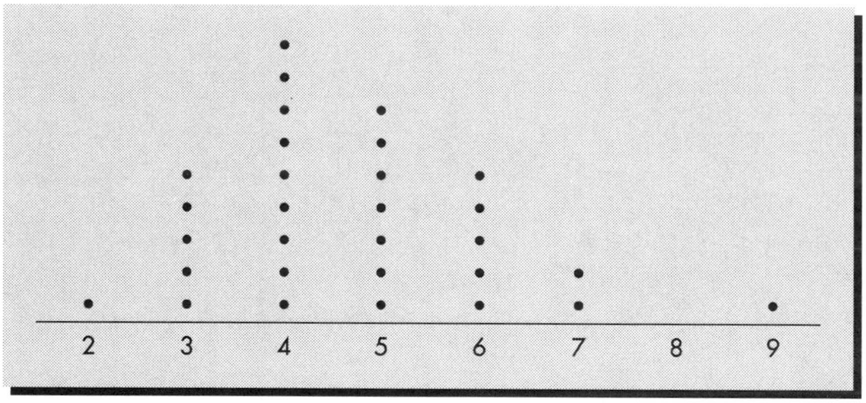

By comparing this distribution with the simulated one in step 5, the students were able to see that the longest length of a head run in their faked sequences tended to be too short.

Sample Results from Wrap-Up

1. As we saw in Activity step 5, the average longest run of heads is 6.6. We would begin to suspect that the player was exhibiting unusually streaky behavior if the player made 11 or more shots in a row. We saw from our histogram that the probability of this is quite small for 200 binomial trials with $p = 0.5$.

2. As we saw in this activity, people don't realize how long streaks generated from a sequence of binomial trials tend to be. Thus they look for explanations of what they perceive as unusually streaky behavior.

3. It should be exactly the same. The probability of heads is the same as the probability of tails. It's all the same to the coin.

Sample Results from Extension

1. Before students begin this activity, remind them that the distribution won't be the same as in Activity step 5 unless the player happens to be a 50% shooter. Students will have to generate a distribution like that of step 5 for the percentage for their player. For that reason, this activity should be done by a group of students.

   Statisticians have examined sequences of field-goal attempts and free-throw attempts and, in general, have found that they look like sequences generated randomly from a binomial distribution. Players do not exhibit streaky behavior. (See the References.)

2. Here are the distributions.

   For $n = 1$:

Longest Run of Heads	Probability
0	1/2
1	1/2

   For $n = 2$:

Longest Run of Heads	Probability
0	1/4
1	2/4
2	1/4

   For $n = 3$:

Longest Run of Heads	Probability
0	1/8
1	4/8
2	2/8
3	1/8

   For $n = 4$:

Longest Run of Heads	Probability
0	1/16
1	7/16
2	5/16
3	2/16
4	1/16

For $n = 5$:

Longest Run of Heads	Probability
0	1/32
1	11/32
2	12/32
3	5/32
4	2/32
5	1/32

Observations will vary, but students may notice that the distributions are becoming more skewed to the right, that the most frequent value is increasing rather slowly, that the probability that the longest run of heads will be either 0 or n is $1/2^n$, and that the probability that the longest run of heads will be $n - 1$ is $2/2^n$.

Sample Assessment Questions

To answer the first two questions, students will need a copy of the histogram of the probability distribution (including the mean) constructed by the class in Activity step 5.

1. Estimate the probability that there will be 8 or more heads in a row when a coin is flipped 200 times.

2. In this activity you constructed a histogram of the longest run of heads in a sequence of 200 coin flips. The probability of a head was 0.5 on each flip. Suppose instead you were constructing a histogram of the longest run of heads in a sequence of 200 coin flips for a weighted coin for which the probability of a head was 0.6 on each flip. How would this new histogram be different from the one you constructed? How would it be the same?

3. A book review in the *Los Angeles Times* of December 29, 1993, about discrimination against upper-class African-Americans contains the following example:

   > A Harvard law students says her professor tended to ignore the raised hands of the black students in class—and then, suddenly, he would call on several black students in a row: "As if," she explains, "the professor had suddenly realized that he was neglecting an important segment of the student body and had resolved to make amends."

   Discuss this quotation in the light of what you have learned from this activity.

References

You may want to ask students to read and report on some of these articles:

1. "Slumps," *Money* (Feb. 1985), p. 12.
2. Amos Tversky and Thomas Gilovich (1989), "The cold facts about the 'hot hand' in basketball," *Chance*, **2**(1):16–21.
3. Patrick D. Larkey et al. (1989), "It's okay to believe in the 'hot hand,' " *Chance*, **2**(4):22–30.
4. Amos Tversky and Thomas Gilovich (1989), "The 'hot hand': Statistical reality or cognitive illusion?" *Chance*, **2**(4):31–34.
5. Robert Hooke (1989), "Basketball, baseball and the null hypothesis," *Chance*, **2**(4):35–37.
6. Mark F. Schilling (1990), "The longest run of heads," *College Mathematics Journal*, **21**:196–207.

    Schilling's article gives the formulas for the probability distributions of the longest run of successes (and failures) for different numbers of trials and probability of success:

7. Christian Albright (1992), "Streaks & Slumps," *OR/MS Today* (April):94–95.

    The author ends, "For now, we simply conclude by stating that if there are perennially streaky hitters, the present analysis was not able to find them."

8. S.C. Albright (1993), "A statistical analysis of hitting streaks in baseball (with discussion)," *The American Statistician*, **88**:1175–1183.

Streaky Behavior: Runs in Binomial Trials

SCENARIO

When they have made several baskets in succession, basketball players are often described as being "hot." When they have been unsuccessful for a while, they are described as being "cold" or "in a slump." Fans and basketball players alike tend to believe that players shoot in streaks. That is, players have long periods when they are shooting better than we would expect followed by long periods when they aren't doing as well as we would expect. In this activity you will begin to evaluate whether it is true that basketball players exhibit streaky behavior in shooting. The first step is to learn to recognize a streak of unusually successful or unsuccessful shooting.

Question

How can we recognize a streak of unusually successful basketball shooting?

Objective

In this activity you will study the distribution of the length of the runs of successes in a sequence of binomial trials.

Activity

1. Without using a coin or a table of random digits, write down a sequence of heads (H) and tails (T) that you think looks like the results that one would get from actually flipping a coin 200 times.

2. In the following two sequences of 200 coin flips, H stands for a head and T for a tail. One of the sequences is the result from actually tossing a fair coin. The other was made up. Which do you think is the real sequence?

    THTTTTHHTHTHTTTTTHHTHTHHHHHHHHHTHTHTHHHTHHHT
    THHTHTHHHHHTHHTHHTTHTHTHTTTTTTHHTHTHTHTTHTTT
    THHTTTTHTTTTHTTTTHHHTHTHHTTTHTTTHTTHHTHHHTHHH
    HHTHHHTTHHHTHHTTTTTTTHTTTHTTTHTTHTTTHTTHHHHTH
    HHTTTHTTTTTTHHTHTTTHH

    THTHTTTHTTTTHHTHTTTHTTHHHTHHTHTHTHTTTTHHTTHHT
    THHHTHHHTTHHHTTTHHHTHHHHTTTHTHTHHHHTHTTHHHTH
    HTHTTTHHTHHHTHHHHTTHTHHTHHHTTTHTHHHTHHTTTHHH
    TTTTHHHTHTHHHHTHTTHHTTTTHTHTHTTHTHHTTHTTTHTTT
    THHHHTHTHHHTTHHTHHTHH

3. How many from your class chose the first sequence? the second? What were the reasons for the choices?

4. Construct a real sequence of 200 coin tosses by tossing a coin or using a random digit table. What is the longest run of heads in your sequence?

5. Get the results for step 4 from all members of your class, and make a frequency table of the lengths of the longest run of heads. From the frequency distribution, make a plot of the lengths. Compute an estimate of the mean length of the longest run of heads in a sequence of 200 coin tosses. Mark this mean on the histogram.

6. Which sequence in step 2 do you now think is the real one? Explain.

7. Does your impostor sequence from step 1 resemble a real sequence of coin tosses? In what way is it different?

Wrap-Up

1. If a basketball player is a 50% field-goal shooter and shoots 200 times in a series of games, what would you expect the longest streak of baskets to be if the player doesn't exhibit unusually streaky behavior? How could you tell if the player was "hot?"

2. What could be some of the reasons why people tend to believe in streak shooting?

3. Would a frequency table of the lengths of the longest run of *tails* be the same as or different from that of the longest run of heads? Explain your reasoning.

Extensions

1. Pick a basketball player whom you can follow for a few games and who has a well-established field-goal percentage. Using this percentage and random digits, make a frequency table of the longest run of heads in a sequence of 200 attempts. After following the player for 200 field-goal attempts, do you see any evidence of an unusually long streak?

2. Construct the probability distribution for the length of the longest run of heads when a coin is tossed one, two, three, four, five, and six times. For example, when a coin is tossed two times, the possible outcomes are HH, HT, TH, and TT. The longest run of heads in these outcomes are 2, 1, 1, and 0, respectively. The probability distribution for the longest run of heads is as follows:

Longest Run of Heads	Probability
0	1/4
1	2/4
2	1/4

After examining your probability distributions, what conjectures can you make?

Reference

Mark F. Schilling (1994), "Long-run predictions," *Math Horizons*, Spring, pp. 10–12.

The Law of Averages

NOTES FOR THE INSTRUCTOR

Statistical Setting

The phrase "according to the law of averages" appears frequently, incorrectly, and inconsistently in literature, sports commentary, and everyday speech. This activity asks students to examine such statements critically in order to prepare them for a correct statement of the law of large numbers.

This activity can serve as an introduction to probability. Concepts that may be explored include independent events, the law of large numbers, and expectation.

Prerequisites

None.

Materials

This activity requires a pair of scissors for each group. For Extension #3, groups may need additional equipment such as dice.

Procedure

This activity is best done in small groups. If you have a computer available, use it to generate a (longer) sequence for Activity step 3 to illustrate that the percentage of heads tends to 50% while the difference between the expected number of heads and the actual number of heads tends to infinity. This is an important point for students to understand. People tend to believe that events are "due" because they don't think that the proportion of successes can get closer to p unless the number of successes gets closer to np.

Sample Results from Activity

1. Students may have heated discussions about this activity. One possible grouping is as follows:

   - Statements that convey the idea that an unlikely event must happen given enough opportunities:

     Colin Dexter, *MacWorld*, *Boy Meets World*, Anne Tyler, Michael Innes

   - Statements that convey the idea that an event is "due" if it hasn't happened in a while:

     Dear Abby, Vin Scully, Pamela Naber Knox, Antonia Fraser

   - Statements that convey the idea that events even out in order to maintain some "normal" average:

     Kiner, Alison Lurie, Thomas Pynchon

   - Miscellaneous statements:

     FM radio program, Los Angeles Times Magazine

2. The only version of the law of averages that is a correct statement is the first one, that even an unlikely event will eventually happen given enough opportunities.

3. a. 5
   b. 4; 40%
   c. 25
   d. 28; 56%

4. 100 times (Questions 4 and 5 are adapted from David Freedman et al. (1991), *Statistics*, second ed., New York: W.W. Norton, p. 249.)

5. 10 times

6. a. They are equally likely.
   b. 2 + 8/2 = 6 heads are expected; 60%. This is also an important idea for students to understand. The coin doesn't remember that the first two tosses were heads, and so we expect one-half of the remaining 8 flips to be heads.
   c. 501; 50.1%

d. The expected number of heads is always one more than half the number of flips. The expected percentage of heads gets closer to 50%, but always stays slightly more than 50%.

Sample Results from Wrap-Up

1. No, it says that the percentage of heads should get closer to 50%, which is not the same thing. For example, looking at the chart below, we see that the percentage of heads is getting closer to 50%, but the number of heads and the number of tails do not even out.

Number of Tosses	Number of Heads	Number of Tails	Percentage Heads
10	6	4	60
100	55	45	55
1,000	540	460	54
10,000	5,300	4,700	53

2. No, it isn't. Chick Hearn believes that Perkins is "due" to miss because he has made six free throws in a row. Statisticians have analyzed sequences of free throws and believe that each free throw is independent of the others. (See the references for the activity "Streaky Behavior: Runs in Binomial Trials.") If Perkins is an 80% free-throw shooter, the probability he makes this seventh free throw is 80%.

Sample Results from Extensions

1. Results from the survey will vary.

2. This question is adapted from Daniel Kahneman et al. (1982), *Judgment Under Uncertainty: Heuristics and Biases*, Cambridge: Cambridge Univ. Press.
   a. 100
   b. $[150 + 49(100)]/50 = 101$

3. Experiments will vary.

Sample Assessment Questions

1. A roulette wheel has 18 black positions and 18 red positions. A gambler observes six consecutive reds and then bets heavily on black because "black is due." Is his reasoning correct? Explain.

2. After hearing your explanation to Question 1, the gambler moves on to a poker game. He is dealt four red cards. He remembers what you said and assumes that the next card dealt to him is equally likely to be red or black. Is the gambler right or wrong? Why?

Answers to Sample Assessment Questions

This question is adapted from one in David S. Moore and George P. McCabe (1993), *Introduction to the Practice of Statistics*, second ed., New York: Freeman.

1. No, it isn't correct. On each spin of the wheel, red and black are equally likely; the roulette wheel has no "memory."

2. He is wrong. The deck of cards does, so to speak, have a memory. Four red cards are gone from the original deck of 52. So the chance he will be dealt a red card next is 22/48. He is more likely to get a black card.

Reference

William Feller (1968), *An Introduction to Probability Theory and Its Applications*, Vol. I, third ed., New York: Wiley.

The Law of Averages

SCENARIO

Have you ever heard someone appeal to the "law of averages"?

- As many times as she's been guilty before, she *has* to be innocent this time—it's the law of averages. You should know that. You want to go to law school. (Tom on *The Tom Arnold Show*, March 9, 1994)
- President Reagan explained why there is corruption in the Pentagon's purchasing process: "The law of averages says that not all 6 million [people involved] are going to turn out to be heroes." (*Pittsburgh Press*, June 18, 1988)
- Disobey the law of averages. Let others take the traditional course. We prefer creativity over conformity. Invention over imitation. Inspired ideas over tired ideas. In short, Audi offers an alternative route. (From a print advertisement for Audi)

Question

What do people mean when they use the term "law of averages"?

Objective

In this activity you will examine how people use the term "law of averages" to make correct and incorrect inferences about probability. Then you will explore the consequences of the law of large numbers, a correct statement about probability.

Activity

1. There are several popular versions of the law of averages. Cut apart the quotations on the pages "Statements About the Law of Averages" and sort them into piles according to what the author or speaker meant when using the phrase the "law of averages." For example, one pile might include those statements that are supposed to convey the point that given enough opportunities, even unlikely events must eventually happen. The Reagan statement would go into this pile. A second pile might include those statements that contend that if an event has not happened on several previous opportunities, it is much more probable that it will happen on the next opportunity. Tom Arnold's statement would go into this pile. A third pile might include those statements that use the following definition:

   law of averages: the proposition that the occurrence of one extreme will be matched by that of the other extreme so as to maintain the normal average. (*Oxford American Dictionary*, 1980)

   Miscellaneous uses, such as the Audi ad, might go into a final pile.

2. Which of the above versions of the "law of averages" are correct interpretations of probability?

3. Here is a sequence of flips of a fair coin:

   T T H H T H T T H T H H T T T H T H H H H T T T H H T H H H H H T
   H T H T T H H H T H T H H T H H T

   a. How many heads would you expect to have after 10 flips of a fair coin? ("Expect" is a technical term that means about the same as "average." This question means, "On the average, how many heads do people get after flipping a coin 10 times?")
   b. In the sequence above, what is the actual number of heads after 10 flips? What is the percentage of heads after 10 flips?
   c. How many heads would you expect to have after 50 flips of a fair coin?
   d. In the sequence above, what is the actual number of heads after 50 flips? What is the percentage of heads after 50 flips?

What many people find counterintuitive about the answers to question #3 is that the percentage of heads gets closer to 50% while the actual number of heads gets further away from half the total number of flips. In fact, the law of large numbers tells us that this is what we can expect.

The law of large numbers says that as a fair coin is flipped more and more times,

- the percentage of heads tends to get closer to 50%.
- the number of heads tends to swing more and more wildly about the expected number of heads (which is half the total number of flips).

4. You will win a prize if you toss a coin a given number of times and get between 40% and 60% heads. Would you rather toss the coin 10 times or 100 times, or is there any difference?

5. You will win a prize if you toss a coin a given number of times and get exactly 50% heads. Would you rather toss the coin 10 times or 100 times, or is there any difference?

6. Suppose you plan to flip a coin indefinitely. The first two flips are heads.
   a. Is a head or a tail more likely on the next flip?
   b. How many heads do you expect to have at the end of 10 flips? What is the expected percentage of heads at the end of 10 flips?
   c. How many heads do you expect to have at the end of 1,000 flips? What is the expected percentage of heads at the end of the 1,000 flips?
   d. As you keep flipping the coin, what happens to the expected number of heads? What happens to the expected percentage of heads?

Wrap-Up

1. Does the law of large numbers imply that if you toss a coin long enough, the number of heads and the number of tails should even out? Explain.

2. As Los Angeles Laker Carl Perkins comes up for a seventh free throw, announcer Chick Hearn notes that Perkins had made the last six out of six free throws and concludes that "the law of averages starts working for Golden State" (December 15, 1990). What does Chick Hearn mean? Is this a correct interpretation of probability?

Extensions

1. Design a survey to determine what a typical person thinks the phrase "law of averages" means.

2. a. The average math achievement test score of the population of eighth graders in a large city is known to be 100. You have selected a child at random. Her score turns out to be 110. You select a second child at random. What do you expect his or her score to be?
   b. The average reading achievement test score of the population of eighth graders in a large city is known to be 100. You have selected 50 children randomly. The first child tested has a score of 150. What do you expect the mean score to be for the whole sample?

3. Design an experiment to test this version of the law of averages: In gambling games, if an event hasn't happened at the last few opportunities, it is more likely to happen at the next opportunity.

References

1. David Freedman, et al. (1991), *Statistics*, second ed., New York: W. W. Norton, p. 249.
2. Myles Hollander and Frank Proschan (1984), *The Statistical Exorcist: Dispelling Statistics Anxiety*, New York: Marcel Dekker, pp. 203–204.
3. David Moore (1985), *Statistics: Concepts and Controversies*, second ed., New York: Freeman, pp. 266–267.
4. Ann E. Watkins (1995), "The law of averages," *Chance Magazine*, Spring.

Counting Successes: A General Simulation Model

NOTES FOR THE INSTRUCTOR

Statistical Setting

This lesson involves quite simple simulations and is designed to introduce the topic of simulation. It should come early in the discussion of probability, and could even serve as an introduction to probability. The concept of a probability distribution should be emphasized, rather than the calculation of probabilities for isolated events. The expected value is then one measure of the center of that distribution.

Prerequisites for Students

None, other than some elementary knowledge of fractions, percents, and proportions.

Materials

A device, such as a random number table or generator, that will produce outcomes with various probabilities is needed.

Procedure

A key to understanding simulations is to have students follow a standard procedure. First, identify the basic random component and find a device that will generate outcomes with this probability. Second, understand how many of these basic components are necessary for one trial of the simulation (one in the case of purchasing one soft drink, 10 in the case of taking a 10-question multiple-choice test). Third, identify what is being counted as a success and accumulate the total number of successes across the trial. This results in one random outcome. The whole procedure must be repeated many times to generate a simulated distribution of outcomes.

Simulation is a very powerful tool, but it must be introduced with simple examples before moving to the more difficult ones. The waiting time (geometric) distribution is introduced in the lesson "Waiting for Reggie Jackson."

Sample Results from the Activity

A typical simulated distribution of the number of correct answers for the five-question true–false test looks something like the one shown below. Here, we get 3 or more correct answers 23 times out of 50, or 46% of the time. The average number of correct answers per trial is 2.4.

```
Stem-and-leaf of Correct   N = 50
Leaf Unit = 0.10
     2   0  00
    13   1  00000000000
  (14)   2  00000000000000
    23   3  00000000000000
     9   4  00000000
     1   5  0
```

A simulation of the multiple-choice test of Wrap-Up #2 may look something like the one below. Here, the chance of 6 or more correct answers is only 1/50, or .02. The average number of correct answers per trial is 2.20

```
Stem-and-leaf of Correct   N = 50
     5   0  00000
    17   1  000000000000
  (11)   2  00000000000
    22   3  000000000000000
     7   4  0000
     3   5  00
     1   6  0
```

Sample Assessment Questions

1. The star free-throw shooter on the girls' basketball team makes 80% of her free throws. She gets about 10 such shots per game.
   a. Set up and conduct a simulation that shows the approximate distribution of the number of successful free throws per game for this player.
   b. What is the approximate probability that she makes more than 80% of her free throws in any one game?

c. How many free throws should she expect to make in a typical game?
d. Over the course of a 15-game season, how many points should this player expect to have from free throws?
e. Are there any assumptions built into the simulation that might not be realistic? Explain.

2. About 33% of the people who come into a blood bank to donate blood have type A+ blood. The blood bank under study gets about 20 donors per day.
a. Set up and conduct a simulation that shows the approximate distribution of the number of A+ donors per day coming into the blood bank. Could you conveniently use some device other than random numbers here?
b. If the blood bank needs 10 A+ donors tomorrow, is it likely to get them?
c. How many A+ donors can the blood bank expect to see each day?
d. Are there any assumptions built into the simulation that seem unrealistic? Explain.

3. Read the attached article, "Love is not blind, and study finds it touching." For the 72 blindfolded people in the forehead test, simulate the distribution of the number of correct decisions that would be made had each of them just been guessing. Where does the observed value of 58 fall on this distribution? Do you agree with the conclusion that most people were not guessing but instead could actually recognize their mates?

Reference

Mrudulla Gnanadesikan, Richard L. Scheaffer, and Jim Swift (1987), *The Art and Techniques of Simulation*, Palo Alto, CA: Dale Seymour Publications.

Love is not blind, and study finds it touching

Associated Press

NEW YORK—How well do lovers know each other? A new study suggests that if blindfolded, they might recognize each other just by feeling their partners' foreheads.

And if he's a man, touching his hand might do.

Seventy-two blindfolded people in the study tried to distinguish their romantic partner from two decoys of similar age, weight and height.

The blindfolded participants stroked the back of each person's right hand in one test, and the forehead in another. Each time, they were asked to pick out the lover.

Random guessing would be right 33 percent of the time. But the blindfolded people were correct 58 percent of the time in the forehead test, and women identified their man's hand 69 percent of the time.

"I think that in real life we could probably do a whole lot better," said researcher Marsha Kaitz. The stress of being in a laboratory experiment and the carefully matched decoys probably hindered the real-world ability of recognition by touch, she said.

Blindfolded couples have feel for relationships.

"I think that probably everyone can do it," Kaitz, a psychologist at Hebrew University in Jerusalem, said in a telephone interview. Touch recognition is "just a skill that has not been tapped before," she said.

Men did not show evidence of recognizing their partner's hands. Kaitz said women probably did better because the hair on men's hands made them more distinctive.

Tiffany Field, director of the Touch Research Institute at the University of Miami School of Medicine, said the finding made sense to her. Touch is an important sense in intimate contact, she said.

The experiment involved 36 heterosexual couples in their 20s who had been in their relationship for an average of two years.

Sixteen of the couples were married, and a total of 25 were living together.

Kaitz said that since the couples were relatively new in their relationship, it is possible that the touch recognition they showed is present only in the "getting to know you" phase of a relationship. She is now studying long-married couples, she said.

Source: *The Gainesville Sun, Monday, June 22, 1992*

Counting Successes: A General Simulation Model

SCENARIO

A basketball player goes to the free-throw line 10 times in a game and makes all 10 shots. A student guesses all the answers on a 20-question true–false test and gets 18 of them wrong. Twelve persons are selected for a jury, and 10 of the 12 are female. Unusual? Some would say that the chances of these events happening are very small. But, as we have seen, calculating the probabilities to evaluate the chances of these events requires some careful thought about a model. The three scenarios outlined above all have certain common traits. They all involve the repetition of the same event over and over. They all have as a goal counting the number of "successes" in a fixed number of repetitions. This activity discusses how to construct a simulation model for events of this type so that we can approximate their probabilities and decide for ourselves whether or not the events are unusual.

Question

How can we simulate the probability distribution of "the number of successes out of n repetitions"?

Objectives

In this lesson you will learn to recognize probability problems with outcomes of the form "number of successes in n repetitions of an event," simulate the distribution of such outcomes, and use simulated distributions for making decisions.

Activity

1. Read the article from the *Milwaukee Journal* (May 1992) entitled "Non-cents: Laws of probability could end need for change," which follows.
   a. Does this seem like a reasonable proposal to eliminate carrying change in your pocket?
   b. Do you think the proposal is fair? Explain your reasoning.

Non-cents: Laws of probability could end need for change

Chicago, Ill.—AP—Michael Rossides has a simple goal: to get rid of that change weighing down pockets and cluttering up purses.

And, he says, his scheme could help the economy.

"The change thing is the cutest aspect of it, but it's not the whole enchilada by any means," Rossides said.

His system, tested Thursday and Friday at Northwestern University in the north Chicago suburb of Evanston, uses the law of probability to round purchase amounts to the nearest dollar.

"I think it's rather ingenious," said John Deighton, an associate professor of marketing at the University of Chicago.

"It certainly simplifies the life of a businessperson and as long as there's no perceived cost to the consumer it's going to be adopted with relish," Deighton said.

Rossides' basic concept works like this:

A customer plunks down a jug of milk at the cash register and agrees to gamble on having the $1.89 price rounded down to $1 or up to $2.

Rossides' system weighs the odds so that over many transactions, the customer would end up paying an average $1.89 for the jug of milk but would not be inconvenienced by change.

That's where a random number generator comes in. With 89 cents the amount to be rounded, the amount is rounded up if the computerized generator produced a number from 1 to 89; from 90 to 100 the amount is rounded down.

Rossides, 29, says his system would cut out small transactions, reducing the cost of individual goods and using resources more efficiently.

The real question is whether people will accept it.

Rossides was delighted when more than 60% of the customers at a Northwestern business school coffee shop tried it Thursday.

Leo Hermacinski, a graduate student at Northwestern's Kellogg School of Management, gambled and won. He paid $1 for a cup of coffee and a muffin that normally would have cost $1.30.

Rossides is seeking financial backing and wants to test his patented system in convenience stores.

But a coffee shop manager said the system might not fare as well there.

"Virtually all of the clientelle at Kellogg are educated in statistics, so the theories are readily grasped," said Craig Witt, also a graduate student. "If it were just to be applied cold to average convenience store customers, I don't know how it would be received."

Source: Milwaukee Journal, May 1992

2. Investigate a single random outcome per trial.
   Suppose the soft drink machine you use charges $.75 per can. The scheme proposed by Mr. Rossides requires you to pay either $0 or $1, depending on your selection of a random number. You select a two-digit random number between 01 and 00 (with 00 representing 100). If the number you select is 75 or less, you pay $1. If the number you select is greater than 75, you pay nothing.
   a. From a random number table, calculator, or computer, choose a random number between 01 and 100. If this represents your selection at the drink machine, how much did you pay for your drink?

b. The article suggests that things will even out in the long run. Suppose that over a period of time you purchase 60 drinks from this machine and use the random mechanism for payment each time. This can be simulated by choosing 60 random numbers between 01 and 100. Make such a selection of 60 random numbers.
   i. How many times did you pay $1? What is the total amount you paid for 60 drinks?
   ii. If you had paid the $.75 for each drink, how much would you have paid for 60 drinks? Does the scheme of random payment seem fair?

c. Now, suppose you are buying a box of cookies that cost $2.43. You pay either $2 or $3, depending on the outcome of a random number selection.
   i. For what values of the random number should you pay $2? For what values of the random numbers should you pay $3?
   ii. Since you will pay $2 in any case, the problem can be reduced to one similar to the soft drink problem by looking only at the excess you must pay over $2. This excess amount will be either $0 or $1, just as in the case of the soft drinks. Using the rule you determined in i, simulate what will happen if you and your friends buy 100 boxes of these cookies. How many times did you have to pay the excess of $1?
   iii. How much did you and your friends pay in total for the 100 boxes of cookies from the simulation? How much would you have paid for the 100 boxes if you had paid $2.43 per box? Does the randomization scheme seem fair?

3. Investigate many random outcomes per trial.
Because you have not studied for the true–false test coming up in history, you will not know any of the answers. You decide to take the test anyway and guess at all the answers. Is this a wise decision? Let's investigate by simulation. There are five questions on the test. The simulation must be designed to give an approximate distribution for the number of correct answers on such a test.

a. The basic random component is to make a selection of a random number correspond to guessing on a true–false question. What is the probability of guessing the correct answer on any one question? How can we define an event involving the selection of a random number that has this same probability?

b. Each trial of the simulation represents one taking of the test; therefore, each trial must have five random selections of question outcomes within it. Select five random numbers, and with outcomes as defined in part a, count the number of correct answers obtained. Record this number.

c. Repeat the procedure for a total of 50 trials (which represents taking the test 50 times). Record the number of correct answers for each trial. Construct a dot plot or stem plot of the results.

d. Approximate the probability that you would get three or more correct answers by guessing.

e. What is the average number of questions you answered correctly per trial? This

is an approximation to your **expected** number of correct answers when you take the test by guessing.

Wrap-Up

1. Suppose you are now guessing your way through a 10-question true–false test. Conduct a simulation for approximating the distribution of the number of correct answers.

a. Does the basic probability per selection of a random number change over what it was in part 3a?

b. Does the number of random selections per trial change? If so, what is it now?

c. Conduct at least 50 trials and record the number of correct answers for each. Make a plot of the results.

d. What is the approximate probability of getting six or more questions correct? How does this compare with the answer to part 3d? If you had to guess at the answers on a true–false test, would you want to take a long test or a short one? Explain.

e. What is your expected number of correct answers when guessing on this test?

2. Suppose you are now guessing your way through a 10-question multiple choice test, where each question has 4 plausible choices, only 1 of which is correct. Conduct a simulation for approximating the distribution of the number of correct answers.

a. Does the basic probability per selection of a random number change over what it was for the true–false test? If so, what is this probability now?

b. What is the number of random selections per trial?

c. Conduct at least 50 trials and record the number of correct answers for each. Make a plot of the results.

d. What is the approximate probability of getting more than half the answers correct? How does this compare with the answer to part 1d here? Explain any differences you might see.

e. What is the average number of correct answers per trial? How many answers would you expect to get correct when guessing your way through this test?

Extensions

1. Pairs of people eating lunch together enter a cafeteria that has square tables, with one chair on each side. Each pair chooses a separate table at which they may sit next to each other or across from each other. Suppose 100 such pairs enter the cafeteria today. Assuming that they choose seats randomly, construct an approximate distribution for the number of pairs that sit next to each other. How many of these pairs would you expect to be sitting across from each other?

2. For each of the simulations run in this lesson, make the following identifications:

n = the number of selections per trial;

p = the probability of obtaining the outcome of interest (a "success") on any one selection;

m = the mean number of successes per trial across the simulation;

sd = the standard deviation of the number of successes per trial across the simulation.

a. Do you see any relationship between m, calculated from the simulation data, and n and p? That is, can you write the expected value for the number of successes in n selections as a function of n and p?

b. Theory suggests that $(\text{sd})^2 = np(1 - p)$. Does this rule seem to hold for your simulation data?

Waiting for Reggie Jackson: The Geometric Distribution

NOTES FOR THE INSTRUCTOR

Statistical Setting

This activity introduces students to the geometric, or discrete waiting time, distribution. It can be used as the student's first experience with simulation, the idea of a probability distribution, and the idea of a rare event. (If Reggie must buy 8 boxes to get his poster, has he experienced a rare event?) It can also be used to motivate the multiplication rule for independent events. (The probability it will take 4 trials to get the first success is $1 - p$ times the probability it will take 3 trials to get the first success.)

You will also have the opportunity to help students confront the common misconception that an event must be "due" even if trials are independent. For example, many students will believe that if Reggie has purchased six boxes of cereal without getting his poster, he is more likely to get it in the next box.

Prerequisites for Students

Students should know how to make a histogram.

Materials

Each student will need one die (or a random digit table that can be used to simulate the rolls of a die) and two Post-It stickers (you may want about 40 extra).

Procedure

This activity works best when done with a large class so that you can have at least 100 simulations in Activity steps 1 and 2. If you have fewer students, have them do several repetitions each. Steps 1 and 2 should be done with the entire class. The remaining questions are best done in small groups.

One nice way to do the simulations in steps 1 and 2 is to use Post-It stickers to make a histogram of the waiting time distribution. Give every student a Post-It. Place other Post-Its numbered 1, 2, 3, . . . , 15 along the bottom of a classroom wall. Have everyone in the class roll their die. In the Reggie Jackson simulation, for example, ask the class what proportion of them should get a Reggie Jackson poster (a one or two on the die) on the first roll. Have those who did get a Reggie Jackson poster on the first roll come to the wall and put their Post-Its above the number 1. Ask the class what proportion of the remaining students should get a one or a two on the second roll. Have the remaining students roll and those who got a one or a two put their Post-Its above the number 2. Continue until all students have placed their Post-Its on the wall.

Sample Results from Activity

1. a. This table for 100 students is close to the theoretical one.

Number of Boxes	Frequency
1	33
2	22
3	15
4	10
5	7
6	4
7	3
8	2
9	1
10	1
11	1
12	1

The histogram for the above table is as follows.

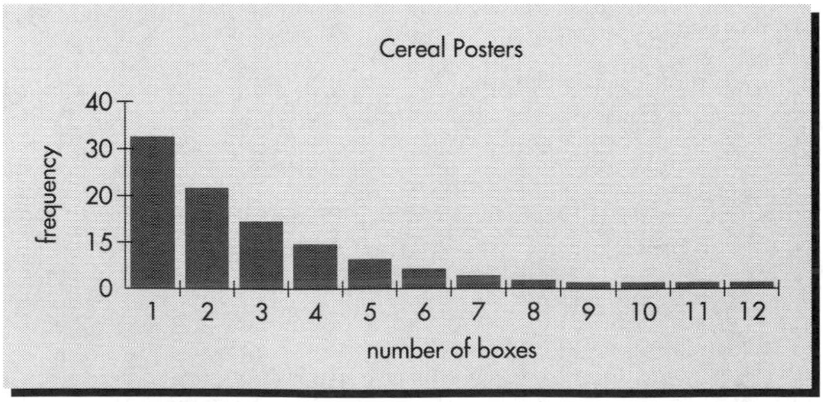

b. The histogram is skewed right. Each bar is about 2/3 the height of the one to the left.

c. Answers will vary depending on the results from your class. The theoretical answer is 3 1/3.
d. About 0.06, not quite a rare event.
e. We are assuming that each time a box is opened, the probability is 1/3 that there is a Reggie Jackson poster inside. This assumption would not be true if, for example, there were more Reggie Jackson posters made than the others, or if all the boxes containing Reggie Jackson posters went to New York, or if all boxes contain a Reggie Jackson poster for a month or so, then all boxes contain a Nolan Ryan poster, etc.

2. a. A table that is close to the theoretical one appears next.

Event	Number of Rolls Required	Frequency
Rolled doubles on first try	1	17
Rolled doubles on second try	2	14
Rolled doubles on third try	3	12
Rolled doubles on fourth try	4	10
etc.		etc.
	Total	100

b. The histogram is skewed right, with each bar about 5/6 the height of the one on its left.

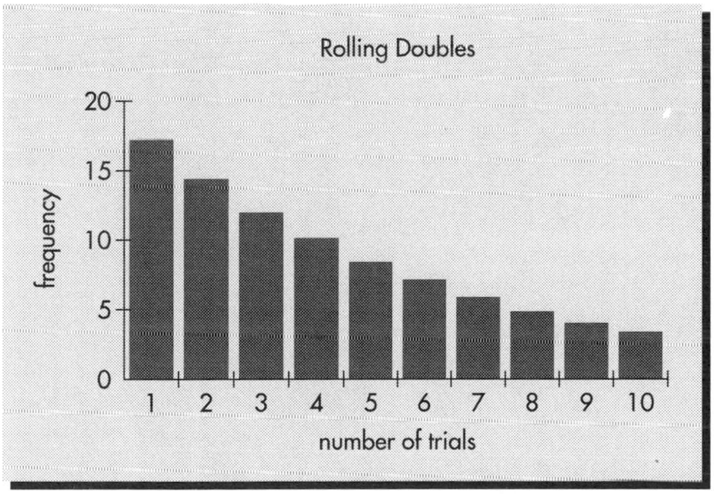

c. Answers will vary depending on the results from your class. The theoretical answer is 6.

3. a. One-fourth of the 4,096 children, or 1,024 children. We are assuming that 1/4 of the boxes contain a parrot sticker and that the boxes are randomly distributed to the stores.

b. 4,096 − 1,024 = 3,072
c. One-fourth of 3,072 children, or 768 children
d. You may need to remind students what we mean by "expect" in statistics. For example, when we flip a coin 5 times, we expect 2.5 heads. That is, if a large number of people did this experiment, results would vary but the average number of heads would be 2.5.

Students should eventually realize that to complete the table quickly, all they have to do is multiply the previous number of children by 0.75.

Number of Boxes Purchased to Get First Parrot Sticker	Number of Children
1	1,024
2	768
3	576
4	432
5	324
6	243
7	182.25
8	136.69
9	102.52
10	76.89
11	57.67
12	43.25
13	32.44
14	24.33
15	18.25
16	13.68
17	10.26
18	7.70
19	5.77
20	4.33

e.

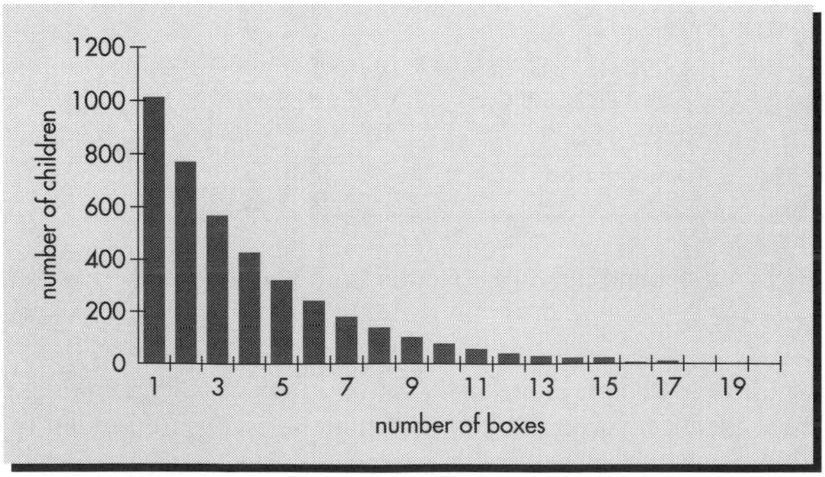

f. Each bar is 0.75 of the height of the bar to its left.
g. The theoretical average is 4. Here, the average is only 3.94 because the table ended at 20 although about 13 children are still trying to get their parrot sticker after buying 20 boxes.

Sample Results from Wrap-Up

1. The leftmost bar in a waiting time distribution is the tallest. We saw why in our experiments in class. All students are available to have a success on the first trial. Those that had a success on the first trial don't participate in the second trial, so there will probably be fewer successes on the second trial than on the first, etc. The waiting time distribution has an infinite number of bars, with the height of each one $1 - p$ times the height of the previous one.

2. Students need to find examples where the probability of a success remains the same on each successive trial, no matter how many trials there have been. Examples include waiting for a winning ticket when buying scratch-off lottery tickets where the probability of winning is the same on each ticket, waiting for a rainy day on days when the weather report says there is a 30% chance of rain, waiting for a child of a given sex when having a family.

Sample Results from Extensions

1. If prizes are not put randomly and in equal numbers into the boxes, say if there is a larger number of one prize than the others, then the expected number of boxes that must be purchased is larger.

2. If p is the probability of a success on each trial, the average waiting time is $1/p$. For example, the average number of times you must roll a die before the first five appears is $1/(1/6) = 6$. Their experiments should convince students that this simple formula is correct.

   There are several ways to prove it. One that avoids infinite series is to let A be the average waiting time for a success. The probability is p that we will succeed on the first trial. If we don't succeed, which happens with probability $(1 - p)$, then it will take an average of A additional trials (a total of $A + 1$ trials) for the first success. (We are given no "credit" for having missed on the first trial as the process has no memory.) Thus,

$$A = p(1) + (1 - p)(A + 1)$$

or

$$A = 1/p.$$

3. The probability the first success occurs on the first trial is p. The probability the first success occurs on the second trial is $(1 - p)p$ since the first trial must be a failure and the second a success. The probability the first success occurs on the third trial is $(1 - p)(1 - p)p = p(1 - p)^2$ since the first two trials must be failures and the third a success. The probability that the first success occurs on the nth trial is then $p(1 - p)^n$. For example, in rolling a die and waiting for a five, the probability the first success occurs on the first roll is 1/6. The probability the first success occurs on the second roll is $(5/6)(1/6) = 5/36$. The probability the first success occurs on the third roll is $(5/6)(5/6)(1/6) = 25/216$. So the probability that the first success occurs on the nth trial is $(1/6)(5/6)^{(n-1)}$.

Sample Assessment Questions

1. In the game of Monopoly, you must roll doubles in order to get out of "jail." If you haven't rolled doubles in three tries, you must pay $50. Out of every 36 people who go to jail, how many would you expect to have to pay the $50?

2. Describe some of the characteristics of a waiting time distribution.

Reference

1. Richard J. Larsen and Morris L. Marx (1986), *An Introduction to Mathematical Statistics and Its Applications*, second ed., Englewood Cliffs, NJ: Prentice-Hall, pp. 218–221.

Waiting for Reggie Jackson: The Geometric Distribution

SCENARIO

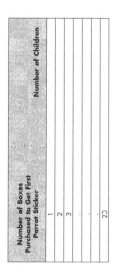

Children's cereals sometimes contain small prizes. For example, not too long ago, boxes of Kellogg's Frosted Flakes contained one of three posters: Ken Griffey Jr., Nolan Ryan, or Reggie Jackson. Reggie wanted to get a Reggie Jackson poster and had to buy eight boxes until getting his poster. Reggie feels especially unlucky.

Question

Should Reggie consider himself especially unlucky? On the average, how many boxes would a person have to buy to get the Reggie Jackson poster? What assumptions would you have to make to answer this question?

Objective

In this activity you will become familiar with the **geometric,** or **waiting time, distribution**, including the shape of the distribution and how to find its mean.

Activity

1. You will need a die or another method of simulating an event with a probability of 1/3. Roll your die. If the side with one or two spots lands on top, this will represent the event of buying a box of Frosted Flakes and getting a Reggie Jackson poster. If one of the other sides lands on top, roll again. Count the number of rolls until you get a one or a two.

a. Make a histogram of the number of rolls the students in your class require to get their first Reggie Jackson poster.
b. Describe the shape of this distribution.
c. What was the average number of "boxes" purchased to get a Reggie Jackson poster?
d. Estimate the chance that Reggie would have to buy eight or more boxes to get his poster.
e. What assumptions are made in this simulation about the distribution of the prizes? Do you think they are reasonable ones?

2. In some games, such as Monopoly, a player must roll doubles before continuing. Use a pair of dice or use random digits to simulate rolling a pair of dice. Count the number of rolls until you get doubles.
a. Make a histogram of the number of rolls the students in your class required to roll doubles.
b. Describe the shape of this distribution.
c. What was the average number of rolls required?

3. In questions 1 and 2, you constructed a waiting time distribution using simulation. In this question, you will construct a theoretical waiting time distribution. Boxes of Post's Cocoa Pebbles recently contained one of four endangered animal stickers: a parrot, an African elephant, a tiger, or a crocodile. Suppose 4,096 children want a sticker of a parrot.
a. How many of them would you expect to get a parrot in the first box of Cocoa Pebbles they buy? What assumptions are you making?
b. How many children do you expect will have to buy a second box?
c. How many of them do you expect will get a parrot in the second box?
d. Fill in the following table.

Number of Boxes Purchased to Get First Parrot Sticker	Number of Children
1	
2	
3	
.	
.	
20	

e. Make a histogram of your theoretical waiting time distribution.
f. The height of each bar of the histogram is what proportion of the height of the bar to its left?
g. What is the average number of boxes purchased?

Wrap-Up

1. Describe the shape of a waiting time (geometric) distribution for a given probability p of a success on each trial. Will the first bar in a waiting time distribution always be the highest? Why or why not? The height of each bar is what proportion of the height of the bar to its left?

2. Find an example of another real-world situation that would be modeled by a geometric distribution.

Extensions

1. There is some evidence that prizes are not put randomly into boxes of cereal. Design an experiment and determine how this would affect the average number of boxes that must be purchased to get a specific prize.

2. Look at the average waiting times in Activity questions 1, 2, and 3. Can you find the simple formula that gives this average in terms of the probability p of getting the desired event on each trial?

3. Find a formula that gives the probability that the first five occurs on the nth roll of a die.

Reference

Frederick Mosteller, Robert E. K. Rourke, and George B. Thomas, Jr. (1970), *Probability with Statistical Applications*, second ed., Reading, MA: Addison-Wesley, pp. 176, 189, and 219.

SAMPLING DISTRIBUTIONS

Spinning Pennies

NOTES FOR THE INSTRUCTOR

Statistical Setting

The idea that a statistic, such as a sample proportion, is itself a random variable is of central importance in statistics but is a very difficult concept. This activity helps by giving a concrete example to refer to as the course develops. The activity can be used early in the semester and referred to many times thereafter, such as when discussing confidence intervals.

Prerequisites for Students

Students should have some familiarity with histograms or dot plots and with the concepts of mean and standard deviation.

Materials

Each student needs a penny and a copy of the Stirling recording sheet. All pennies should be minted in the same year. (Be prepared for some students to lose their pennies during the activity—have a few extra pennies minted in the proper year with you.) Students should have a hard, *flat* surface (e.g., a table top) available on which to spin the pennies.

Procedure

Distribute pennies to the class, making sure that all of the pennies were minted in the same year. Note that pennies minted in different years have different values of p = Pr(heads). For example, for 1961 pennies p is near .1, but for 1990 pennies p if close to .4. Thus it matters that each student have a penny from the same year.

Demonstrate to the students how to spin the pennies. Holding the penny with one hand and flicking with a finger of the other hand works best. Some will want to hold it with both thumbs (one on each side) and try to get it spinning by pulling their hands apart; this tends not to work as well. Having Lincoln's profile right side up and facing you each time reduces between-trial variability, which may affect the probability of getting heads on a given spin.

A hard, flat surface works best for the spinning, which takes 20 to 25 minutes. You might want to have students do the spinning outside of class, especially if they only have slanted desktops available during class time.

After all data have been added to the dot plot, a number of features can be discussed. For example, it will be close to bell-shaped. The combined class data, with 50*(number of students) trials, will give a good estimate of p if the class is reasonably large. One can see how $\hat{p}$ varies about p and relate this to the variability in sample percentages from polls and from other sources.

Sample Results from Activity

Here is a dot plot of values obtained by one class when using pennies minted in 1993:

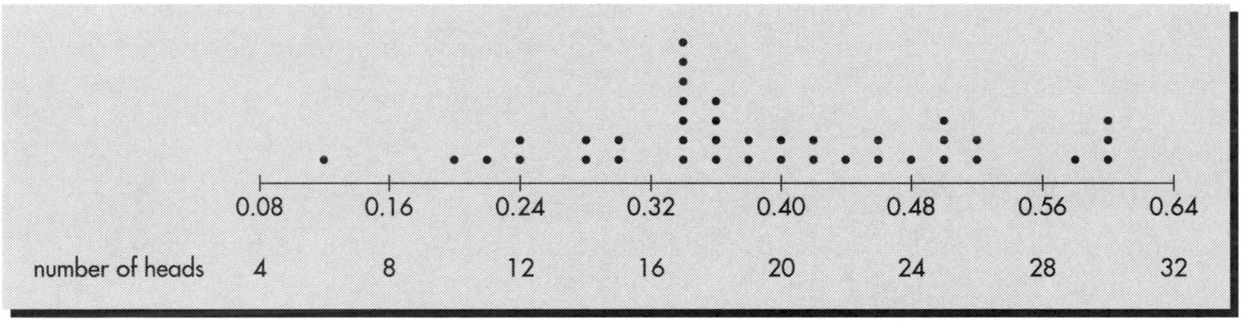

Warning: In an ideal sense, the distribution of $\hat{p}$ should be binomial with $n = 50$ and should be the same for each student. However, depending on the penny-spinning skills of the students and the conditions of the various pennies used, the class distribution of $\hat{p}$ may be more variable than one would expect from a true binomial. That is, the standard deviation of the data in the dot plot may exceed $\sqrt{p(1-p)/50}$ (as in the dot plot above). This provides a good opportunity for a class discussion of the assumptions that underlie a binomial.

Extensions

At the appropriate time, one can talk about the standard deviation (SD) of $\hat{p}$, comparing theory to experience. Indeed, entering the students' values of $\hat{p}$ into a data file and then finding the SD helps students see that the formula for the SD of $\hat{p}$ *is* related to something real. When discussing confidence intervals, you may have each student construct a CI for p based on his or her value of $\hat{p}$ and then see how many CIs include p (using the combined class data percentage as p). Of course, if the class is small, there will not be all that many observations in the combined class data set, so the estimate of p will be off by a bit.

The second extension in the student notes involves the longest run in the 50 trials. There are other sampling distributions that you could consider, such as the longest run of heads or the number of spins needed in order to get heads for the first time.

The Stirling recording sheet gives a record of the order in which the data were recorded, so that one can ask questions like "Does the probability of heads depend on whether the previous result was heads? What does the distribution of the longest run of consecutive heads in 50 trials look like? What is the probability that two paths (traced out on the recording sheet by the record of heads and tails in 50 trials) will cross?" This last question may not hold intrinsic interest, but it is hard to answer, which gives one the chance to introduce simulation ("Find a partner, hold your papers together and up to a light, and see whether or not your paths cross. What is the sample fraction in class today?") as a tool in tackling difficult probability problems.

You can tie this activity to "the law of large numbers and the central limit theorem" activity. One extension of that activity is to use values other than .5 in the computer simulation of coin tossing. You could use the value of p determined from the combined class data in place of .5.

Sample Assessment Questions

1. Consider the results from the class dot plot. Suppose someone who was absent made up the activity by spinning 50 pennies. What percentage, $\hat{p}$ would you expect this person to get? What is the interval in which you expect $\hat{p}$ to fall?

2. National opinion polls are often reported in the newspaper. Suppose the Gallup organization takes a random sample of 1,000 voters and asks them whether or not they approve of the job performance of the president. What sampling distribution is related to this process?

3. When one opinion poll of 1,249 adults was reported in the newspaper, the following statement was included: "Results have a margin of sampling error of plus or minus 3 percentage points." How is this statement related to a sampling distribution?

Spinning Pennies

SCENARIO

When an organization conducts an opinion poll, it reports the percentage of the people sampled who favor some issue, such as the percentage who favor the death penalty. If the poll were repeated many times, the resulting sample percentages—one from each poll—would form a sampling distribution. Sampling distributions are very important in statistics, but we usually must imagine what the sampling distribution looks like because the people who conduct a poll don't repeat it. (If they *do* ask the same question in another poll, it is only after time has elapsed so that people's opinions may have changed.)

Question

If you toss a penny, you have a 50–50 chance of getting heads. What happens if you spin a penny and wait for it to fall? Do you still get heads 50% of the time? Suppose you spin the penny 50 times and record the sample percentage of heads. How does this sample percentage vary from one sample of 50 spins to another?

Objective

The goal of this activity is to teach you what a sampling distribution is. By the end of the activity you should have developed a feel for sampling distributions in general and distributions of sample proportions in particular.

Activity

Your instructor will loan you a penny; you will also need the Stirling recording sheet. Place your penny on its edge, with Lincoln's profile right side up and facing you, holding the penny lightly with one finger of one hand. Flick the edge of the penny sharply with a finger of your other hand to set it spinning. Let the penny spin freely until it falls. If it hits something while spinning, do not count that trial. When the penny is at rest, record whether it is showing heads or tails. Use the recording sheet to record, in order, the results of 50 penny-spinning trials. Record the trial as a WIN if the penny lands showing "heads" and as a LOSS if it lands showing "tails."

Repeat this 50 times; keep track of the number of heads, y, the number of tails, and the order in which heads and tails occur.

Compute the sample percentage of heads for the 50 observations and call this $\hat{p}$, where $\hat{p} = y/50$. This is representative of the process of estimating a population proportion, such as the percentage of all persons who favor the death penalty, using a sample proportion based on a sample size of 50. (Of course, most opinion polls use sample sizes of several hundred or more, but spinning the penny more than 50 times would be excessively tiring.)

After you have computed your value of $\hat{p}$, go up to the chalkboard and add your sample percentage to the dot plot that the instructor has started. When all class data are collected, the dot plot will resemble the sampling distribution for the sample percentage of heads in 50 penny spins.

Wrap-Up

1. Write a brief summary of what you learned in this activity about sampling distributions.

2. Suppose each student in the class conducted an opinion poll in which they asked 50 randomly chosen persons whether or not they approve of the job performance of the U.S. president. If they each reported a sample percentage, what do you expect that the graph of those percentages would look like?

Extensions

1. Rather than considering the sampling distribution of the sample percentage, $\hat{p}$, you could consider the sampling distribution of the number of heads in 50 penny spins.

2. Suppose you had spun the penny only 10 times and had obtained the sequence TTHTTTHHHT. We can break this into groups as TT H TTT HHH T and say that the sequence of 10 trials contains a run of 2 tails, then a run of 1 head, a run of 3 tails, a run of 3 heads, and a run of 1 tail. There are 5 runs here, the longest of which has length 3.

Look at the sequence of 50 heads and tails you obtained and divide the 50 trials into runs of heads and tails. Find the length of the longest run in your 50 trials. Combine your value with those of the other students in your class in a dot plot.

Reference

G. Giles (1986), "The Stirling recording sheet for experiments in probability," in P. Holmes (ed.), *The Best of Teaching Statistics*, Sheffield, England: Teaching Statistics Trust, pp. 8–14.

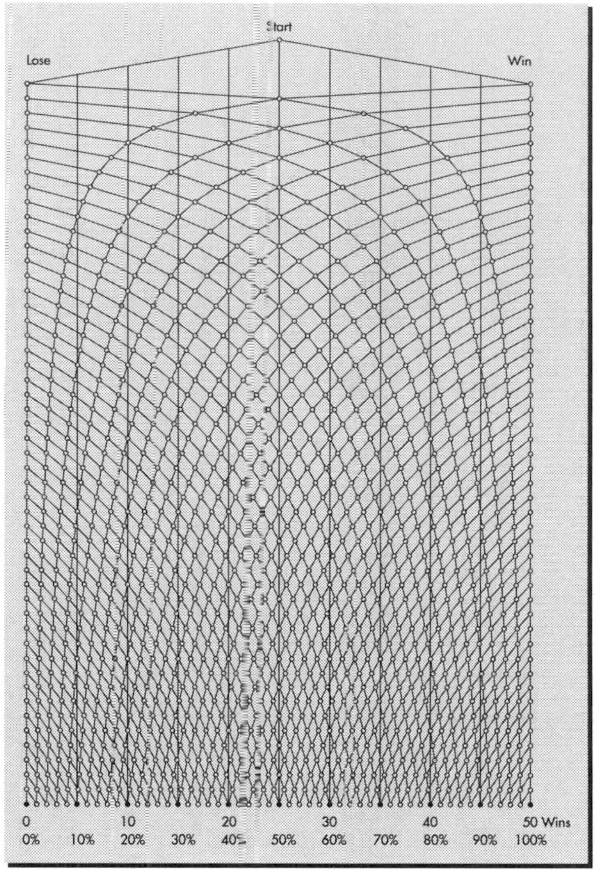

Cents and the Central Limit Theorem

NOTES FOR THE INSTRUCTOR

Statistical Setting

In this activity students construct a distribution of the population of the ages of a large number of pennies. This distribution will have roughly the shape of a geometric distribution. Students then discover the central limit theorem by computing the mean age of pennies in samples of size 5, 10, and 25 from that distribution, making a histogram of the resulting sampling distribution of the mean, and finding the shape, mean, and standard error of the sampling distribution.

Prerequisites for Students

Students should know that the mean of a distribution is its "balance point." Also, students should understand the standard deviation as a measure of spread.

Materials

Each student will need 25 pennies. During the previous week, have each student collect the first 25 pennies that he or she receives in change. Students should bring in the pennies and a list of the 25 dates on the pennies. Throughout this activity, be sensitive to the fact that some students may have difficulty reading the dates on the pennies.

In addition, each student will need at least one nickel, one dime, and one quarter.

A computer or calculator will also be needed to compute means and standard deviations.

Procedure

This activity works well with a class of any size. If your class is small, each student should be responsible for several samples; if your class is very large, you may not want to use all of the pennies when constructing the histogram for the population. It is an activity for the whole class to do together.

Sample Results from Activity

1. Few students realize that the shape of the distribution of the ages of all pennies will be roughly geometric (see #2 below). Many will believe that it should be normal ("A few pennies are new, a few pennies are old, most are lumped in the middle.").

2. Students love to make the histogram by placing the pennies themselves above a number line on a large table or on the floor of the classroom. As this can take some time, you could construct the histogram during a break in the class or else collect the pennies the class session before and ask for volunteers (with sharp eyes) to come early next time to construct the histogram. Other ways of making the histogram are to collect a list of the ages during the previous class session, type the data into the computer, and bring copies of the histogram to class.

   One class's histogram for 648 pennies is as follows:

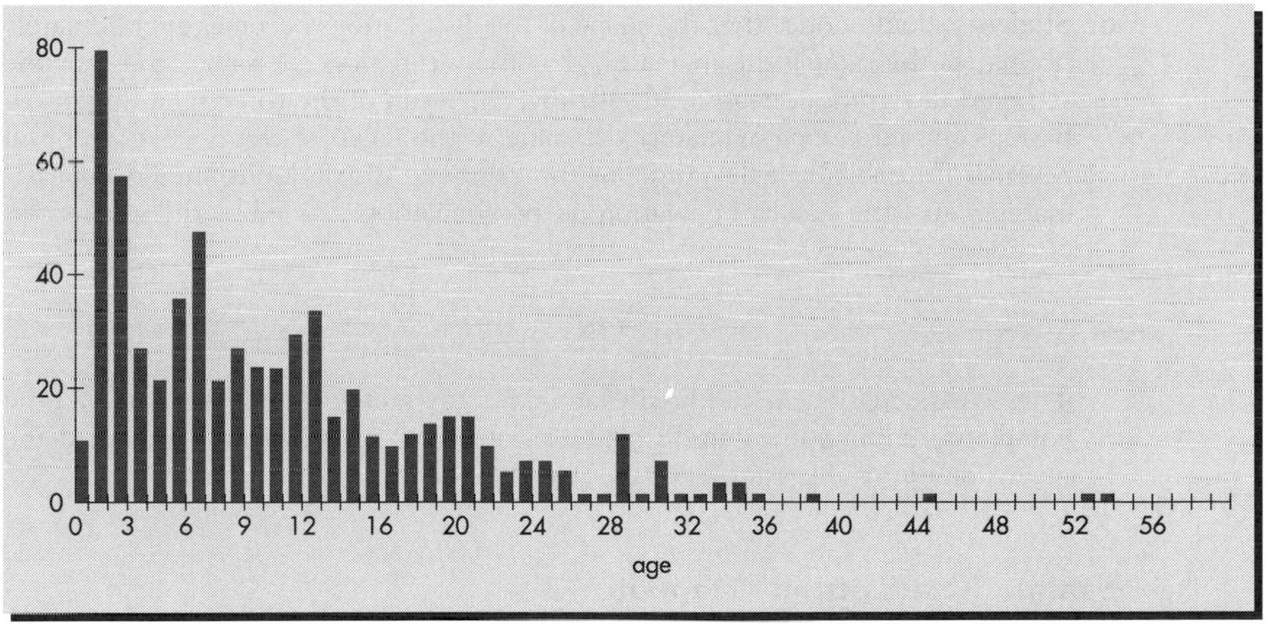

3. Answers will vary. It is best if you have all of the ages entered in a computer or calculator and can get the exact mean and standard deviation. The mean age for the pennies in the histogram above, 10.4, is larger than usual. One suspects that a student found the older pennies in the back of a drawer. The mean age tends to be between 7 and 8 years, with a standard deviation around 8 years.

4. Students very much like for this distribution to be displayed in the following way. Each student places a nickel above a number line on the floor or table to repre-

sent the mean age of their five pennies. It's best if there are 100 or so nickels in the distribution. Thus, if you have a small class, you may want to have students compute the means of several samples of size 5.

5. The mean will be the same. Most students will not realize this. A typical answer is to say it will be smaller. Most students will realize that the standard deviation will be smaller.

6. If the mean of the population of ages of the pennies was, say 7.5 years with a standard deviation of 8 years, then the mean of the sampling distribution should be about 7.5 years with a standard deviation of about $\frac{8}{\sqrt{5}} \approx 3.6$ years.

7. Make a display of the histograms using the dimes and quarters. Students will then be able to visualize that the numbers graphed represent the means of samples, not individual values.

8. Students should notice that the shape of the distributions becomes approximately normal as the sample size increases, that the mean stays the same, and that the standard deviation decreases. Specifically, the mean of the three sampling distributions should be approximately the same as the mean of the population of all pennies. The standard deviation of the sampling distributions should approximately equal the standard deviation of the population divided by the square root of the sample size:

$$\mu_{\bar{x}} = \mu, \qquad \sigma_{\bar{x}} = \frac{\sigma}{\sqrt{n}}.$$

These two equations would hold exactly, for any sample size, if students had the complete sampling distribution of the mean; that is, if they had computed the means of all possible samples of size n.

Sample Results from Wrap-Up

1. The central limit theorem says the following:

> If samples of size n are taken from a distribution with mean μ and standard deviation σ, then the sampling distribution of the mean will have mean μ and standard deviation $\frac{\sigma}{\sqrt{n}}$. As n increases, the sampling distribution becomes approximately normal.

2. The sampling distribution for $n = 36$ would be approximately normal and would have the same mean as the distribution of the ages of all pennies, and its standard deviation would be the standard deviation of the ages of all pennies divided by 6.

Sample Results from Extensions

1. The number of pennies minted each year has been increasing slowly.

2. The distribution probably will be reasonably geometric. Estimates of r will vary. The value of r tells you what percentage of the pennies minted in a given year that are still in circulation will remain in circulation next year. The first bar, for pennies 0 years old, is short because these are pennies for the current year and pennies for the current year are still being put into circulation.

3. You would need to quadruple the sample size.

Sample Assessment Questions

(For this exercise, give your students the dates from a sample of 30 nickels.)

1. What is your best guess of the mean age of all nickels in circulation? Explain.

2. Approximately how far off might this guess be? Explain. List the assumptions you are making.

Cents and the Central Limit Theorem

SCENARIO

Many of the variables that you have studied so far in your statistics class have had a normal distribution. You may have used a table of the normal distribution to answer questions about a randomly selected individual or a random sample taken from a normal distribution. Many distributions, however, are not normal or any other standard shape.

Question

If the shape of a distribution isn't normal, can we make any inferences about the mean of a random sample from that distribution?

Objective

In this activity you will discover the central limit theorem by observing the shape, mean, and standard deviation of the sampling distribution of the mean for samples taken from a distribution that is decidedly not normal.

Activity

1. You should have a list of the dates on a random sample of 25 pennies. Next to each date, write the age of the penny by subtracting the date from the current year. What do you think the shape of the distribution of all the ages of the pennies from students in your class will look like?

2. Make a histogram of the ages of all the pennies in the class.
3. Estimate the mean and the standard deviation of the distribution. Confirm these estimates by actual computation.
4. Take a random sample of size 5 from the ages of your pennies, and compute the mean age of your sample. Three or four students in your class should place their sample means on a number line.
5. Do you think the mean of the values in this histogram (once it is completed) will be larger than, smaller than, or the same size as the one for the population of all pennies? Regardless of which you choose, try to make an argument to support each choice. Estimate what the standard deviation of this distribution will be.
6. Complete the histogram, and determine its mean and standard deviation. Which of the three choices in part 5 appears to be correct?
7. Repeat this experiment for samples of size 10 and size 25.
8. Look at the four histograms that your class has constructed. What can you say about the shape of the histogram as n increases? What can you say about the center of the histogram as n increases? What can you say about the spread of the histogram as n increases?

Wrap-Up

1. The three characteristics you examined in Activity question 8 (shape, center, and spread of the sampling distribution) make up the **central limit theorem**. Without looking in a textbook, write a statement of what you think the central limit theorem says.
2. The distributions you constructed for samples of size 1, 5, 10, and 25 are called sampling distributions of the sample mean. Sketch the sampling distribution of the sample mean for samples of size 36.

Extensions

1. Get a copy of the current *Handbook of United States Coins: Official Blue Book of United States Coins*, and graph the distribution of the number of pennies minted in each year. What are the interesting features of this distribution? Compare this distribution with the distribution of the population of ages of the pennies in the class. How are they different? Why? How can you estimate the percentage of coins from year x that are out of circulation?

2. In a geometric distribution, the height of each bar of the histogram is a fixed fraction r of the height of the bar to the left of it. Except for the first bar, is the distribution of the ages of the pennies approximately geometric? Estimate the value of r for this distribution. What does r tell you about how pennies go out of circulation? Why is the height of the first bar shorter than one would expect in a geometric distribution?

3. To cut the standard deviation of the sampling distribution of the sample mean in half, what sample size would you need?

The Law of Large Numbers and the Central Limit Theorem

NOTES FOR THE INSTRUCTOR

Statistical Setting

This activity can be used to reinforce ideas about probability as a limiting relative frequency.

Prerequisites for Students

Students need to know something about probability as a limiting relative frequency. They should also have conducted a simulation "by hand" (e.g., with dice, cards, or a random number table) so that they understand the nature of a simulation and what it is that the computer is doing during the activity.

Materials

None.

Procedure

This activity uses the program DataDesk. The instructions given below are for use with that program, so you should have some familiarity with DataDesk. (Other statistics software could be used, but detailed instructions are not given here for other software.) The students need to know how to use the software before starting the activity—do not underestimate the number of "simple" problems that can arise when students use the computer.

If the students are not comfortable with using the computer, they may tend to see the entire exercise as button pushing. That is, they might be able to follow the steps outlined, but not understand what is being simulated. In any event, it is a good idea to stop them periodically and review what they have done so far, using statements such as, "We have simulated having each of 25 students conduct an opinion poll of 1,000 voters."

You might consider using $n = 500$, rather than 1,000, or 15 "coin tossers" rather than 25, if computing speed is a concern.

Sample Results from Activity

Step g should produce a graph similar to the following:

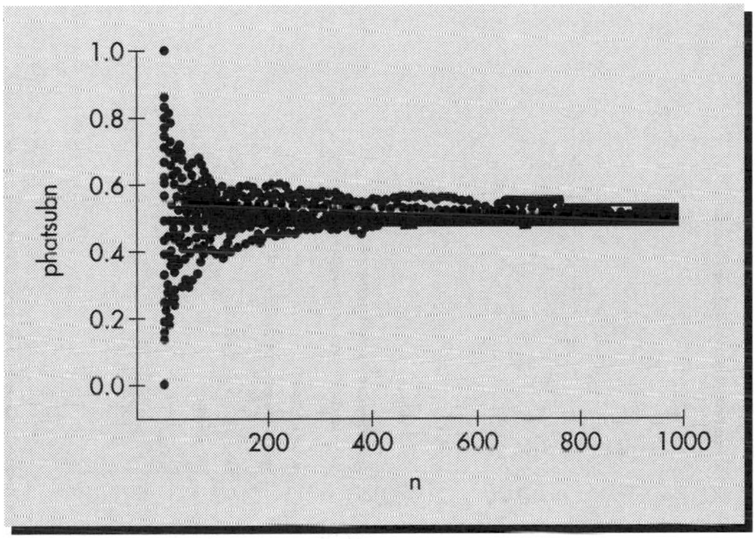

When embedded in a Microsoft Word document, this graph is a single object that contains 25,000 points. It takes the typical computer a bit of time to create such an object on the screen, so when the graph first appears it is created dynamically: The points are added in 1,000 at a time, in order, so that one can see the 25 paths being traced on the screen. (This is one case in which it is advantageous to have slower equipment. A very fast computer would flash the 25,000 points onto the screen almost all at once, taking away one's ability to visually follow the changes in phatsubn as n goes from 1 to 1,000.)

The following scatter plot presents four slices of the scatter plot taken at $n = 5$, 20, 100, and 1,000:

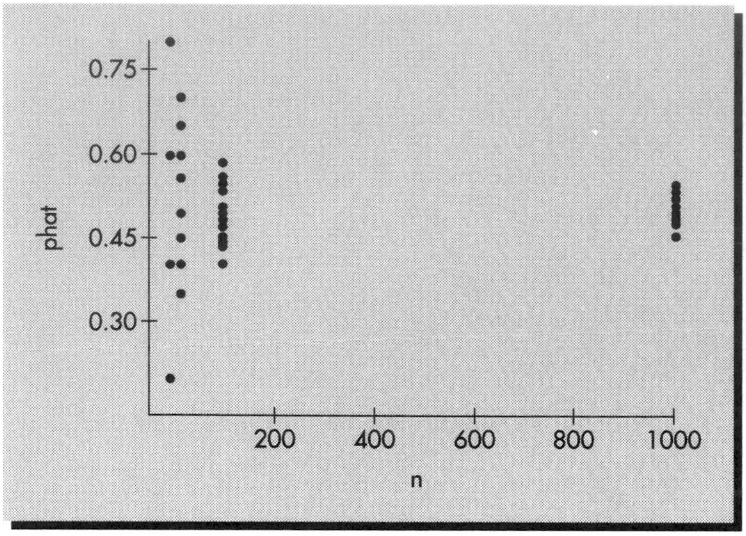

The following are histograms for 25 sample percentages when $n = 5$, 20, 100, or 1,000.

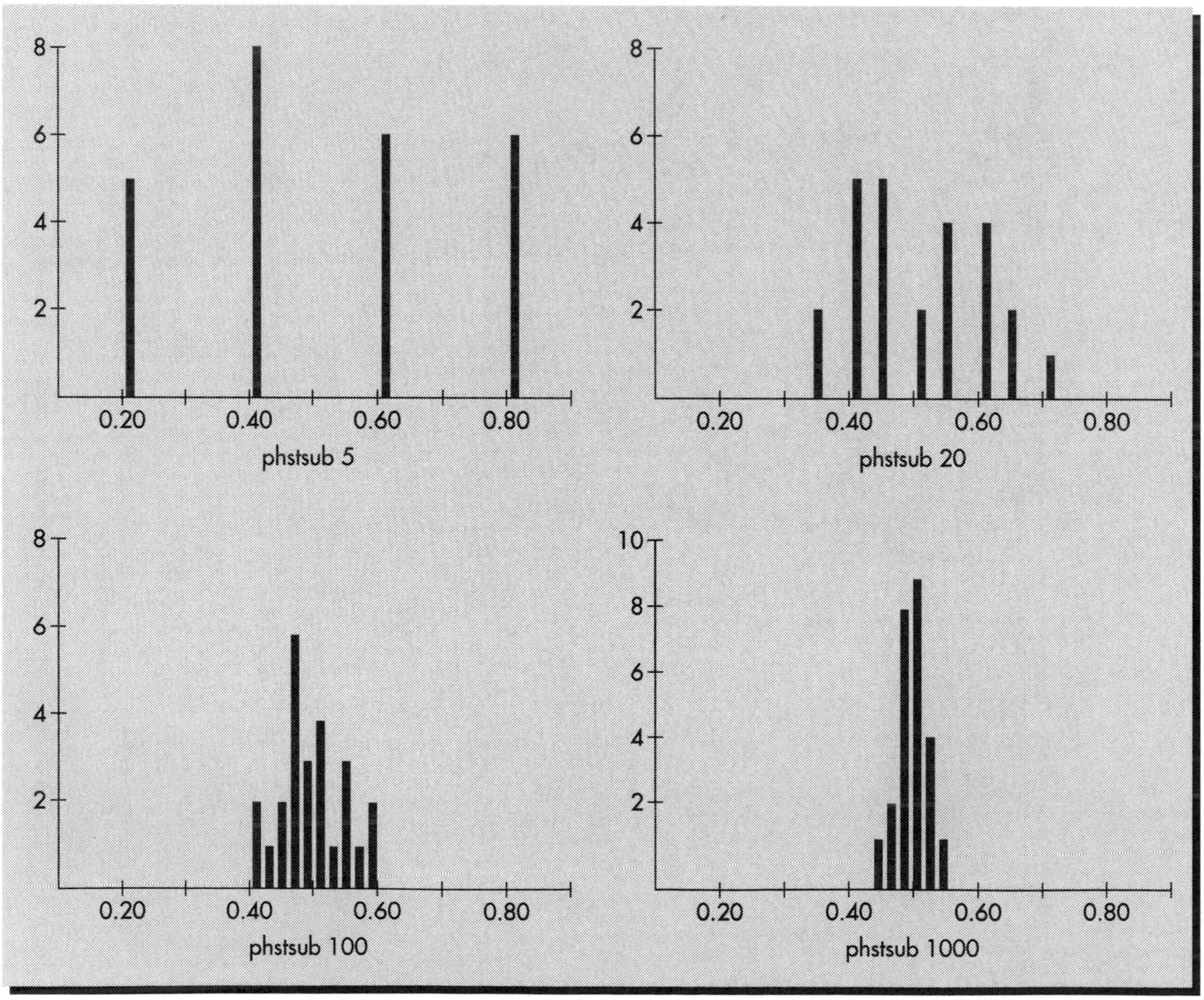

Sample Assessment Question

We used 25 variables in part 2 of the activity. What would the results have been like had we used 50 variables instead of 25?

The Law of Large Numbers and the Central Limit Theorem

SCENARIO

When pollsters ask a question such as "Do you approve of the job performance of the president?" they usually take large samples. They expect the sample percentage to be close to the population percentage, but they are never certain if their results are accurate. Likewise, suppose you toss a coin over and over again and keep track of the percentage of heads obtained along the way. You expect to get heads half of the time, but that doesn't mean that you'll get exactly 50 heads in the first 100 tosses. As the number of tosses goes up, you expect the sample percentage to approach 50%, but there will be variability.

Question

Suppose you were to toss a coin 1,000 times. How close to having 50% heads do you expect to get?

Objective

In this activity you will learn how a sample proportion converges to the corresponding population percentage as the sample size increases (the law of large numbers), and you will see how this behavior is related to the central limit theorem.

Note: This activity makes use of *Data Desk* software for Macintosh computers. If this is not available, the details of the lesson can not be completed.

Activity

1. Generate a sample path
   a. Use the **"Generate Random Numbers . . ."** command (under the **Manip** menu) to generate 1 variable with 1,000 cases from a Bernoulli distribution with $p = .5$. This simulates tossing a coin 1,000 times.
   The result of step a is an icon with a string of zeroes and ones. We want to know what the sample percentage of ones is as the number of trials increases. (Think of "one" as representing "heads" in a coin toss.) To compute this requires several steps.
   b. Select the icon from step a, go to the **Manip** menu, and choose **Transform > . . . Summary > . . . CumSum(•).** (This calculates the cumulative number of heads for each number of tosses; i.e., it keeps track of "the number of heads obtained so far.")
   c. Use the **"Generate Patterned Data . . ."** command (under the **Manip** menu) to generate a variable with the numbers 1 to 1,000. (Generate numbers from 1 to 1,000 in increments of 1.) Rename this variable "n".
   d. Divide the variable from step b by the variable from step c. To do this, first select the icon from b, then select the icon from c. Then go to the **Manip** menu, and choose **Transform > . . . Arithmetic > . . . y/x.** This calculates the percentage of heads after n tosses, as n goes from 1 to 1,000. You might want to rename this variable "phatsubn."
   e. Create a scatter plot of phatsubn versus n. This should produce a graph similar to the following:

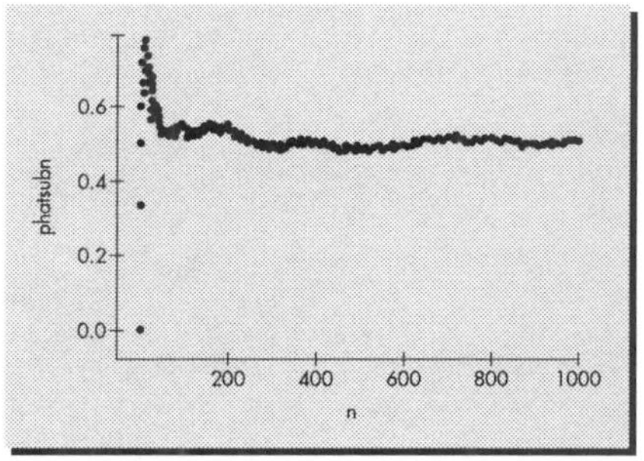

2. Generate many sample paths

Next we want to find out what happens if we repeat part 1 many times. Before proceeding it is a good idea to put everything from part 1 in the trash and then empty the trash (under the **Special** menu).

a. Use the **"Generate Random Numbers . . ."** command (under the **Manip** menu) to generate 25 variables with 1,000 cases each from a Bernoulli distribution with $p = .5$. This simulates 25 persons each tossing a coin 1,000 times.

The result of step 2a is a folder with 25 icons, each of which holds a string of zeroes and ones.

b. Select all 25 of the icons from step 2a, go to the **Manip** menu, and choose **Transform > . . . Summary > . . . CumSum(•).** This should create 25 new icons, each with the cumulative number of heads for one of the 25 variables from 2.

c. Use the **"Generate Patterned Data . . ."** command (under the **Manip** menu) to generate a variable with the number 1 to 1,000. (Generate numbers from 1 to 1,000 in increments of 1.) Rename this variable "n".

d. Divide *each* of the variables from 2b by the variable "n" from 2c. To do this, first hold down the Option key and select the icons from 2b (you can do this by dragging the mouse from left to right), then hold down the Shift key and select the icon "n". Then go to the **Manip** menu and choose **Transform > . . . Arithmetic > . . . y/x.**

Next, we want to combine the results of these 25 simulated coin tossers.

e. Select the 25 icons from 2d, go to the **Manip** menu, and choose **Append & Make Group Variable.** This will create an icon called "Data" that holds 25,000 values (the 25 icons from 2d stacked on top of one another). Rename this icon "phatsubn." (If you are using Version 3.0 or Version 4.0 of DataDesk, you'll need to move this icon into the same relation as the icon from step 2e before going on to step g.)

f. Use the **"Generate Patterned Data . . ."** command (under the **Manip** menu) to generate a variable with the number 1 to 1,000 with the entire sequence repeated 25 times. (Generate numbers from 1 to 1,000 in increments of 1; repeat each number in the sequence 1 time; repeat the entire sequence 25 times.) Rename this icon "n".

g. Create a scatter plot of "phatsubn" from step e versus "n" from step f. This should produce a graph that shows the "history" of 25 sample percentages as n goes from 1 to 1,000.

The law of large numbers says that the sample percentage of heads should approach 50% as n increases. What do you notice in the scatter plot about the behavior of your 25 sample percentages as n increases?

The central limit theorem says that if n is large, then the distribution of the sample percentages should approximate a normal distribution. To investigate this, we want to consider a vertical "slice" of the scatter plot from step g. That is, we want to fix the sample size, n, and see what the distribution of the 25 sample percentages looks like.

h. Go to the **Data** menu and choose **New > . . . Derived Variable.** Name the variable "n = 5". In the window that DataDesk opens, type in

If 'n' = 5   Then 1 Else 0.

Then select the "n = 5" icon, go to the **Special** menu, and choose **Selector > Assign.** This will create a button near the bottom of the screen that will say "Selector Variable: n = 5". As long as this button is dark, only the data that correspond to a sample size of $n = 5$ will be included in any analysis you do. (If you click on the button, you turn it off, which makes it white; click on it again to turn it on.)

i. With the "Selector Variable: n = 5" button on (dark), select "phatsubn" and create a histogram. This will produce a histogram of the 25 sample percentages after 5 coin tosses. Does this histogram look like a normal curve?

j. Repeat steps 2h and 2i using a different value of n. For example, if you want to look at the histogram of the 25 sample percentages after 50 coin tosses, you need to create a derived variable that contains the command

If 'n' = 50   Then 1 Else 0.

(You might want to rename this variable "n = 50".) After selecting this icon and then choosing the **Selector > Assign** command, you will have selected only the sample percentages after 50 coin tosses. A histogram of the "phatsubn" data will now show the distribution of the 25 sample percentages after 50 coin tosses. Does this histogram look like a normal curve when n is large? How large does n need to be for this to happen?

Wrap-Up

1. Suppose we had used 10,000 in places of 1,000 in the simulations. What would the final graph have looked like?

2. Write a brief summary of what you learned in this activity about the law of large numbers and the central limit theorem.

Extensions

The steps outlined in this activity are for simulating the tossing of (25) fair coins. If you want to simulate the results of opinion polls of a population in which 80% answer "yes" to the survey question you can repeat the steps given here but use .8 as the Bernoulli parameter rather than .5. You could also extend the number of trials to a number larger than 1,000, although choosing a value like 10,000 may tax the capacity and speed of your computer.

If you have a color monitor, you can select one of the 25 sequences in the scatter plot from step 2h and plot it in blue, then choose another one and plot it in red, etc. To do this, create a bar chart of the Groups icon from step 2e, then use a plot tool to select one of the groups (e.g., select the first bar in the bar chart) and choose a color from the color palette.

SAMPLING

Random Rectangles

NOTES FOR THE INSTRUCTOR

Statistical Setting

This lesson should be introduced early in the course so that students can see the importance of randomization in collecting data. Discussion of sampling bias should be paired with discussion of measurement bias so that students see how these two forms of bias differ and how important each is to the measurement and data collection process.

The concepts introduced here can be extended when one gets to the notions of sampling distributions for means and the construction of confidence intervals.

Prerequisites for Students

Students should have some knowledge of how to use random numbers to select random samples. In addition, they should have an understanding of the difference between a population and a sample and of why sampling is essential in most statistics problems.

Materials

The sheet of rectangles that is included in this lesson and a random number table or generator are all that is needed for the lesson.

Procedure

1. For the guessed value of the average area, make sure that the students have only a few seconds to study the sheet of rectangles. Under these conditions, most guesses will be on the high side of the true average due to the number of small rectangles that the eye does not readily see. The same phenomenon occurs in the subjective, or judgmental, samples; the averages here are usually on the high side of truth.

Comparing the guesses to the subjective samples often shows that the two are very similar; if you are going to take subjective samples, you might as well just guess.

2. Random samples have averages that center close to the true mean area. Thus, randomization reduces the bias. Students must understand that sampling bias is a function of the sampling procedure, not of the outcomes that might have been obtained from one or two students.

3. In the data analysis, pay some attention to the shape and variability of the distributions of sample means, even though the main focus here is on center. This example can lead nicely into a discussion of sampling distributions, the central limit theorem, and confidence intervals.

Sample Results from Activity

The actual distribution of rectangle areas and the numerical summaries are provided below.

VALUES	PROB
1	0.16
2	0.02
3	0.06
4	0.16
5	0.08
6	0.06
8	0.08
9	0.05
10	0.07
12	0.10
15	0.01
16	0.10
18	0.05

	N	MEAN	MEDIAN	TRMEAN	STDEV
AREAS	100	7.420	6.000	7.189	5.228

	MIN	MAX	Q1	Q3
AREAS	1.000	18.000	4.000	12.000

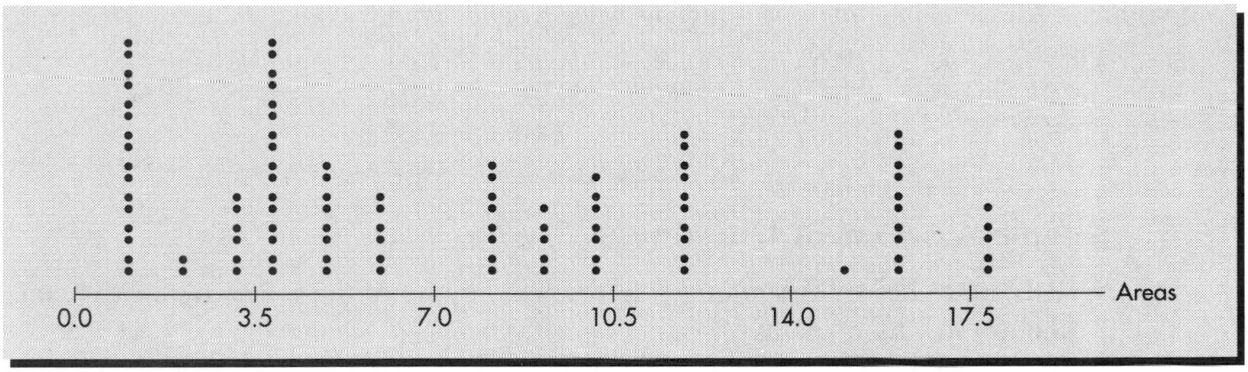

The following data on this activity were obtained from a class of 48 students. The three methods are the guesses, the averages of subjective samples of size 5, and the averages of random samples of size 5, respectively.

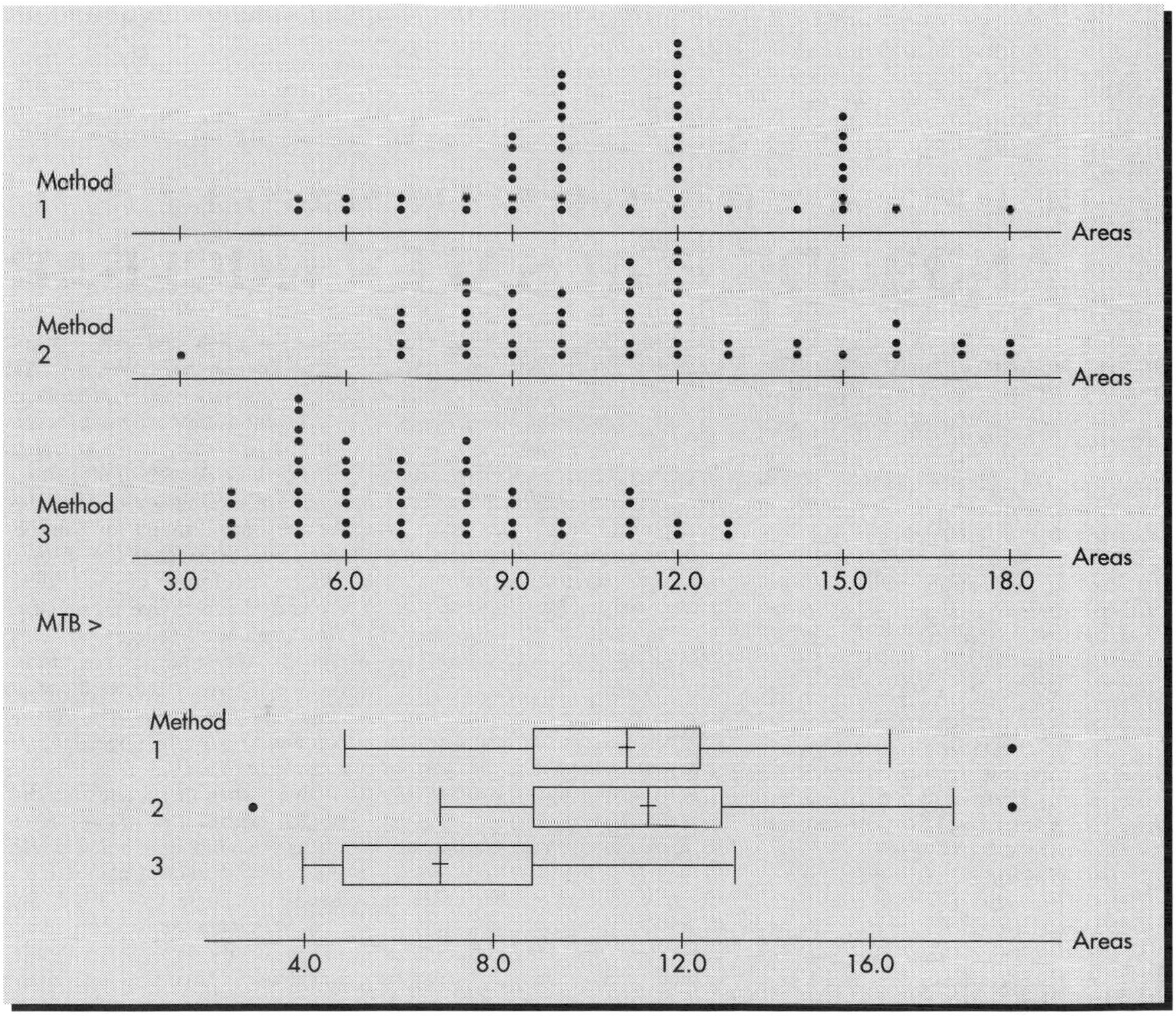

	N	MEAN	MEDIAN	TRMEAN	STDEV
GUESS	48	10.979	10.500	10.977	2.979
EYEBALL	48	11.125	11.000	11.091	3.266
RANDOM	48	7.417	7.000	7.318	2.533

Sample Assessment Questions

1. You are to design a sampling plan to measure one of the following for the community in which you live:
   a. average number of TV sets per household
   b. average number of pets per student
   c. percent of foreign autos driven by students at your school
   Choose one of the items, and discuss how you would obtain the data. Mention the role of randomization and the possibility of bias in your data collection plan.

2. Critique the following article with respect to the sampling issues involved and the conclusions reached.

THE POPULAR SCIENCE NUCLEAR POWER POLL

LANDSLIDE FOR SAFER NUCLEAR

By a stunning majority, voters in the POPULAR SCIENCE poll [Cast Your Vote On Next Generation Nuclear Power, April '90] favor building a new generation of nuclear power plants. Responding to our detailed report on nuclear plant designs and concepts claimed to be safer than existing nuclear generating stations, voters from 38 states took part in the poll and more than 5,600 persons cast electronically recorded ballots on the main issue: Should the United States build more fossil-fuel power plants or the new so-called safe nuclear generators to meet the energy crisis of the '90s? The pro-nuke voters prevailed by a margin of more than six to one.

Why did the vote turn out in this fashion? In the follow-up interviews, voters said that nuclear power in a new package could meet energy needs in a safe, environmentally friendly way. Said one voter in Oklahoma City: "It's also better than acid rain." The Oklahoma reader also said he was in favor of building "little ones rather than mega ones." A New Jersey minister told us that "more people have gotten sick from Jane Fonda movies than from nuclear power."

A huge majority of voters also indicated a preference for nuclear power research over alternate energy. In addition, the tally showed disapproval of the government's energy policies.

The vote was clear-cut despite an attempt on the part of Westinghouse, one of the world's major nuclear power plant designers, to stuff the electronic ballot box. Westinghouse set up a special speed dialing number, enabling its Pittsburgh headquarters employees to call our 900 poll number (which cost voters 75 cents a minute) for free, and then publicized the article and poll in an internal newsletter. Westinghouse employees responded with a total of 20 percent of the total votes cast in our poll, according to our research.

(POPULAR SCIENCE protested the actions of Westinghouse, criticizing the nuke manufacturer for a sinister and cynical disregard for an open and honest forum of free ideas. Richard Slember, vice president of Westhinghouse's energy systems, replied: "We in no way tried to influence any votes, but we did want our employees to know about the poll and to have the opportunity to participate.")

Even when these votes are discarded, the result is a landslide favoring the building of a new generation of safer nuclear plants. If we assume that all the votes originating from Westinghouse were pro nukes and disqualify them, the margin is still about five to one in favor of the new nuclear plants.

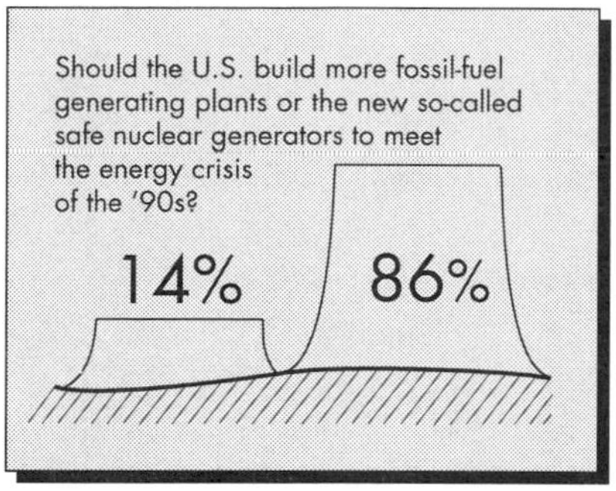

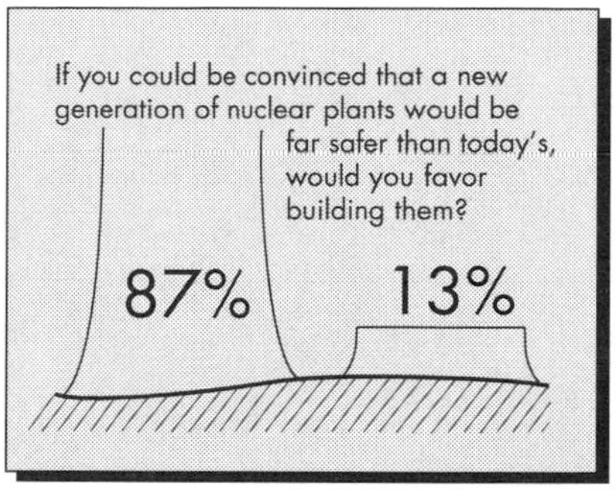

If you could be convinced that a new generation of nuclear plants would be far safer than today's, would you favor building them?
87%
13%

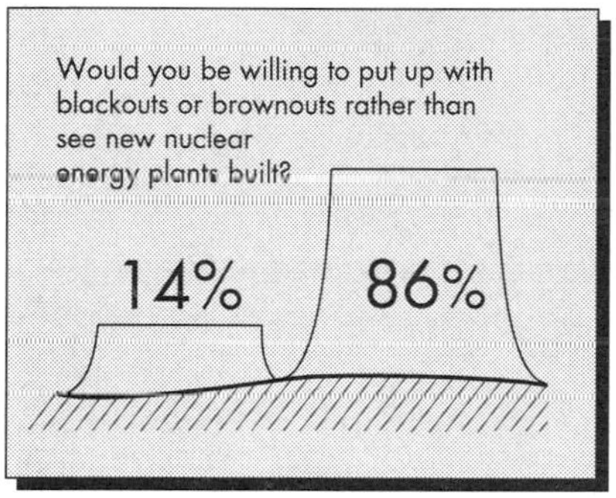

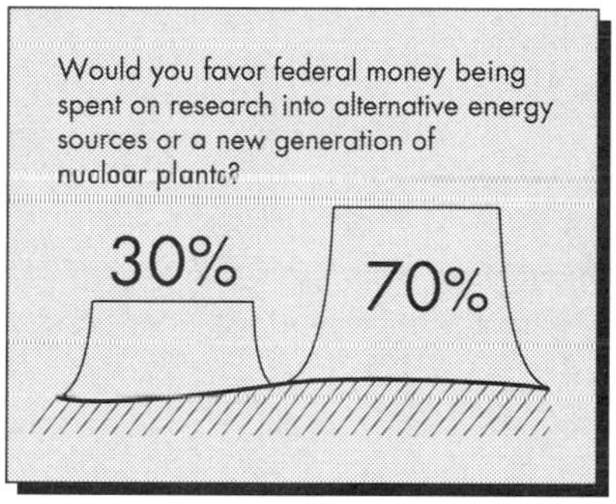

Would you favor federal money being spent on research into alternative energy sources or a new generation of nuclear plants?
30%
70%

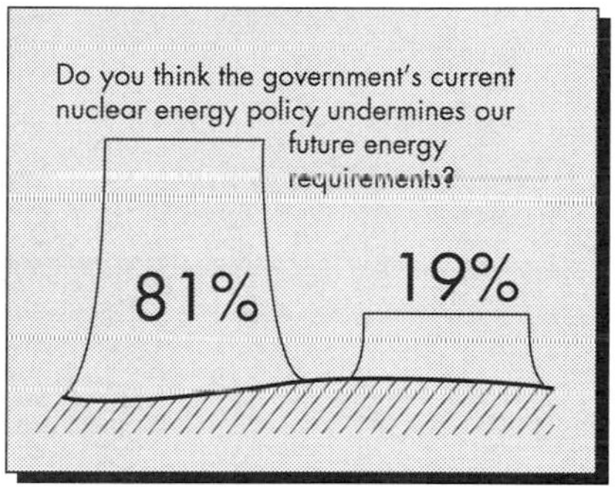

Voting on the second question, which posed the issue of safety, and on the question that asked if respondents would rather put up with blackouts than see new nukes built, revealed that there is less concern about safety than might be supposed. And there's definitely no strong sentiment for cutting back on energy consumption.

A California student explained his pro vote this way: "You're more likely to be struck by lightning twice than hurt by nuclear power." And a Coloradan simply explained his pro vote with: "It's a necessary evil, we have to have it."

Our April cover article also provoked the greatest volume of letters we've received for one story in recent memory. In these, there was a distinct trend *against* the proposed new generation of nuclear power plants, although the ratio was less pronounced: 60 percent to 40 percent (a sampling appears in the box at right).

Voters who opposed the new nukes also had strong feelings on the subject. One negative vote was cast by a Reno, Nev. cook who declared, "Scientists are destroying our society. I've lost all confidence in these people."

Some voters simply told us that there are no easy solutions to the energy problem—that fossil and nuclear fuels both posed threats to mankind. But virtually all of these comments came from people who voted in favor of the new nukes.

The voting also revealed a preference for spending more money on nuclear-energy research rather than on alternate-energy research. However, in our exit poll calls, there were a number of respondents who indicated that alternate energy is a viable alternative to nuclear and fossil fuels. And, by a margin of four to one, the voters also expressed a deep distrust of the government's energy policies. The vote indicates that there is a strong feeling that the government's energy policy undermines future needs.

Copies of the April article, the results of our poll, and the letters will be forwarded to leading policy makers, including President Bush.

Source: POPULAR SCIENCE AUGUST 1990

Random Rectangles

SCENARIO

Results from polls and other statistical studies reported in a newspaper or magazine often emphasize that the samples were randomly selected. Why the emphasis on randomization? Couldn't a good investigator do better by carefully choosing respondents to a poll so that various interest groups were represented? Perhaps, but samples selected without objective randomization tend to favor one part of the population over another. For example, polls conducted by sports writers tend to favor the opinions of sports fans. This leaning toward one side of an issue is called **sampling bias.** In the long run, random samples seem to do a good job of producing samples that fairly represent the population. In other words, randomization *reduces* sampling bias.

Question

How do random samples compare to subjective samples in terms of sampling bias?

Objectives

In this lesson subjective (or judgmental) samples will be compared to random samples in terms of sampling bias. The goal is to learn why randomization is an important part of data collection.

Activity

1. Judgmental samples
   a. Keep the accompanying sheet of rectangles covered until the instructor gives the signal to begin. Then look at the sheet for a few seconds and write down your guess as to the average area of the rectangles on the sheet. (The unit of measure is the background square. Thus, rectangle 33 has area $4 \times 3 = 12$.)
   b. Select five rectangles that, in your judgment, are representative of the rectangles on the page. Write down the area for each of the five. Compute the average of the five areas, and compare it to your guess. Are the two numbers close?
   c. The instructor will provide you with all class members' guesses and the averages of their subjective samples of five rectangles. Display the two sets of data on separate dot plots. Comment on the shape of these distributions and where they center. Why is the center an important point to consider?

2. Random samples
   a. Use a random number table (or a random number generator in a computer or calculator) to select 5 distinct random numbers between 00 and 99. Find the 5 rectangles with these numbers, using 00 to represent rectangle number 100, and mark them on the sheet. This is your random sample of five rectangles.
   b. Compute the areas of these five sampled rectangles, and find the average. How does the average of the random sample compare with your guess in step 1a? How does it compare with your average for the subjective sample?
   c. The instructor will collect the averages from the random samples of size 5 and construct a dot plot. How does this plot compare with the plots of the guessed values and the averages from the subjective samples in terms of center? In terms of spread?

3. Data analysis
   a. From the data the instructor has provided for the whole class, calculate the mean of the sample averages for the subjective samples and for the random samples. How do the centers of the distributions of means compare?
   b. Calculate the standard deviation of the averages for the subjective samples and for the random samples. How do the spreads of the distributions of means compare?
   c. Having studied two types of sampling, subjective and random, which method do you think is doing the better job? Why?

4. Sampling bias
   Your instructor will now place the true average area of the rectangles on each of the plots.
   a. Does either of the plots have a center that is very close to the true average?
   b. Does either of the plots have a center that is larger than the true average?

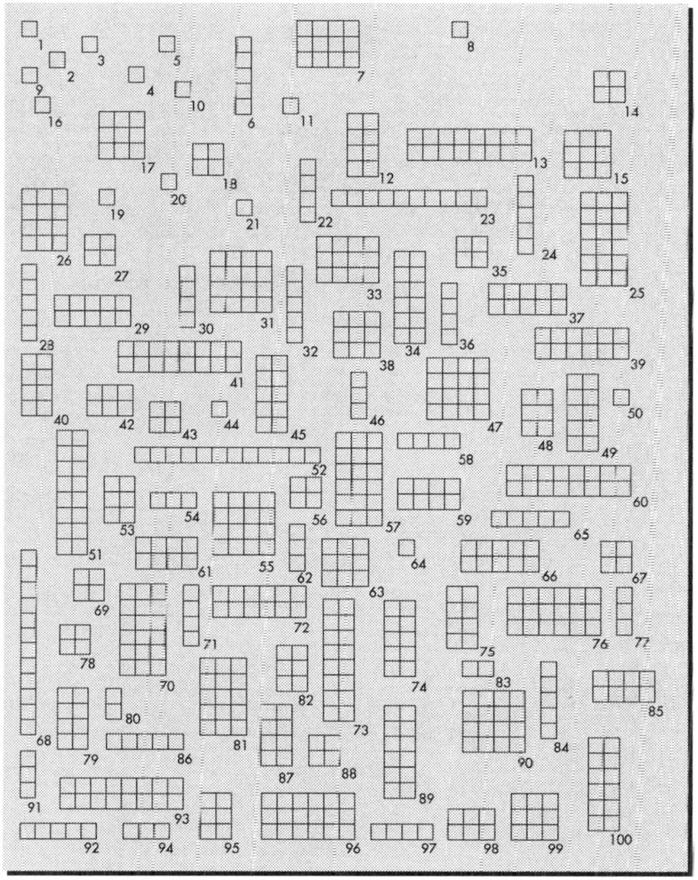

RANDOM NUMBER TABLE

39634	62349	74088	65564	15579	19713	39153	69459	17986	24537
14595	35050	40469	27478	44526	67331	93365	54526	22356	93208
30734	71571	83722	79712	25775	65178	07763	82928	31131	30196
64628	89126	91254	24090	25752	03091	39411	73146	06089	15630
42831	95113	43511	42082	15140	34733	68076	18292	69486	80468
80583	70361	41047	26792	78466	03395	17635	09697	82447	31405
00209	90404	99457	72570	42194	49043	24330	14939	09865	45906
05409	20830	01911	60767	55248	79253	12317	84120	77772	50103
95836	22530	91735	80210	34361	52228	33869	94332	83868	61672
65358	70469	87149	89509	72176	18103	55169	79954	72002	20582
72249	04037	36192	40221	14918	53437	60571	40995	55006	10694
41692	40581	93050	48734	34652	41577	04631	49184	39295	81776
61885	50796	96822	82002	07973	52925	75467	86013	98072	91942
48917	48129	48624	48248	91465	54898	61220	18721	67387	66575
88378	84299	12193	03785	49314	39761	99132	28775	45276	91816
77800	25734	09301	92087	02955	12872	89848	48579	06028	13827
24028	03405	01178	06316	81916	40170	53665	87202	88638	47121
86558	84750	43994	01760	96205	27937	45416	71964	52261	30781
78545	49201	05329	14182	10971	90472	44682	39304	19819	55799
14969	64623	82780	35686	30941	14622	04126	25498	95452	63937
58697	31973	06303	94202	62287	56164	79157	98375	24558	99241
38449	46438	91579	01907	72146	05764	22400	94490	49833	09258
62134	87244	73348	80114	78490	64735	31010	66975	28652	36166
72749	13347	65030	26128	49067	27904	49953	74674	94617	13317
81638	36566	42709	33717	59943	12027	46547	61303	46699	76243
46574	79670	10342	89543	75030	23428	29541	32501	89422	87474
11873	57195	32209	67663	07990	12238	59245	83638	23642	61715
13862	72778	09949	23096	01791	19472	14634	31690	36602	62943
08312	27886	82321	28666	72998	22514	51054	22940	31842	54245
11071	44430	94664	91294	35163	05494	32882	23904	41340	61185
82509	11842	86963	50307	07510	32545	90717	46856	86079	13769
07426	67341	80314	58910	93948	85738	69444	09370	58194	28207
57696	25592	91221	95386	15857	84645	89659	80535	93233	82798
08074	89810	48521	90740	02687	83117	74920	25954	99629	78978
20128	53721	01518	40699	20849	04710	38989	91322	56057	58573
00190	27157	83208	79446	92987	61357	38752	55424	94518	45205
23798	55425	32454	34611	39605	39981	74691	40836	30812	38563
85306	57995	68222	39055	43890	36956	84861	63624	04961	55439
99719	36036	74274	53901	34643	06157	89500	57514	93977	42403
95970	81452	48873	00784	58347	40269	11880	43395	28249	38743
56651	91460	92462	98565	72062	18556	55052	47614	80044	60015
71499	80220	35750	67337	47556	55272	55249	79100	34014	17037
66660	78443	47545	70736	65419	77489	70831	73237	14970	23129
35483	84563	79956	88618	54619	24853	59783	47537	88822	47227
09262	25041	57862	19203	86103	02800	23198	70639	43757	52064

c. Discuss the concept of bias in sampling and how it relates to the two sampling methods (subjective and random) you just used.

Wrap-Up

1. Discuss the difference between sampling bias and measurement bias. Give examples of statistical studies in which each is an important consideration. Is either of these biases reduced appreciably by increasing the sample size?

2. Find an article printed in the media that reflects sampling bias (The article below and on p. 93 is an example.). Discuss how the sampling bias could have affected the conclusions reported in the article.

Extensions

1. Using the same sheet of rectangles as used earlier, select multiple random samples of 10 rectangles each and compute the average area for each sample. Plot these averages, and compare the plot to the one for the random samples of size 5 with regard to
   a. center,
   b. spread, and
   c. shape.

2. Obtain a map of your state that shows the county boundaries. Your job is to show the geography department how to select a random sample of counties for purposes of studying land use. How would you select the sample? What might cause bias in the sampling process?

THE POPULAR SCIENCE NUCLEAR POWER POLL

LANDSLIDE FOR SAFER NUCLEAR

By a stunning majority, voters in the POPULAR SCIENCE poll [Cast Your Vote On Next Generation Nuclear Power, April '90] favor building a new generation of nuclear power plants. Responding to our detailed report on nuclear plant designs and concepts claimed to be safer than existing nuclear generating stations, voters from 38 states took part in the poll and more than 5,500 persons cast electronically recorded ballots on the main issue: Should the United States build more fossil-fuel power plants or the new so-called safe nuclear generators to meet the energy crisis of the '90s? The pro-nuke voters prevailed by a margin of more than six to one.

Why did the vote turn out in this fashion? In the follow-up interviews, voters said that nuclear power in a new package could meet energy needs in a safe, environmentally friendly way. Said one voter in Oklahoma City: "It's also better than acid rain." The Oklahoma reader also said he was in favor of building "little ones rather than mega ones." A New Jersey minister told us that "more people have gotten sick from Jane Fonda movies than from nuclear power."

A huge majority of voters also indicated a preference for nuclear power research over alternate energy. In addition, the tally showed disapproval of the government's energy policies.

The vote was clear-cut despite an attempt on the part of Westinghouse, one of the world's major nuclear power plant

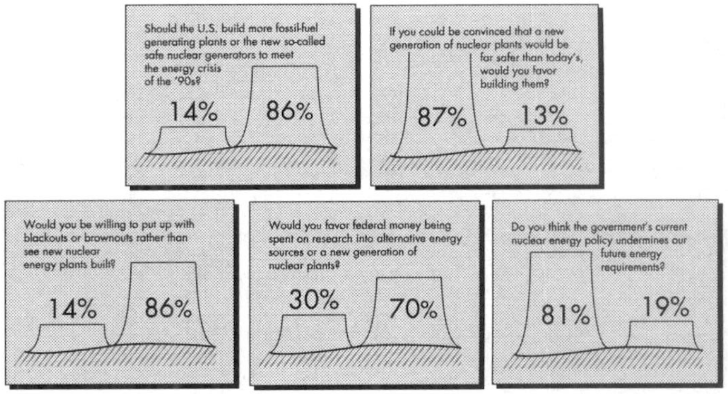
Should the U.S. build more fossil-fuel generating plants or the new so-called safe nuclear generators to meet the energy crisis of the '90s?
14%
86%
If you could be convinced that a new generation of nuclear plants would be far safer than today's, would you favor building them?
87%
13%
Would you be willing to put up with blackouts or brownouts rather than see new nuclear energy plants built?
14%
86%
Would you favor federal money being spent on research into alternative energy sources or a new generation of nuclear plants?
30%
70%
Do you think the government's current nuclear energy policy undermines our future energy requirements?
81%
19%

designers, to stuff the electronic ballot box. Westinghouse set up a special speed dialing number, enabling its Pittsburgh headquarters employees to call our 900 poll number (which cost voters 75 cents a minute) for free, and then publicized the article and poll in an internal newsletter. Westinghouse employees responded with a total of 20 percent of the total votes cast in our poll, according to our research. (POPULAR SCIENCE protested the actions of Westinghouse, criticizing the nuke manufacturer for a sinister and cynical disregard for an open and honest forum of free ideas. Richard Slember, vice president of Westinghouse's energy systems, replied: "We in no way tried to influence any votes, but we did want our employees to know about the poll and to have the opportunity to participate.")

Even when these votes are discarded, the result is a landslide favoring the building of a new generation of safer nuclear plants. If we assume that all the votes originating from Westinghouse were pro nukes and disqualify them, the margin is still about five to one in favor of the new nuclear plants.

Voting on the second question, which posed the issue of safety, and on the question that asked if respondents would rather put up with blackouts than see new nukes built, revealed that there is less concern about safety than might be supposed. And there's definitely no strong sentiment for cutting back on energy consumption.

A California student explained his pro vote this way: "You're more likely to be struck by lightning twice than hurt by nuclear power." And a Coloradan simply explained his pro vote with: "It's a necessary evil, we have to have it."

Our April cover article also provoked the greatest volume of letters we've received for one story in recent memory. In these, there was a distinct trend *against* the proposed new generation of nuclear power plants, although the ratio was less pronounced: 60 percent to 40 percent (a sampling appears in the box at right).

Voters who opposed the new nukes also had strong feelings on the subject. One negative vote was cast by a Reno, Nev. cook who declared, "Scientists are destroying our society. I've lost all confidence in these people."

Some voters simply told us that there are no easy solutions to the energy problem—that fossil and nuclear fuels both posed threats to mankind. But virtually all of these comments came from people who voted in favor of the new nukes.

The voting also revealed a preference for spending more money on nuclear-energy research rather than on alternate-energy research. However, in our exit poll calls, there were a number of respondents who indicated that alternate energy is a viable alternative to nuclear and fossil fuels. And, by a margin of four to one, the voters also expressed a deep distrust of the government's energy policies. The vote indicates that there is a strong feeling that the government's energy policy undermines future needs.

Copies of the April article, the results of our poll, and the letters will be forwarded to leading policy makers, including President Bush.

Source: POPULAR SCIENCE AUGUST 1990

Stringing Students Along: Selection Bias

NOTES FOR THE INSTRUCTOR

Statistical Setting

This lesson illustrates a sampling plan that looks random but still produces bias. The lesson should be done in close proximity to the "Random Rectangles" activity, as both illustrate similar notions of sampling bias. Since sampling by touch is size biased, you might want to guide more advanced students to think about how the size-biased sample could be adjusted to obtain a good estimate of the mean length of the strings.

Prerequisites for Students

Students should have some idea about the uses of sample data to estimate population parameters, and they should be familiar with the use of a random number table or generator. This should not be their first exposure to sampling.

Materials

Strings cut according to the procedure described below are needed for this activity.

Procedure

The activity works best within groups of about 10 students, so the class should be divided accordingly.

Select N random numbers between a and b. These represent service times of the customers at the bank window. $N = 25$, $a = 4$, and $b = 12$ work satisfactorily. For each random number, cut a string of that length in inches. These strings now represent the lengths of the service times.

For the first sampling, place the collection of N strings in a bag. Mix the strings thoroughly, and then have students reach in and draw out the first string touched. Record the string length, and return the string to the bag. Repeat the process until 10 string lengths are recorded.

For the second sampling, number the strings from 1 to N. An easy way to do this is to mark the numbers 1 to N on a table and place each string next to its number. Using random numbers, have students select a random sample of 10 strings (with replacement) and record the lengths of the sampled strings.

Provide the sample means for each group in the class, being careful to keep the two distributions separated. Students should plot these two distributions of sample means and comment on the differences.

Sample Results from the Activity

The stem plots and numerical summaries here show the results of 25 samples by each of the methods presented in the activity. The bag used here contained 35 strings, 5 each of lengths 2, 4, 6, 8, 10, 12, and 14 inches. Notice that the first method (touch) produces a sampling distribution of means that centers at nearly 10, while the second method (random numbers) produces a sampling distribution centering at 8, as it should. Also, notice that the variation is smaller for the first method. It is useful to have students explain why the variation is smaller for the touch method, in addition to discussing why the mean is larger.

```
Stem-and-leaf of Means    N = 25
Leaf Unit = 0.10

  1     7 8
  1     8
  3     8 68
  9     9 022444
 (8)    9 68888888
  8    10 44
  6    10 668
  3    11 2
  2    11 66
```

	N	MEAN	MEDIAN	TRMEAN	STDEV	SEMEAN
Means	25	9.848	9.800	9.861	0.906	0.181

	MIN	MAX	Q1	Q3
Means	7.800	11.600	9.300	10.500

```
Stem-and-leaf of Mean      N = 25
Leaf Unit = 0.10

   1    5 4
   2    5 8
   2    6
   5    6 668
  10    7 02244
 (3)    7 688
  12    8 24
  10    8 6688
   6    9 02
   4    9 668
   1   10
   1   10
   1   11 0
```

	N	MEAN	MEDIAN	TRMEAN	STDEV	SEMEAN
Mean	25	8.008	7.800	7.991	1.326	0.265

	MIN	MAX	Q1	Q3
Mean	5.400	11.000	7.100	8.900

Sample Assessment Questions

1. You are called upon to help design a sampling plan for the purpose of estimating the average length of stay for patients in a certain hospital. Two options are presented. The first involves randomly sampling patients who happen to be in the hospital the day the sampling is to be done. The second involves randomly sampling names from a list of hospital patients over the past month, and calling them to ask about the length of their stay. Which plan would you choose, and why?

2. Design a sampling plan to estimate the average size of farms in your county.

3. To estimate the average size of classes on a college campus, you could go to the campus, randomly sample some students, and ask them about the sizes of their classes. Or, you could go to the registrar's office and sample from a list of classes being offered. Which method will most likely produce the higher sample mean? Explain why this is true.

Stringing Students Along: Selection Bias

SCENARIO

People entering a bank queue need different lengths of time to complete their transactions. For example, depositing a check usually requires less time than obtaining foreign currency. We are interested in looking at ways to sample these times so that we might describe their distribution and estimate the average transaction time.

Question

Can I obtain a good estimate of mean transaction time at my bank by averaging the times of all the customers who are at the windows when I arrive?

Objective

In this lesson you will see a common form of selection bias in a sampling plan that appears to be random.

Activity

The goal is to estimate the average length of the N strings in the bag the instructor is holding. The sampling will be done by two different methods, both involving randomization.

1. Selection by touch
   a. Without looking into the bag reach into it and select a string. Record its length (to the nearest inch) and return the string to the bag. Repeat this process until 10 strings have been sampled.

   b. Calculate the average of your sampled string lengths.
   c. Report your average to the instructor, who will provide you with all the averages recorded in the class.

2. Selection by random numbers
   a. Lay the strings out on a table, and number them from 1 to N.
   b. Select 10 strings by choosing 10 random numbers between 1 and N and matching those numbers to the strings.
   c. Measure the length of the sampled strings, and calculate the average.
   d. Report the average to the instructor, who will provide you with all the averages recorded in the class.

3. Data analysis
   a. Construct a dot plot or stem plot of the averages from the sampling by touch.
   b. Construct a dot plot or stem plot of the averages from the sampling by random numbers.
   c. Compare the two plots. Do they have the same centers?
   d. Which sampling plan do you think produces the better estimate of the true mean length of the strings? Explain your reasoning.

Wrap-Up

1. a With regard to the service times in the bank, what method of sampling is represented by the sampling that involves reaching in the bag and selecting 10 strings.
   b. With regard to the service times in the bank, what method of sampling is represented by the sampling by random numbers?
   c. What is the key difference between the observed results for the two different sampling methods?

2. Look at a map of your state, or a neighboring state, that has the lakes marked in a visible way. An environmental scientist wants to estimate the average area of the lakes in the state. She decides to sample lakes by randomly dropping grains of rice on the map and measuring the area of the lakes hit by the rice. Comment on whether or not you think this is a good sampling plan.

Extensions

1. The sampling by touch method is said to suffer from selection bias, in that the longer strings get chosen more often than the shorter ones. What are the implications of this phenomenon in the case of sampling service times at the bank window?

2. Suggest other situations in which selection bias could be important. Discuss how one might get unbiased data in these situations.

ESTIMATION

Sampling Error and Estimation

NOTES FOR THE INSTRUCTOR

Statistical Setting

This activity can be used as a follow-up activity to "Random Rectangles" or by itself. It dramatically illustrates the central limit theorem although that is not its focus. It also provides an introduction to the concept of sampling error and can easily lead into a discussion of confidence intervals. (Note however, that unlike the "Random Rectangles" activity, the samples in this activity are drawn with replacement.)

Prerequisites for Students

Students should be familiar with simple random sampling and sampling distributions. They should know how to drawn random samples, computer sample means, and get stem and leaf plots of data.

Materials

The sheet of rectangles from the "Random Rectangles" activity and a random number table are needed for this activity.

Procedure and Sample Results from Activity

The activity is suitable for both large and small classes and can be used as an assignment or in-class exercise. Examples of stem and leaf plots of average areas of random samples of 5 rectangles (Figure 1) and 10 rectangles (Figure 2) appear next.

```
Stem-and-leaf of avgar5      N = 100
Leaf Unit = 0.10
     1     1  6
     1     2
     7     3  044666
    13     4  226668
    29     5  0000222444668888
    47     6  000022444666668888
  (18)     7  002222444666666888
    35     8  000222246668888
    20     9  000024468
    11    10  046688
     5    11  0448
     1    12
     1    13
     1    14
     1    15
     1    16  1
```

Figure 1

```
Stem-and-leaf of avgar10     N = 100
Leaf Unit = 0.10
     8     4  56667778
    19     5  22333667899
    41     6  0011223355566777899999
  (27)     7  000011122334456667777789999
    32     8  011223334455577889
    14     9  001122457
     5    10  1245
     1    11  5
```

Figure 2

For the above data in Figure 1, we get the following Table 1:

Table 1 SAMPLES OF SIZE 5

Sampling Error Between	# of Sample Means	Proportion of Means
−1 and +1	38	.38
−2 and +2	67	.67
−3 and +3	82	.82
−4 and +4	96	.96
−5 and +5	98	.98

Thus, for this set one of the answers to statement A will be

A. The proportion of sample means (for sample size 5) that are within 3 units of the population mean is .82.

This proportion increases as we allow the sampling error to increase, or as our sample size increases.

The margin of sampling error is 4. This means that one can expect approximately 95 out of 100 means to be within 4 units of the population mean. Obviously, your answers will depend on your simulation. Similar results can be obtained for samples of size 10.

In order to use this activity as a lead-in to confidence intervals, you can remind the students that in the real world the population mean is not known, and one is forced to use the theoretical expression for the margin of error, namely, $z_{\alpha/2}\sigma/n$.

Sample Results from Extensions

The Extension asks students to calculate these margins of error for a probability of .90 and .95 and compare them to the simulated data. For example, the population standard deviation is 5.2, and the theoretical 90% margin of error is given by $1.645(5.2\sqrt{5})$, which is 3.8. In the data set 92% of the sample means lie within 3.8 units of the population mean, which compares favorably to the theoretical 90%. The confidence interval then can be derived as the sample mean plus the appropriate margin of error.

Sampling Error and Estimation

SCENARIO

How far can an estimate of a parameter be from the true value of the parameter? In particular, how far can a sample mean be from a population mean? For example, all car manufacturers are required to post the average gas mileage of their cars. Typically, an estimate is obtained by taking a sample of cars, driving them under different driving conditions, and calculating the average gas mileage for this sample. It is also assumed that if everything stays the same, this sample mean can be used as an estimate of the mean of the gas mileage of all cars of that model. But how accurate is that estimate? How far is it from the true value? Can we get some understanding of the difference between the estimate and the true value, called the **sampling error**? How often will we observe large sampling errors? If we made certain assumptions, there are theoretical results that will give us a **margin of sampling error.** For example, based on our sample standard deviation and sample size, we may be able to state that the sampling error should be between −5 and +5 in, say, 90 out of 100 cases. So we can assume that there is a 90% chance that our sample mean is within 5 units of the unknown population mean. Another way of saying it is that the margin of sampling error is ±5 with a probability of 90%.

Question

How does an **estimate** of a measure of location, the population mean, differ from the **true value** of the parameter?

Objective

This activity illustrates the sampling distribution of the sampling mean, the sampling error, the margin of error, and the effect of the sample size on these quantities.

Activity

1. Collecting the data
   Using the rectangle sheet from the earlier "Random Rectangles" activity and a random number table, draw as many samples of size 5 as required in order to have 100 samples for the class. Each student will calculate the average area of the sample(s) of rectangles. The instructor will collect these averages and pass out a list of them to the class. Repeat the data collection procedure, using samples of size 10.

2. Studying sampling procedures
   a. Construct stem and leaf plots of the sample means for each sample size. These plots give you approximate sampling distributions of the sample means. They are approximate because they are based on only 100 samples.
   b. Guess the value of the population mean based on the stem and leaf plots. Show the population mean provided by your instructor on the plots. How does this value compare to your guess?
   c. Count the sample means that lie within one unit of the population mean. We say that these means have **sampling errors** between −1 and +1, or that these means are at most 1 unit away from the population mean. Complete the following table for different values of the sampling error.

Table 1

Sampling Error Between	Number of Sample Means	Proportion of Means
−1 and +1		
−2 and +2		
−3 and +3		
−4 and +4		
−5 and +5		

   d. Repeat steps 2a–2c using the second sampling distribution with samples of size 10. Label the table Table 2.

3. Analyzing the sampling error
   a. Compare the sampling distributions for the two sample sizes with respect to the shape and spread of the distributions. Are both distributions centered at about the same point? Do the sample means have the same range of values for the two sample sizes?
   b. Using Table 1, complete the following statement for different values of the sampling error:

      A. The proportion of sample means (for sample size 5) that are within ______ units of the population mean is ______.

      What happens to this proportion of sample means as we allow the sampling error to increase; that is, what happens to the proportion as we look at different intervals of sampling errors?
   c. What happens to this proportion of sample means as we allow the sample size to increase to 10 in Table 2?
   d. The term "margin of sampling error" or "margin of error" is often used to indicate the size of the sampling error that produces a proportion of .95 in the answer to statement A.

   Find an approximate "margin of sampling error" for the rectangle area data with $n = 5$. Repeat for $n = 10$.

Wrap-Up

Often in public surveys and opinion polls, the results of the poll give the value of an estimate as well as the margin of error for that estimate, generally associated with a chance of .95. This is supposed to help the readers better interpret the value of the estimate. Find a newspaper article that is reporting results of a poll, including the sampling error or the margin of error. Write a paragraph explaining what the sampling error or the margin of error mean in the context of the poll.

Extensions

1. If you have a choice between two estimates, the sample mean from a sample of size 5 and a sample mean from a sample of size 10, which one would generally give a "better" estimate? Clearly state your definition of "better."

2. In practice, people do not know the population mean like you did in step 2 and do not have several estimates to get a sampling distribution of a sample mean. Suppose you have only one sample. In that case you would need to use theoretical results. If you have studied the central limit theorem, find the margins of error associated with a chance of .90 and a chance of .95. Compare these values with those calculated in step 2.

Estimating Proportions: How Accurate Are the Polls?

NOTES FOR THE INSTRUCTOR

Statistical Setting

The problem of estimating a population proportion is the simplest realistic setting available for introducing the ideas of sampling distribution and margin of error. This lesson should precede any discussion of confidence intervals, and it forms a nice bridge to confidence intervals. Emphasize the critical role of random sampling in producing predictable sampling distributions.

The notion of a sampling distribution is fundamental to the understanding of statistical inference. Therefore, it is important to spend enough time on the subject for students to grasp the idea.

Prerequisites for Students

Students should have some experience with the notion of random sampling, some understanding of the empirical rule (95% rule) for mound-shaped, symmetric distributions, and facility at plotting distributions of data (for which a computer is helpful). Also, they should be able to calculate sample statistics such as the standard deviation.

Materials

In this activity students will randomly sample discrete items of at least three different types and generate sets of sample proportions. This could be done with bead boxes (with at least three colors of beads), boxes of marbles, candies (such as M&M's), or random number tables (although this is, perhaps, least interesting). In each case, the instructor should know the true proportion of items in the population before sampling is begun.

Procedure

Allow students to examine the contents of the "bead box" so that they see beads of different colors and can select a color that occurs reasonably often. The procedure does not work well for population proportions close to zero or one.

Have students discuss various ways of selecting random samples. The idea of "thorough mixing" of the beads should be mentioned.

Work is simplified and the activity speeded up by teamwork. Actually, the teams can be assigned different sample sizes; not all teams have to use all sample sizes, but five or six different sample sizes should be used overall to see the effect.

Sample Results from Activity

The attached display shows computer-selected samples of sizes 10, 20, 40, 80, and 100 from a population with 40% red balls. Each display contains 100 randomly selected samples. The mound shapedness of the distributions and the centering around .4 are readily apparent.

The variances of the sample proportions within each display are, respectively,

$$.0241,\ .0112,\ .0069,\ .0031,\ .0021.$$

A plot of these values against n shows a decreasing curve. A plot of these values against $(1/n)$ shows an increasing straight line with slope close to .24.

Sample Assessment Questions

1. Assign students, individually or in groups, to write a critique of a newspaper or magazine article that includes the results of a sample survey designed to estimate a proportion. The critique might be arranged according to the following outline:
   a. What are the **objectives** of the study?
   b. What is the **target population** of the study?
   c. How was the **sample** selected; was **randomization** used?
   d. What was the **method of measurement;** is **bias** a potential problem?
   e. How are the results **summarized?**
   f. Is the **data analysis** complete, or should more have been reported?
   g. What are the **conclusions** of the study?
   h. Given your knowledge of **margin of sampling error,** do you agree with the reported conclusions?
   i. If you were to do a **follow-up study,** how would you design it and what data would you collect?

   Be sure that students develop the connection between margin of error and the notion of randomization in the sampling. Does it make sense to report a margin of error for a poll that was not based on a random sample?

2. Have students interpret sample proportions and their margins of error by relating them to decision making in the world around them. Two scenarios for such discussion are presented below.
   a. There is a true–false test coming tomorrow and you expect to know none of the answers. That is, you will have to guess at the answers to all of the questions. You must get 70% correct to pass the test. Would you rather have a 10-question test or a 20-question test? Why?
   b. The National Football League (NFL) plays a 16-game schedule with the best team in the league usually having around a .875 winning percentage and the worst team in the league having around a .125 winning percentage. The National Basketball Association (NBA) plays an 82-game schedule with the best team usually having around a .700 winning percentage and the worst team having around a .300 winning percentage. Major League Baseball (MLB) plays a 162-game schedule with the best team usually having around a .600 winning percentage and the worst team having a .400 winning percentage. It might appear that the NFL is less competitive than the NBA, which, in turn, is less competitive than MLB. Do you agree with this assessment? Why or why not?

3. Each issue of *The Gallup Poll Monthly* contains a statement similar to the one presented below. After reading the statement, comment upon the accuracy of their explanation of sampling error. Do you agree with this statement?

> Readers are cautioned that all sample surveys are subject to the potential effects of sampling error, a divergence between the survey results based on a selected sample and the results that would be obtained by interviewing the entire population in the same way. The risk of this kind of divergence is necessary if probability sampling is used, and probability sampling is the basis for confidence in the representativeness of sample survey results.
>
> The chance that sampling error will affect a percentage based on survey results is mainly dependent upon the number of interviews on which the percentage is based. In ninety-five out of 100 cases, results based on national samples of 1000 interviews can be expected to vary by no more than 4 percentage points (plus or minus the figure obtained) from the results that would be obtained if all qualified adults were interviewed in the same way. For results based on smaller national samples or on sub-samples (such as men or persons over the age of fifty), the chance of sampling error is greater and therefore larger margins of sampling error are necessary in order to be equally confident of survey conclusions.
>
> In addition to sampling error, readers should bear in mind that question wording, and practical difficulties encountered in conducting surveys can introduce additional systematic error or "bias" into the results of opinion polls. Unlike sampling error, it is not possible to estimate the risk of this kind of error in a direct way, but survey organizations can protect against the effects of bias on survey conclusions by focusing careful attention on sampling, questionnaire construction, and data collection procedures and by allowing an adequate of time for the completion of data collection.

In addition to sampling error, readers should bear in mind that question wording and practical difficulties encountered in conducting surveys can introduce error or bias into the findings of opinion polls.

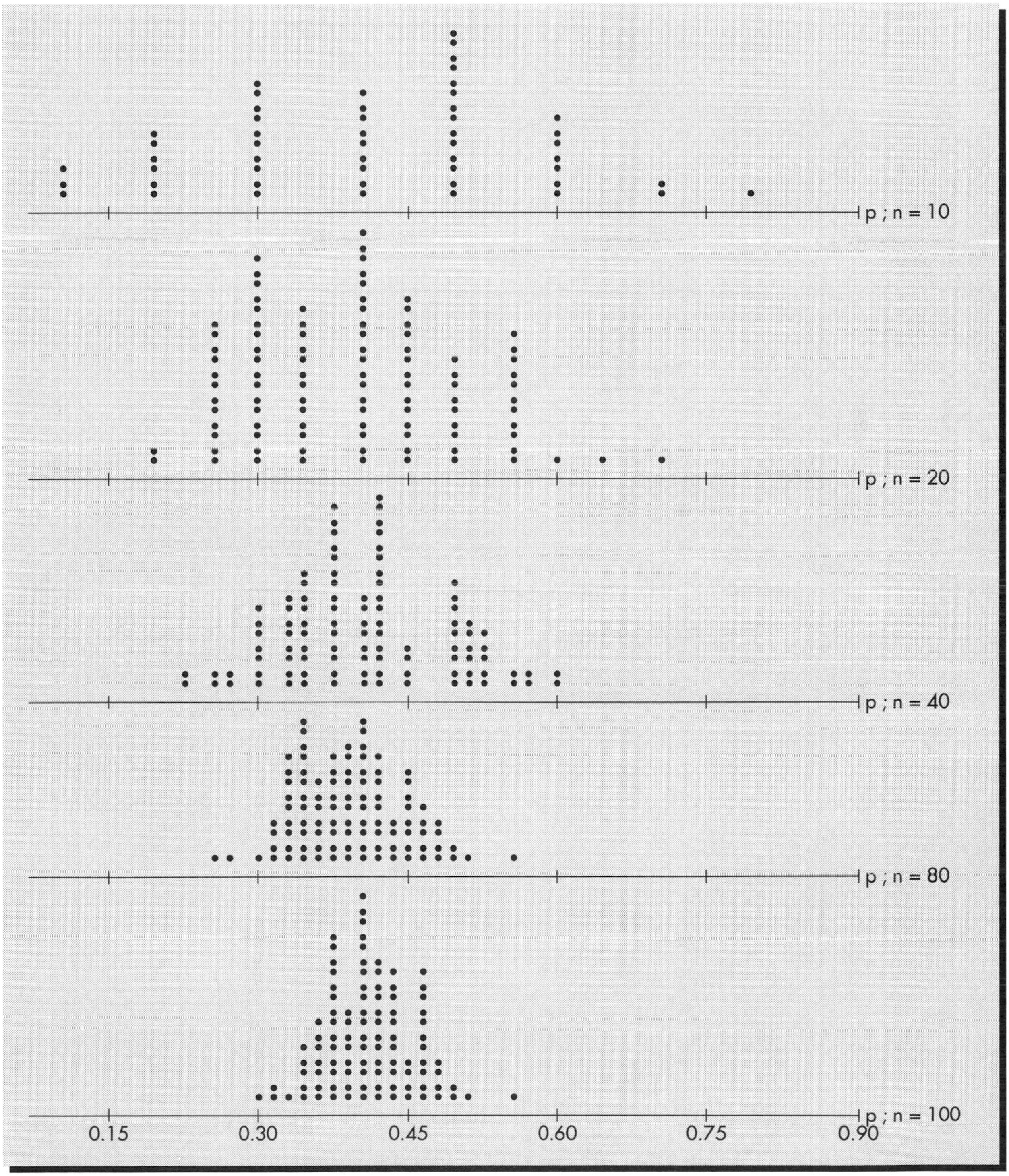

	N	MEAN	MEDIAN	STDEV
p;n = 10	100	0.4100	0.4000	0.1554
p;n = 20	100	0.3945	0.4000	0.1061
p;n = 40	100	0.40475	0.40000	0.08285
p;n = 80	100	0.39263	0.38750	0.05554
p;n = 100	100	0.41120	0.41000	0.04549

Estimating Proportions: How Accurate Are the Polls?

SCENARIO

A recent Gallup Poll says that 60% of the public favor stricter gun control laws and that this result was based on a survey of 1,200 people. The sampling error is reported to be 3%. What does this mean? In the Clinton–Bush–Perot presidential race of 1992, the major polls on the day before the election all estimated Clinton would receive around 43% or 44% of the vote, with a margin of error of about 2.5%. Clinton actually collected 43% of the popular vote. The polls did well! How can polls do so well in estimating proportions? How is the sampling error calculated, and what does it mean? These questions will be investigated in this activity.

Questions

Why do opinion polls work, and how can their accuracy be measured?

Objectives

In this lesson you will generate **sampling distributions** of sample proportions and study their patterns for the purpose of understanding the concept of the **margin of sampling error.**

Activity

1. Working with your team members, examine the box of "beads" given to you for this activity.
   a. Choose one color of bead, say red, which seems to occur relatively often in the box.
   b. Select a random sample of 10 beads from the box, recording the sample proportion of red beads.
   c. Repeat the sampling procedure four more times, for a total of five sample proportions for red beads. Each sample should be returned to the box before the next is selected, since all samples are supposed to come from the same population.
   d. Record the five sample proportions, and give the results to the instructor.

2. Using the same bead color, repeat the procedure of step 1 for samples of size

   20, 40, 80, 100

   Make sure you record five sample proportions for each sample size; give the results to your instructor.

3. Construct dot plots of the five sample proportions for each sample size on a single real number line. You may want to use different symbols for the different sample sizes. Do you see any pattern emerging?

4. Your instructor will now provide the data from all teams. Using a graphing calculator or computer, construct dot plots of the sample proportions for each sample size. (You should have five dot plots, one each for samples of size 10, 20, 40, 80, and 100.) These distributions are approximations to the sampling distributions of these sample proportions.

5. Describe the patterns displayed by the individual sampling distributions. Then describe how the pattern changes as the sample size increases. In particular, where do the dot plots appear to center? What do you think is the true proportion of red beads in the box?

6. Calculate the variance for the sample proportions recorded in each dot plot.

7. Plot the sample variances against the sample sizes, with sample size on the horizontal axis. What pattern do you see? Does this look like the points should fall on a straight line?

8. It appears that the variances are related to the sample size, but the relationship is not linear. Plot the variances against (1/sample size), the reciprocal of the sample size. Observe the pattern that appears. "Fit" a straight line through this pattern either by eye or by using linear regression. Does the straight line appear to fit well?

9. What is the approximate slope of the line fit through the plot of variance versus reciprocal of sample size? Can you see any relationship between this slope and the true proportion of red beads in the box, as estimated in step 5?

10. The above analysis should suggest that the slope of the line relating variance to (1/sample size) is approximately

$$[p(1-p)],$$

where p represents the true proportion of red beads in the box.
   a. Write a formula for the variance of a sample proportion as a function of p and the sample size, n.
   b. Write a formula for the standard deviation of a sample proportion as a function of p and n.

11. A two-standard-deviation interval is called the margin of sampling error (or the margin of error or the sampling error) by most pollsters.
   a. In repeated sampling, how often will the distance between a sample proportion and the true proportion be less than the margin of error?
   b. Using the class dot plots constructed earlier in this lesson, count the number of times a sample proportion is less than two standard deviations from the true proportion for each plot. Do the results agree with your answer to part a?

Wrap-Up

1. Revisit the Scenario at the beginning of this lesson. For a Gallup poll, a typical sample size is 1,000. For the Clinton–Bush election polls, the sample sizes were around 1,500.
   a. Are the reported margins of error correct?
   b. Interpret the results of these polls in light of the margin of error.

2. Does the concept of margin of error make sense if the data in a poll do not come from a random sample? Explain.

3. Write a brief report on what you learned about sampling distributions and margins of error for proportions.

Extensions

1. Find examples of polls published in the media. If the margin of error is given, verify that it is correct. If the margin of error is not given, calculate it. Discuss how the margin of error helps you to interpret the results of the poll.

2. The approximation to the standard deviation of a proportion was developed in step 10 for a single true value of p. How do we know it will work for other values of p? To see that it does, work through the steps outlined earlier for samples selected from a population with a different value of p. (The approximation does not work well for values of p very close to 0 or 1; so choose your new value of p between .1 and .9.)

What Is a Confidence Interval Anyway?

NOTES FOR THE INSTRUCTOR

Statistical Setting

In this activity students build a model that can be used to visualize confidence intervals. From this model, students can understand how a sample can give a good indication of the results that would occur had the whole population been observed. The activity has been used successfully as an introduction to confidence intervals.

Prerequisites for Students

Students should know how to simulate samples from a given binomial population by using a random digit table.

Materials

Students will need a random digit table or access to a computer program such as Minitab that can draw random samples from a binomial population.

Procedure and Sample Results from Activity

For step 1, you should have exactly 40 people. If there are fewer than 40 students in your class, either bring enough results from another class to bring the total up to 40 or ask students during the previous class session to test roommates, parents, etc., and bring the results to class.

1. Proportions will vary. For illustration, we will suppose that 24 out of 40, or 0.60, of the sample are right-eye dominant.

2. a. and b. Make sure your students use different portions of the random digit table. A typical completed frequency table might look like the following one, which was generated using the Random Data function in Minitab.

Number of Successes	Frequency
9	2
10	1
11	5
12	6
13	9
14	9
15	11
16	9
17	10
18	12
19	12
20	2
21	4
22	5
23	2
24	1
Total	100

You may want to mention to your students that the fact that the percentage in this question is 40 and the sample size is also 40 is a coincidence.

c. For our illustrative proportion of 0.60, no. It's possible, but not plausible. Only 1 time in a 100 will a sample from a population with 40% right-eye dominant have 24 or more people right-eye dominant.

d. For our typical frequency table in step b, the answers are 10 or fewer and 23 or more.

e. For our typical frequency table in step b, the chart should look like Figure 1 here.

f. Yes, we would be surprised. Looking at our typical frequency table, there is less than a 1% chance of getting a random sample with 25 or more who make this claim. We would suspect either that the sample wasn't random or that the 40% number is wrong.

g. Yes, this result is quite plausible. The typical frequency table shows that there is about a 14% chance of getting 12 or fewer women who are not working.

3. Assign the percentages 10, 20, 30, 50, 60, 70, 80, and 90 to different groups. If you have enough groups, assign also 5, 15, 25, etc. Do the 0% and 100% cases as a class. When the groups are finished, they should have a chart roughly like Figure 2.

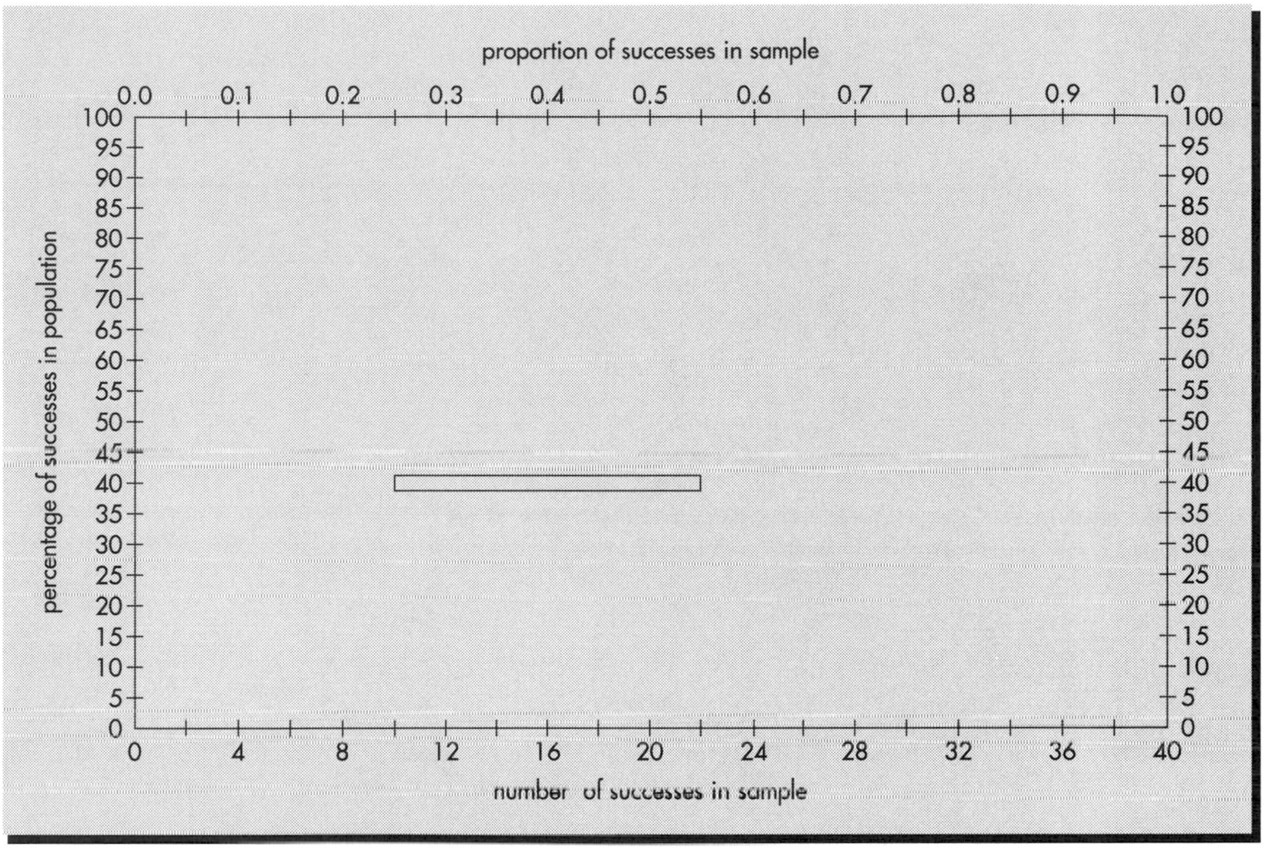

Figure 1 90% Box Plots from Samples of Size 40

4. a. No. From a population with 30% successes, we would usually get between 7 and 17 successes in a random sample of size 40.
   b. No. From a population with 75% successes, we would usually get proportions of between about 0.625 and 0.85 successes.
   c. Looking at the 80% box, 28 to 36; 0.70 to 0.90.
   d. Looking at the 50% box, about 15 to 25.
   e. We'll use the 45% box, since that is the closest to 43%: 0.325 to 0.575.

5. Using our illustrative proportion of 0.60, the answers are yes and 50% to 70%. Have students draw a vertical line from the proportion of successes in the sample to the number of successes in the sample. The boxes that are crossed by the line represent the "plausible" population percentages.

   Remind students that the validity of the confidence interval procedure is based on the assumption of a random sample. People who enroll in a statistics class may for some reason not be representative of the population as far as eye dominance is concerned.

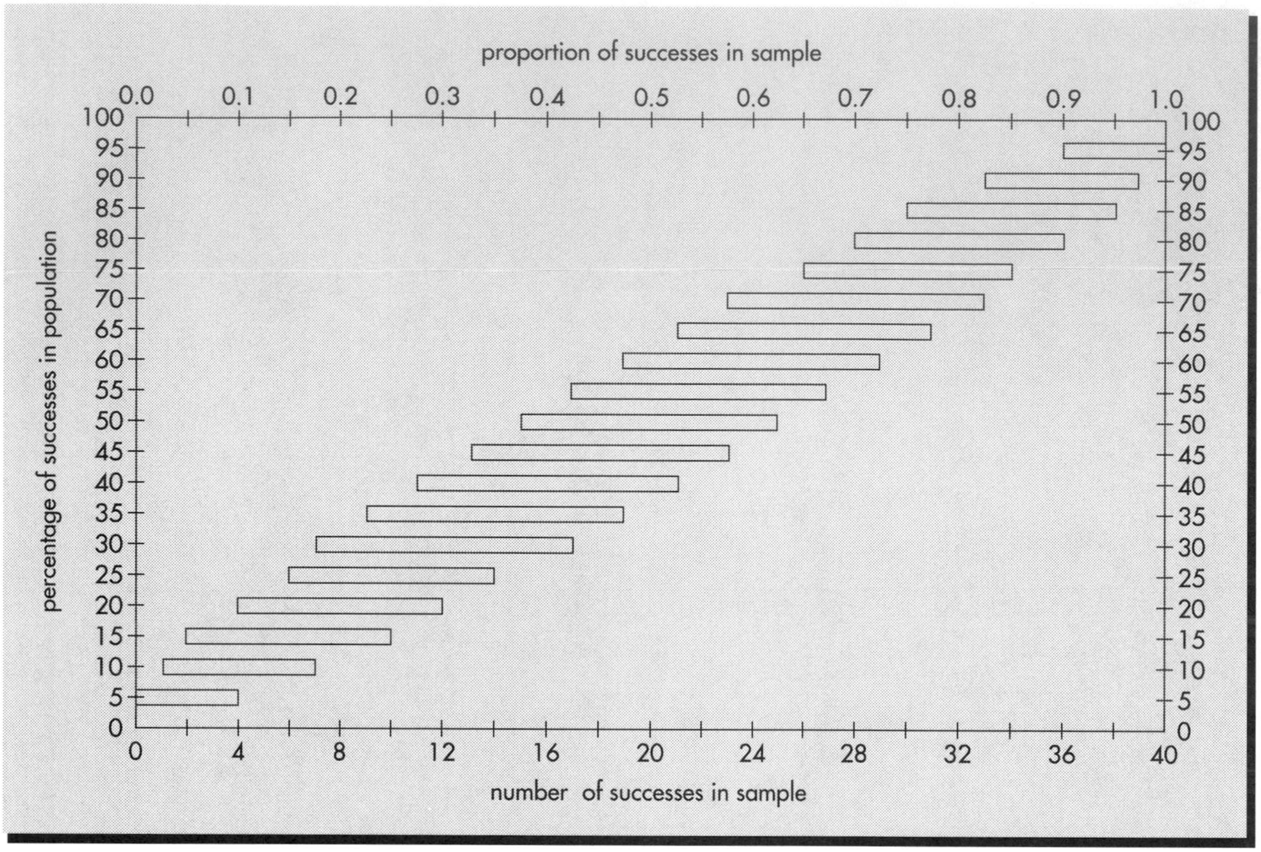

Figure 2 90% Box Plots from Samples of Size 40

6. a. 75% to 90%. (A percentage is included if the vertical line drawn upward from 34 hits the edge of the box.)
   b. 15% to 35%.
   c. Answers will vary.

Sample Results from Wrap-Up

1. Saying that the percentage is 50% with a margin of error of 10% is the same as saying that the confidence interval is 40% to 60%. This is the correct confidence interval. Looking at the chart, we see that getting 20 females in a random sample is a plausible result for the populations with percentages female of 40%, 45%, 50%, 55%, and 60%.

2. As long as the sample is random, it doesn't take a very large sample size until we are sure that we know to a reasonable degree of accuracy what the population percentage is. For example, even with a sample size of only 40, we are pretty sure that

our sample proportion will be within 10% to 12% of the percentage in the population. For example, look on the chart at the population with 50% "successes." Only 10% of the sample proportions were more than 12.5% away from 50%. So if we take a random sample from a population with 50% successes, there is a 90% chance that the sample proportion will be within 12.5% of the true population percentage. Look at another population on the chart, say 20%. There is about a 90% chance that the sample proportion will be within 10% or 20%. Roughly the same is true for all of the populations on the chart. Thus, even with a sample size as small as 40, there is a 90% chance that the proportion we get in our sample is within 10% or 12% of the actual percentage in the population.

Sample Results from Extensions

1. 95% confidence intervals will be longer. To get 95% of the sample proportions in the horizontal box rather than only 90%, the box will have to be longer. When boxes are longer, more of the population proportions will be crossed by the verti-

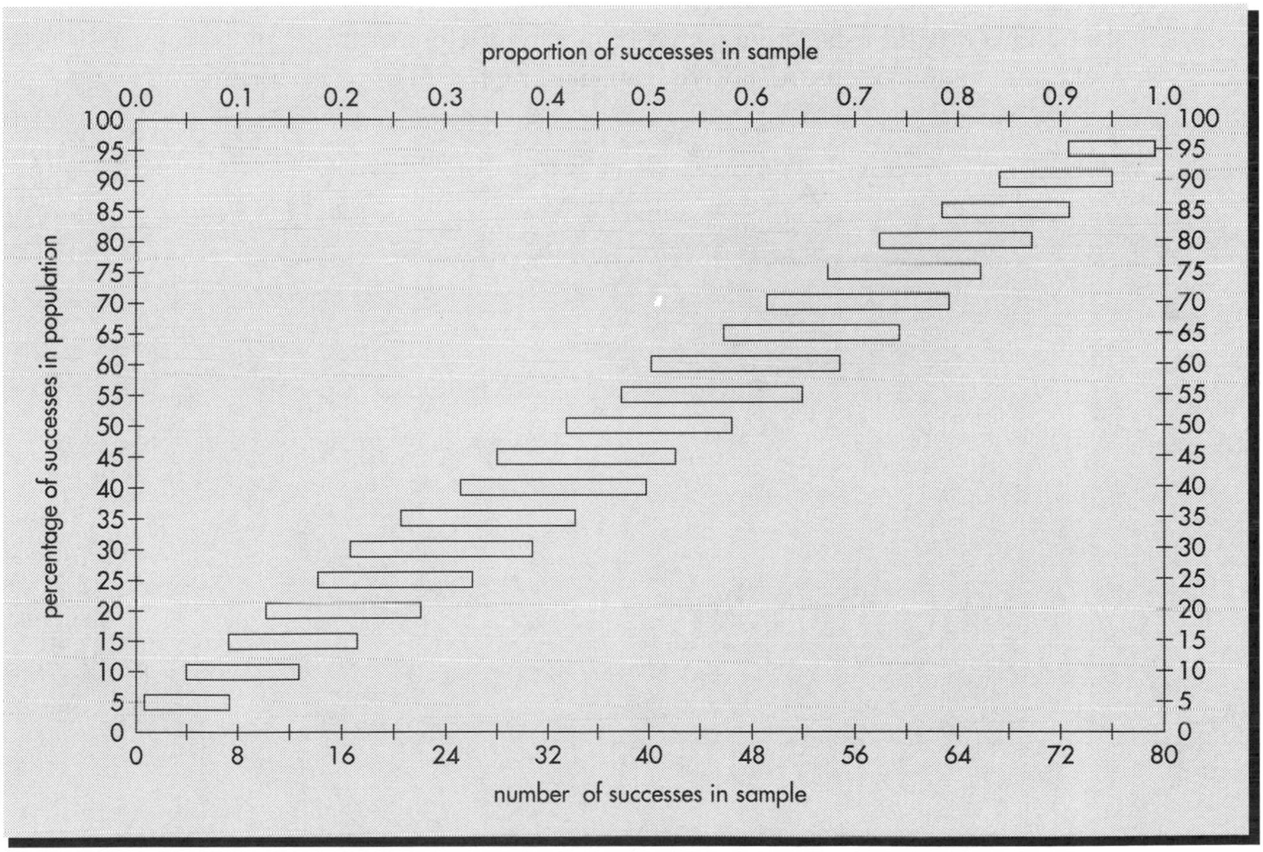

Figure 3 90% Box Plots from Sample Size 80

cal line drawn down from the proportion in the sample. This is the same as saying that the confidence interval will be longer.

2. They will be shorter. The simulation is exactly the same as carried out earlier for the case of a sample size of 40. The completed chart appears here as Figure 3.

Sample Assessment Questions

1. A random sample of 40 statistics students at a large university finds that 27 think the textbook is inscrutable. If you were to ask all statistics students at that university if they think their textbook is inscrutable, explain what you would expect to find.

2. The most difficult idea you will probably encounter in an introductory statistics course is that of a confidence interval. What makes this idea difficult to understand?

Acknowledgment

The idea behind this activity, that of making the chart of box plots, is over 20 years old and came from an inventive Canadian high school teacher, Jim Swift.

What Is a Confidence Interval Anyway?

SCENARIO

Forty people are selected at random and given a test to identify their dominant eye. The person holds a piece of paper about $8^1/_2$" by 11" with a 1" by 1" square cut in the middle at arm's length with both hands. The person looks through the square at a relatively small object across the room. The person then closes one eye. If he or she can still see the object, the open eye is the dominant eye. If not, the closed eye is the dominant eye.

Question

Is this sample of only 40 people large enough for us to come to any conclusion about what percentage of people have a dominant right eye?

Objective

In this activity you will learn how to construct confidence intervals using simulation, and you will learn how to interpret confidence intervals.

Activity

1. Conduct the experiment described in the Scenario with 40 students from your class, adding other people if necessary to bring the total up to 40. What proportion of your sample was right-eye dominant?

2. In this activity, you will be taking samples from a population in which 40% have some characteristic in order to see how close the proportions in the samples tend to come to 40%.
   a. Use a random digit table to simulate taking a sample of size 40 from a population with 40% "successes." Let the digits 0, 1, 2, 3 represent a success and 4, 5, 6, 7, 8, 9 represent a failure. Place a tally mark in a table like the one shown here to represent your result.

Number of Successes	Frequency
0	
1	
2	
.	
.	
.	
40	
Total	100

   b. Combine your results with other members of your class, repeating the simulation until your class has placed tally marks from 100 different samples in the frequency column of the table.
   c. Comparing your proportion from step 1 to the frequency table from step b, is it plausible (has a reasonable chance of being the case) that 40% of the population are right-eye dominant? Explain.
   d. Complete the following sentences based on your frequency table from step b: Less than 5% of the time, there were ______ successes or fewer. Less than 5% of the time, there were ______ successes or more.
   e. On the chart "90% Box Plots from Samples of Size 40" draw a thin horizontal box aligned with the 40% "Percentage of Successes in Population" on each side. Using "Number of Successes in Sample" on the bottom as a guide, the box should stretch in length between your two answers to step d.
   f. About 40% of Americans aged 19 to 28 claim that they have used an illicit drug other than marijuana. If a random sample of 40 Americans aged 19 to 28 finds 25 who claim to have used an illicit drug other than marijuana, would you be surprised? Explain.
   g. According to the U.S. Bureau of Labor Statistics, about 40% of women with children under the age of 6 do not participate in the labor force. Would it be plausible for a survey of 40 randomly chosen mothers of children under the age of 6 to find that 12 are not working? Explain.
3. The class should now divide into groups of about five students each. Your instructor will give each group one of the percentages on the sides of the chart "90%

Box Plots from Samples of Size 40." With your group, repeat steps a, b, d, and e for your new percentage.

4. Use the completed chart to answer these questions.
   a. According to the 1990 U.S. Census, about 30% of people aged 25 to 44 live alone. In a random sample of 40 people aged 25 to 44, would it be plausible to get 20 who live alone?
   b. We select a random sample of size 40 and get a sample proportion of 0.90 successes. Is this sample proportion plausible if the population has 75% successes?
   c. According to the 1990 U.S. Census, about 80% of men aged 20 to 24 have never been married. In a random sample of 40 men aged 20 to 24, how many unmarried men is it plausible to get? What proportion of unmarried men is it plausible to get?
   d. Suppose you flip a coin 40 times; how many heads is it plausible for you to get?
   e. In the 1992 presidential election, Bill Clinton got 43% of the vote. In a random sample of 40 voters, what is the largest proportion of people who voted for Clinton that it is plausible to get? the smallest?

5. Based on the chart your class has completed, it is plausible that 60% is the percentage of the population that are right-eye dominant? What percentages are plausible? These percentages are called the **90% confidence interval.**

6. Use your completed chart to find these confidence intervals.
   a. Suppose that a random sample of 40 toddlers finds that 34 know what color Barney is. What is the 90% confidence interval for the percentage of toddlers who know what color Barney is?
   b. Suppose that a random sample of 40 adults finds that 10 know what color Barney is. What is the 90% confidence interval for the percentage of adults who know what color Barney is?
   c. Observe 40 students on your campus. Find the 90% confidence interval for the percentage of students who carry backpacks.

Wrap-Up

1. Polls usually report a "margin of error." Suppose a poll of 40 randomly selected statistics majors finds that 20 are female. The poll reports that 50% of statistics majors are female, with a margin of error of 10%. Use your completed chart to explain where the 10% came from.

2. People often complain that election polls cannot be right because they personally were not asked how they were going to vote. Write an explanation to such a person about how polls can get a good idea of how the entire population will vote by asking a relatively small number of voters.

Extensions

1. This activity used 90% confidence intervals because it is easy computationally to find the bottom 5% and the top 5% of a distribution. Usually, 95% confidence intervals are reported. Will 95% confidence intervals be longer or shorter than 90% confidence intervals? Explain.

2. Will the confidence intervals for samples of size 80 be longer or shorter than those for samples of size 40? Design and carry out the simulation needed to answer this question (Figure 1).

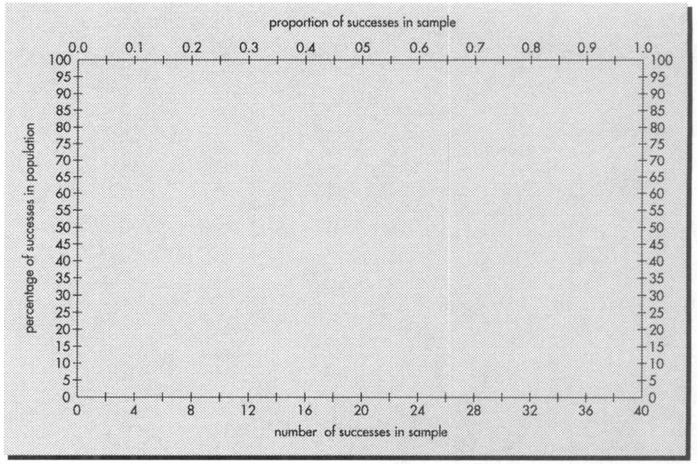

Figure 1 90% box plots from samples of size 40.

References

James M. Landwehr, Jim Swift, and Ann E. Watkins (1987), *Exploring Surveys and Information from Samples*, Palo Alto, CA: Dale Seymour Publications.

Confidence Intervals for the Percentage of Even Digits

NOTES FOR THE INSTRUCTOR

Statistical Setting

Students will take a simple survey: counting the number of even digits in a sample of 200 random digits. If each student constructs a 95% confidence interval for the percentage of even digits, is it true that approximately 95% of the students' 95% confidence intervals will contain 50%?

This activity helps students understand that any statement about "95% confidence" includes sampling error only. If there are errors in measurement or other problems with survey design, we can't expect that 95% of all confidence intervals will contain the population parameter.

Prerequisites for Students

Students should be able to compute a 95% confidence interval for a proportion using the following formula:

$$\hat{p} \pm 1.96\sqrt{\frac{\hat{p}(1-\hat{p})}{n}}\,.$$

Materials

Each student will need a random digit table with 200 random digits (different for each student) and 2 copies of the attached transparency (or it could be copied onto the blackboard).

Procedure

Students should not look at the student pages until after they have counted their samples of 200 random digits and computed their confidence interval.

This activity works best in a large lecture class, but it can be used with any class with at least 25 students. If your class is smaller, have each student do Activity steps 1 to 4 several times.

Sample Results from Activity

1. Each student should be given a different set of 200 random digits.

2. Tell students to count the number of even digits in their sample. Give students the minimum amount of time to do this. Try to discourage questions so that students won't ask whether 0 is an even digit or an odd digit. If you are asked, tell them to use their best judgment on how to perform this task.

3. Each student now constructs a 95% confidence interval based on the number of even digits counted and the sample size of 200. Answers will vary.

4. Using the attached transparency master, or creating it on the blackboard, display your students' confidence intervals as horizontal lines. Students like to do this themselves, each student drawing a horizontal line to represent his or her own confidence interval.

5. 50%

6. Count the number of students whose confidence intervals contain 50%, are below 50%, and are above 50%. Students should note that the confidence intervals tend to be too low. There are two reasons for this. First, some students do not include 0 as an even digit. Second, most students will miss some of the even digits as they are counting. (*Note*: It's possible that your students will be very careful counters. If so, bail out at this point and use the lesson as a demonstration that about 95% of all 95% confidence intervals contain the true percentage (50%) of even digits, if the survey has no design flaws.) If your class is typical and the confidence intervals tend to be too low, put the students into small groups and ask each group to submit a paragraph explaining why the 95% confidence intervals don't seem to be "working."

Sample Assessment Questions

1. When the results of political polls are reported in the newspaper, sometimes a "margin of error" is given. Exactly what kind(s) of "error" are included in this margin of error?

2. Explain under what conditions this statement is true: Ninety-five percent of the time, the true population percentage will be in the 95% confidence interval given in political polls.

95% CONFIDENCE INTERVALS FOR THE PERCENTAGE OF EVEN DIGITS

0.0	0.1	0.2	0.3	0.4	0.5	0.6	0.7	0.8	0.9	1.0	

Confidence Intervals for the Percentage of Even Digits

SCENARIO

Suppose a political poll says that 56% of voters approve of the job the president is doing and that this poll has a margin of error of 3%. The 3% margin of error results from the fact that the poll was taken from a sample of voters. Since not all voters were included, there is some *error due to sampling*, or *sampling error*. If all voters had been asked, the polling organization predicts that the percentage would have been in the **confidence interval** of 53% to 59%. For every 100 polls that report a 95% confidence interval, the polling organization expects that 95 of the confidence intervals will contain the true population percentage. In this activity you will take a "poll" of random digits in order to estimate the percentage that are even.

Question

What percentage of 95% confidence intervals will contain the true proportion of random digits that are even?

Objective

You have learned that we expect that 95 out of every 100 of our 95% confidence intervals will contain the true population proportion. In this activity you will test this statement by constructing confidence intervals for the percentage of random digits that are even.

Activity

1. Your instructor will assign you a group of 200 random digits. Don't look at it yet.

2. When your instructor tells you to begin, count the number of even digits in your sample of 200.

3. Use the following formula to construct a 95% confidence interval for the percentage of random digits that are even:

$$\hat{p} \pm 1.96\sqrt{\frac{\hat{p}(1-\hat{p})}{n}}.$$

4. Place your confidence interval on the chart your instructor gives you.

5. What is the true percentage of all random digits that are even?

6. What percentage of the confidence intervals contained this percentage? Is this what you expected? Explain.

Wrap-Up

1. **Bias** is defined as systematic error in carrying out a survey that results in the sample proportion's tending to be too big (or too small). What were the sources of bias in your class's first survey of random digits? In which direction were the results biased?

2. Devise a method of counting the even digits accurately. Recount the number of even digits in your sample of 200 random digits, and compute a new 95% confidence interval. Put these confidence intervals on a second chart. How many of the confidence intervals now contain 50%? Is this about what you expect? Explain.

Extensions

1. Investigate the kinds of systematic error that might result in bias in a political poll.

2. Investigate how major polling organizations try to minimize bias in political polls.

References

1. American Statistical Association. "What Is a Survey?" Reprinted in Jonathan D. Cryer and Robert B. Miller (1994), *Statistics for Business: Data Analysis and Modeling*, second ed., Belmont, CA: Duxbury, pp. 388–399.

2. David Freedman et al. (1991), *Statistics*, second ed., New York: Norton.
3. James M. Landwehr, Jim Swift, and Ann E. Watkins (1986), *Exploring Surveys: Information from Samples*, Menlo Park, CA: Dale Seymour Publications.
4. David S. Moore (1985), *Statistics: Concepts and Controversies*, second ed., New York: Freeman.

The Rating Game

NOTES FOR THE INSTRUCTOR

Statistical Setting

This activity can be used any time after students have had experience with data displays and summary statistics. It provides valuable experience in designing a questionnaire to measure public opinion. Students learn first hand how the design of the questionnaire influences the validity and reliability of the responses and the usefulness of the subsequent analysis.

Prerequisites

Students need to know how to construct data displays such as the dotplot, stemplot and the boxplot. They need to be able to summarize data using means and medians, as well as the range and the standard deviation.

Materials

This activity will need the student-designed questionnaires.

Procedure

The activity is suitable for both large and small classes and is particularly appropriate as an out-of-class group activity. The final report on the analysis can be done individually or in groups, depending on the class size. The group work forces students to talk with each other about statistical concepts and has been found to be an important learning experience.

The number of questions in each category should probably be the same since that will make it easier to compare the summary ratings for the categories as measures of the degree of satisfaction. If it is not possible for the students to do a pilot of their questionnaires, you may be able to do a pilot in class with a select number of questions. Either way students should be made aware of the importance of the wording of their questions before they collect their responses. You may wish to use the publication "What Is a Survey?" published by the American Statistical Association. A portion of the discussion is given here.

No sample results are included, since issues concerning students tend to be local. The length of the written report will depend on the level of the students and the timing of the assignment. Assessing a group report can be difficult since invariably some of the students will not contribute to a group project. One way around this is to assign a group grade in points and let the students negotiate among themselves how to divide the points among the members of the group. The students can then inform you in writing of the agreed-upon division of the points.

"What Is a Survey?" (Published by the American Statistical Association)

> Designing the questionnaire represents one of the most critical stages in the survey development process, and social sciences have given a great deal of thought to issues involved in questionnaire design. The questionnaire links the information needed to the realized measurement.
>
> Unless the concepts are clearly defined and the questions unambiguously phrased, the resulting data are apt to contain serious biases. In a survey to estimate the incidence of robbery victimization, for example, one might want to ask, "Were you robbed during the last six months?" Though apparently straight straight-forward and clearcut, the question does present an ambiguous stimulus. Many respondents are unaware of the legal distinction between robbery (involving personal confrontation of the victim by the offender) and burglary (involving breaking and entering but no confrontation), and confuse the two in the survey. In the National Crime Survey, conducted by the Bureau of the Census, the questions on robbery victimization do not mention "robbery". Instead, several questions are used which, together, seek to capture the desired responses by using more universally understood phrases that are consistent with the operational definition of robbery.
>
> Designing a suitable questionnaire entails more than well-defined concepts and distinct phraseology. Attention must also be given to its length, for unduly long questionnaires are burdensome to the respondent, are apt to induce respondent fatigue and hence response errors, refusals and incomplete questionnaires, and may contribute to higher nonresponse rates in subsequent surveys involving the same respondents. Several other factors must be taken into account when designing a questionnaire to minimize or prevent biasing the results and to facilitate its use both in the field and in the processing center. They include such diverse considerations as the sequencing of sections or individual questions in the document, the inclusion of check boxes or precoded answer categories versus open-ended questions, the questionnaire's physical size and format, and instructions to the respondent or to the interviewer on whether certain questions are to be skipped depending on response patterns to prior questions.

The Rating Game

SCENARIO

As consumers, we are often faced with making decisions based on ratings, whether they are ratings of movies, the Nielson ratings of TV shows, or the car ratings of *Consumer's Report*. *U.S. News and World Report* annually rates colleges and universities. To come up with the ratings, the magazine use data such as average SAT/ACT scores of their freshmen, the acceptance rate, financial resources, as well as the ratings of schools by the presidents, deans, and admissions directors. The elite schools compete with each other for the top ratings and use them to attract the best students. Do you know how your school did in this rating scheme? How others rate your school may not affect your day-to-day experiences at the school. This activity asks you to find out the level of satisfaction of your fellow students with three different aspects of your school that really matter to you: academic programs, physical plant, and extracurricular activities.

Question

How satisfied is the student body with the academic programs, physical plant, and extracurricular activities?

Objective

The objective of this activity is to give you some experience in designing questionnaires and summarizing and analyzing score data.

Activity

1. This activity is designed to be done in a group. Each group will design a questionnaire that covers the three categories: academic programs, physical plant, and extracurricular activities. Each category should have at least three questions asking students to give a score between 0 and 100, where 0 indicates the lowest level of satisfaction and 100 the highest level of satisfaction. You therefore have to set up the questions such that the response is a score and not a "yes" or "no." For example, if you want to find out how students feel about the scheduling of the courses, a sample question is

   How satisfied are you with the times at which your courses are scheduled?

   The following is an example of issues you could consider in each category.

   Academic programs
   Quality of instruction
   Quality and variety of course offerings
   Availability of required courses
   Scheduling of the courses
   Flexibility in designing your program

   Extracurricular activities
   Recreational/cultural programs on campus
   Recreational/cultural programs off campus
   Student clubs and organizations

   Physical plant
   Condition of classrooms
   Maintenance of dormitories
   Access to libraries/gym
   Parking

2. After you have designed the questionnaire, do a pilot survey. Ask some of your friends to respond to the questionnaire. Change the wording of a question if their response did not give you the information you want.

3. Administer your questionnaire to the rest of the class or to a randomly chosen sample of students outside the class.

4. Analyze the responses using graphs such as the dot plot, stem and leaf plot, and the box plot. Summarize the responses with means, medians, range, and standard deviations. Combine the summaries within each category to get an overall rating for the category. Be prepared to justify your method of combining the response summaries.

5. Write a report that includes a discussion of your questions, the method of analyzing the responses, and conclusions.

Wrap-Up

In any questionnaire, the wording of the questions, the scale used for the responses, the length of the questionnaire, and the order in which the questions are asked can all influence the responses.

1. Select a question from your questionnaire where the responses were not at either extreme of the rating scale and change the wording to draw a mostly negative or mostly positive response.

2. The activity asks you to use a rating scale from 0 to 100. From observing your data, would a scale of 0 to 7 or 0 to 5 have been sufficient?

3. Field tests of this activity have shown that students are most dissatisfied with facilities and most satisfied with the academics at their institution. Suppose you had switched the order of the categories with facilities being rated first. Do you think that would affect the ratings of the academics? How do you suggest the questions should be ordered?

Extension

Get a copy of the *U.S. News and World Report* issue where the schools are rated. The methodology used to rate the schools is included in the article. You are the consumer, so what variables do you believe the magazine should have used to make the ratings meaningful to you? Write a short recommendation with your suggestions to the magazine.

Capture/Recapture

NOTES FOR THE INSTRUCTOR

Statistical Setting

This activity can be used at any point in a course, including on the first day, as an introduction to statistical thinking, to the use of models, and to how the dependence of the model on assumptions differentiates statistics from mathematics. At the end of this activity the students should be able to explain how the violation of an assumption in a statistical model can bias the results, and they should be able to use the capture/recapture methodology to construct population estimates.

Prerequisites for Students

None.

Materials

This activity calls for 2 6-ounce bags of Pepperidge Farm Goldfish (in different flavors, such as Parmesan Cheese flavor and Pizza flavor) and a bag or box to serve as the "lake."

Procedure

This activity can be done with the entire class and takes approximately 20 to 30 minutes. You might prefer to have students work in groups, using the set of estimates from the various groups in place of the replication described in Extensions.

Solving the question in step 6 is trivial, but the solution makes sense as an estimate of the population size only if it is reasonable to set the sample percentage of tagged fish equal to the population percentage of tagged fish. This depends on many assumptions. Ask the class what might go wrong *and how each problem would affect the resulting estimate of* N. Here is a list of assumptions:

(a) Tagged animals do not lose their tags.

(b) In each sample, every animal has the same chance of being captured. In particular, this means that tagged animals are *not* less likely than others to be caught in the second sample, nor is it the case that some animals are "trap happy" (very easy to capture) while others are "trap shy."

(c) The mortality rate among tagged animals is the same as the mortality rate for untagged animals. (Note that we need not assume that tagged animals don't die, only that they die at the same rate as untagged animals.)

(d) New animals do not enter the population in the study.

(e) The tagged animals become randomly mixed with the others between the first and second phases.

Technology Extension

Use the computer to simulate the capture/recapture process and to replicate it many times. To do this you will want to construct a variable of $N - M$ zeroes and M ones, with the ones corresponding to the tagged animals from the first sample. Then sample without replacement n times from this variable and count the number, R, of ones in the group (i.e., calculate the sum of the n zeroes and ones). Now calculate an estimate of N from the formula in Activity step 6. Repeat these steps many times, and graph the results.

Using $N = 400$, $M = 50$, and $n = 40$ yields the following box plot of 100 estimates of N:

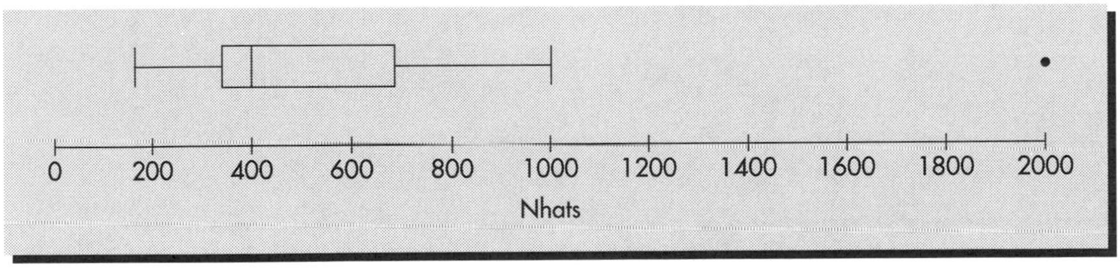

Note that this provides a visual display of the distribution of estimates; it does not give us a confidence interval. (Your students might confuse the distribution of estimates of N, derived from the sampling distribution of the sample sum of ones and zeroes, with the idea of a confidence interval. A confidence interval, as explained in the Extensions section, is an interval estimate of the population size; it is not the interval in which we expect the next estimate to fall.)

EXPERIMENTAL DESIGN

The recapture phase can be repeated many times and the resulting estimates of N compared, perhaps in a stem–leaf plot. One could then vary the number of tagged animals and repeat the process to see how the variability of the estimates depends on M. One could also vary the size of the second capture to see how the variability of the estimates depends on n. Ask the class, "Is it better to mark more fish? Is it better to take a larger sample in the recapture phase?"

There are two factors, M and n, that influence the variability of the estimate of N when using capture/recapture. You could have students use simulations to conduct a factorial experiment to study this. For example, here are box plots of 100 estimates of N when M is either 100 or 200 and n is either 60 or 120. In each case, N was 1,000. Note that if M and n are small, then it is possible for R to be zero (no tagged members in the recapture phase), in which case the estimate of N is undefined.

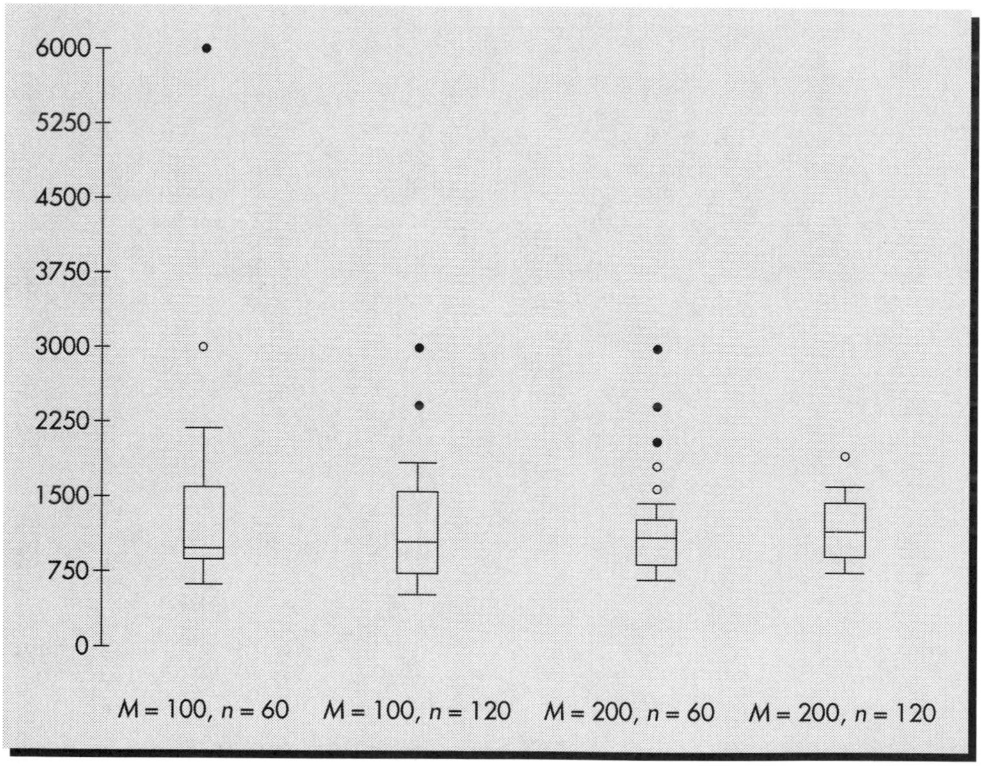

Sample Assessment Questions

1. Consider using the capture/recapture method to estimate the number of fish in a lake. Suppose 200 fish are captured and tagged in the first phase. If 348 fish are caught in the second phase and 32 of them have tags, what is the estimate of the number of fish in the lake?

2. Consider question 1 again. Suppose the tags were not attached very well, so that some of them fell off. How does this affect your estimate? Is it too high, too low, or about right? Why?

3. Use the data from question 1 to construct a 95% confidence interval for the number of fish in the lake.

Capture/Recapture

SCENARIO

Naturalists often want estimates of population sizes that are difficult to measure directly. The capture/recapture method allows one to estimate, for example, the number of fish in a lake. This idea is also the basis of methods that the Census Bureau has developed that can be used in adjusting population figures obtained in the decennial census (although using sampling techniques to adjust the census is controversial).

Question

What is the population of the United States, including those who are not counted in the official census?

Objectives

The purpose of this activity is to introduce the use of statistical models and to learn how assumptions affect statistical analyses. You will also learn about the capture/recapture method and how it is used.

Activity

1. How would you estimate the number of fish in a lake or the number of bald eagles in the United States? Just going out and counting the animals that you see will not work.

2. We will demonstrate how the capture/recapture method works by trying to determine the number of goldfish in a bag of Pepperidge Farm Goldfish crackers. Open one of the bags of goldfish, and pour the goldfish into the "lake."

3. A volunteer should take a large sample from the lake. It is a good idea to get at least 40 or so goldfish in this sample, which often requires taking two handfuls. Then someone needs to count the number, M, of goldfish in the sample; these are the "captured" goldfish.

4. The next step is to put tags on the goldfish. If these were real fish they could be marked with the use of physical tags, but with cracker goldfish we mark them by making them change color. Set aside the captured goldfish and open the second bag. For each goldfish in the captured set, put a goldfish of the other (second) flavor in the lake. (The new goldfish replace the old ones; don't put the old ones back into the lake.)

5. Shake the lake for a while in order to mix the two flavors of fish. Then a volunteer (it need not be the same person who took the first sample) should take a new sample of goldfish; let n denote this sample size, which should again be reasonably large. Count the number in the second sample that are tagged (call this R) and the number that are not tagged ($n - R$).

6. Now consider the percentage of fish in the second sample that are tagged (R/n). A reasonable idea is to set this percentage equal to the population percentage of tagged fish at the time of the second sample, M/N, where N denotes the unknown population size:

$$\frac{R}{n} = \frac{M}{N}.$$

Solving for N, we see that the estimate of the population size is

$$N = \frac{M{*}n}{R}.$$

Another way to view this is to think of a 2×2 table that gives a cross classification according to whether or not each member of the population was captured in the first phase and in the second phase.

		In First Capture? Yes (Tagged)	No (Not Tagged)	Total
In second capture?	Yes	R	$n - R$	n
	No	$M - R$		
	Total	M	$N - M$	N

Discussion: Solving the equation in step 6 is trivial, but the solution makes sense as an estimate of the population size only if it is reasonable to set the sample percentage of tagged fish equal to the population percentage of tagged fish. This depends on many assumptions—for example, that the tags do not fall off. What might go wrong? How would each problem affect the resulting estimate of N? Make a list of assumptions that might be violated and how the estimate of N would be affected in each case.

Wrap-Up

Suppose you wanted to estimate the number of bald eagles in the United States. Explain how you could use the capture/recapture method to do this. What assumptions might be violated? How would this affect your estimate?

Extensions

1. We have now used capture/recapture once and we have an estimate of N. However, this estimate is subject to the uncertainty that arises from the process of taking random samples (rather than counting the entire population). If we repeat the process several times, we will get different estimates of N each time. Repeat the process at least once to see how the estimate of N varies.

2. We can construct a confidence interval for the population size based on a confidence interval for the population percentage tagged. Let p denote the proportion of tagged animals in the population after the first sample (i.e., after step 4); that is, $p = M/N$. Let $\hat{p} = R/n$, the sample percentage tagged. A 95% confidence interval for p is given by $\hat{p} \pm 1.96^* \sqrt{\hat{p}(1 - \hat{p})/n}$. (Here we are assuming that n is small relative to N so that we can ignore the finite population correction factor.) Setting M/N equal to the lower limit of the confidence interval for p and solving for N gives the upper limit of a confidence interval for N. Setting M/N equal to the upper limit of the confidence interval for p and solving for N gives the lower limit of a confidence interval for N.

3. The Census Bureau uses the capture/recapture idea in providing adjustments to the decennial census. The capture phase is the census (persons are "tagged" by being recorded in the Census Bureau computer), and the recapture phase is called the Post Enumerative Survey (P.E.S.) conducted after the census. In this setting the appropriate 2 × 2 table would be as follows:

		Recorded in the Official Census? Yes	No	Total
"Captured" in the P.E.S.?	Yes	R	$n - R$	n
	No	$M - R$		
	Total	M	$N - M$	N

Only about 96% of all persons in the United States are "captured" in the official census. The percentage of persons in the P.E.S. who were missed in the census, around 4% or so, can be used to adjust original census figures, although whether or not to use the adjusted figures is a hotly debated political issue. For more discussion, see the series of articles that Stephen Fienberg has written for *Chance* magazine. (*Note:* The actual adjustment procedure proposed for the census is based on the capture/recapture idea but involves smoothing of estimates and becomes rather complicated.)

4. Another option is to sample fish one at a time during the recapture phase, stopping when the number of tagged fish, R, is 15, say. Thus, n, the size of the second sample, is random. This sampling method ensures that R will not be too small; that is, it eliminates the possibility that R will be very small by chance, which would lead to a very large estimate of N in Activity step 6.

Technology Extension

We could use a computer to simulate the capture/recapture process. Suppose there are actually 400 animals in the population and the first sample captures 50 of them. Then we begin the second capture phase with a population of 50 tagged animals and 350 untagged animals. If the sample size for the recapture phase is $n = 40$, then we need to simulate drawing a sample of size 40 without replacement from a population of 50 tagged and 350 untagged animals.

One way to do this is to construct a variable of 50 ones and 350 zeroes, with the ones corresponding to the tagged animals from the first sample. Then we sample without replacement 40 times from this variable and count the number, R, of ones in the group (i.e., we calculate the sum of the 40 zeroes and ones). Next we calculate an estimate of N from the formula in Activity step 6. By repeating this many times, we can see how the estimates of N vary from sample to sample.

Here is a box plot of 100 estimates of N from this type of simulation:

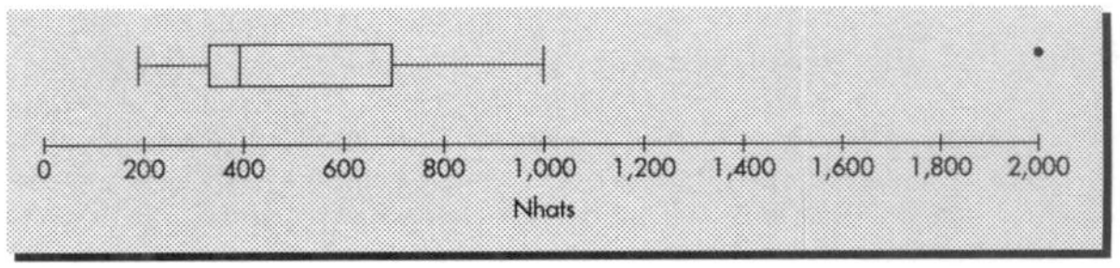

References

1. B. Bailar (1988), "Who counts in America?" *Chance*, **1**:9, 17.
2. D.G. Chapman (1989), "The plight of the whales," in J. M. Tanur, et al. (eds.), *Statistics: A Guide to the Unknown*, third ed., Belmont, CA: Wadsworth, pp. 60–67.
3. E.P. Erickson and J.B. Kadane (1985), "Estimating the population in a census year: 1980 and beyond" (with discussion), *J. Amer. Statist. Assoc.*, **80**:98–131.
4. Stephen Fienberg (1992), "An adjusted census for 1990? The trial," *Chance*, **5**:28–38.
5. Howard Hogan (1992), "The 1990 post-enumerative survey: an overview," *The American Statistician*, **46**:261–269.
6. W.E. Ricker (1975), *Computation and Interpretation of Biological Statistics of Fish Populations*, Bulletin of the Fisheries Board of Canada 191, Ottawa, Canada, pp. 83–86.

Randomized Response Sampling: How to Ask Sensitive Questions

NOTES FOR THE INSTRUCTOR

Statistical Setting

This activity can be used when discussing surveys and questionnaire wording. You might want to incorporate this into a discussion of experimental design.

Students know that some people are not willing to answer sensitive questions truthfully. The randomized response technique provides a way to ask sensitive questions while preserving the anonymity of the respondents.

Prerequisites for Students

Students should understand how to use probability trees. They should also have had some exposure to the use of surveys.

Materials

Each student will need a coin. You need slips of paper to hand out to the students.

Procedure

This activity is done with the entire class and takes approximately 20 minutes plus time for discussion, which will vary from class to class.

You might want to have the students do the necessary coin tossing in advance, so that no one sees the result of someone else's coin toss. You can tell the class, "In preparation for tomorrow's class, go home and toss a coin twice, in private. Make a mental note of the results, in order (first toss, second toss). Don't tell anyone what results you get."

The question "Have you ever shoplifted?" can be replaced by any other sensitive question. Indeed, it is a good idea to *let the class suggest a sensitive question to use in the activity*. If they cannot come up with an acceptable question to use, you could consider one of these: "Are you strictly heterosexual?" (so that homosexual, bisexual, or questioning students would answer "no"); "Have you been the victim of sexual abuse?" (this may require a definition of sexual abuse, although you could let each student define it for himself or herself); "Have you cheated on any exams or quizzes in this course?" (although asking this question may convey the sense that cheating is an accepted part of higher education, so you may not want to use it).

Another option that some instructors like is to ask a question that is not really very sensitive, so that the estimate can later be compared to an estimated based on a regular survey of the class (in which everyone answers the same question and it is assumed that they are all willing to tell the truth). However, see the Extensions section here for another way to handle this approach.

The following instructions assume that the sensitive question is "Have you ever shoplifted?" Note that you need to give a clear and unambiguous definition of "shoplifted" before proceeding.

1. Point out to the students that *if* you were to ask them to raise their hands if they have ever shoplifted, you could not use the responses as a reliable estimate—many students who have shoplifted would not admit this publicly.

2. Tell the class that you intend to collect data that will allow you to estimate the percentage who have shoplifted by asking two questions. One of them, the "real question," is "Have you ever shoplifted?" The other question is a "decoy question," to be explained in a moment.

3. Tell the students that they will each toss a coin twice. The first toss will randomly determine who is to answer the real question and who is to answer the decoy question. The second toss will disguise for one group the fact that they are answering the real question, while for the other group it will provide the answer to the decoy question, which is "Did you get heads on your second toss?" (It helps to write the two questions on the board, with labels "real" and "decoy.")

4. Hand out the slips of paper, one per student. Tell the students to *privately* flip a coin twice, remembering the results (heads/tails) for each toss *in order*. It is imperative that the student is the only person who knows the results of his or her tosses. Tell them that they are never to reveal to anyone the results of their tosses

and that they should not write these down. (It is important to be painfully careful and clear in giving instructions during this activity and to proceed slowly.)

5. Those who get heads on the first toss are in group 1 and are to answer the real question. Those who get tails on the first toss are in group 2 and are to answer the decoy question. Repeat that they are *not* to reveal which group they are in.

6. Tell them that the only thing they are to write is the answer, yes or no, to the appropriate question; they are *not* to indicate which question they are answering.

7. Point out that because only they know which question they are answering, there is no way that a written answer of "yes" can be used against them—someone charged with having shoplifted, based on having written "yes" on his or her slip of paper, can claim that he or she was answering question number 2. Thus, they should feel free to answer truthfully.

8. Collect the papers and have someone tally the results. Present the following probability tree and explain how, *without knowing who answered which question*, you can estimate the proportion who have used illegal drugs this semester. Here θ denotes the proportion of persons in the class who have shoplifted.

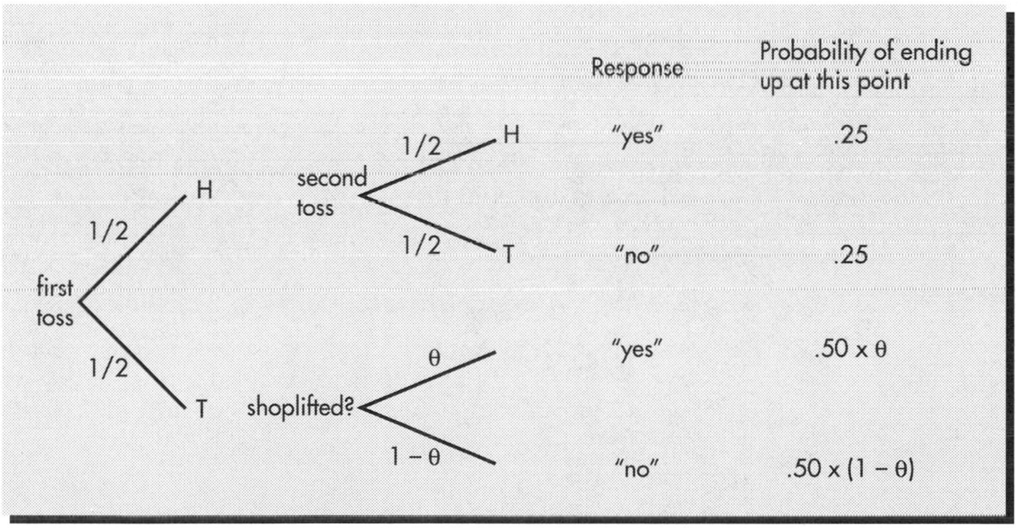

Let p denote the probability of a "yes" response under the randomized response technique, and let $\hat{p}$ denote the proportion of "yes" responses in the sample. The expected value of the sample proportion of "yes" responses is $.25 + .5 \times \theta$. Setting this equal to the sample proportion of "yes" responses, $\hat{p}$, yields

$$.25 + .5 \times \theta = \hat{p}.$$

Solving for θ, we have

$$\theta = 2 \times (\hat{p} - .25).$$

Note: Many students find this probability calculation difficult to understand. Present the calculations slowly, and be prepared for questions as to how this calculation allows one to get a good estimate of the percentage who have shoplifted. You might find it helpful to add to the probability tree the expected number of students at each node (see the Sample Results from Activity section ahead).

Using the randomized response technique effectively cuts the sample size in half (see the Extensions section here), so the estimate of θ is subject to a large standard error. Thus, having a large class helps. For example, if you only have 30 students in the sample and θ is .2, then the standard error of the estimate is around .17, so negative estimates are to be expected from time to time!

Sample Results from Activity

In one class of college students the randomized response technique was used to estimate the proportion who had used illegal drugs since the beginning of the semester. Out of $n = 70$ students, 26 wrote "yes" on their slips of paper. Thus, $\hat{p}$ is $26/70 = .37$, and the estimate of θ is $2 \times (.37 - .25) = .24$. (The standard error of this estimate is .116; see the Extensions section ahead.)

One way to think about this sample is to consider that we would expect 35 of the 70 students to answer the decoy question and 35 to answer the real question. If 35 students answer the decoy question, we expect 17.5 to answer "yes." Thus, we expect $26 - 17.5 = 8.5$ of 35 students to have answered "yes" to the real question. This gives an estimate of θ of $8.5/35 = .24$. Here is a probability tree for this setting:

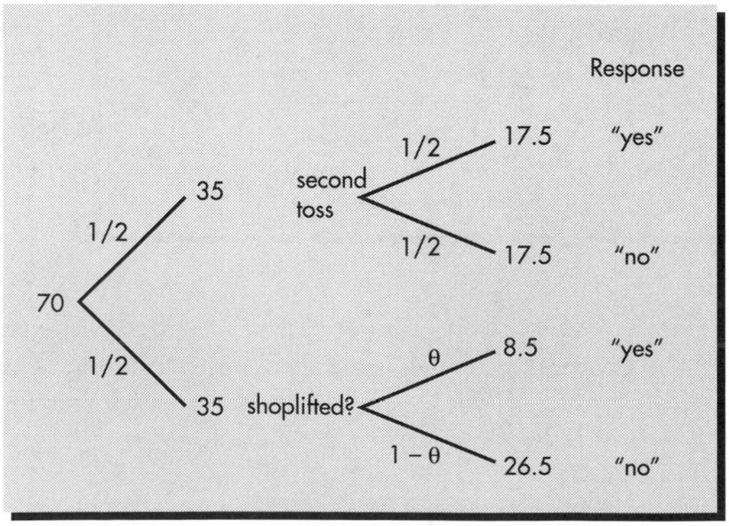

Another way to view these results is to think of a 2 × 2 table. Here is a table of expected cell counts.

	Yes	No	Total
Decoy Q	17.5	17.5	35
Real Q	$35 \times \theta$	$35 \times (1 - \theta)$	35
Total	$17.5 + 35 \times \theta$	$17.5 + 35 \times (1 - \theta)$	70

Setting the observed number of "yes" responses, 26, equal to the expected number of $17.5 + 35 \times \theta$ and solving for θ yields an estimate of .24 for θ.

Extensions

If you used a truly sensitive question and your students doubt that the method can really work (despite your showing them the probability tree calculations, etc.), you could repeat the activity using a benign question like "Are you a female?" The randomized response answer can be compared to the true percentage (presumably you can determine the percentage of females in the class without conducting a poll) as a way of verifying that the method does work (within the bounds of variability measured by the standard error (below).

The randomized response technique effectively cuts the sample size in half, so the estimate is subject to a large standard error. The standard error of the estimate of θ is $2 \times \sqrt{\hat{p}(1 - \hat{p})/n}$, which can be used to construct a confidence interval for θ.

You might want to preface the presentation of randomized response with a general discussion of how it is that controlled randomness can be useful. Many students think that having no randomness is always better than having some randomness; this is not true. For example, a random sample can be *better* than a census in cases in which a census is difficult, tiring, and likely to lead to errors. Thus, an auditor may prefer accurately analyzing a sample of financial records over an attempt to analyze all records, a task that would likely lead to many careless errors as tedium set in. Controlled randomness lets us remove effects of confounding variables in experiments, which is one reason that subjects are randomly assigned to treatments in clinical trials.

Options

Another way to conduct the activity is as follows: Hand out slips of paper and have each student toss a coin once (privately). If the coin lands heads, they are to write "yes" on their paper. If the coin lands tails, they are to write the honest answer to the question "Have you ever shoplifted?" (or whatever sensitive question you have chosen). As before, they should be reassured that they can answer honestly.

In this setting the probability tree is as follows:

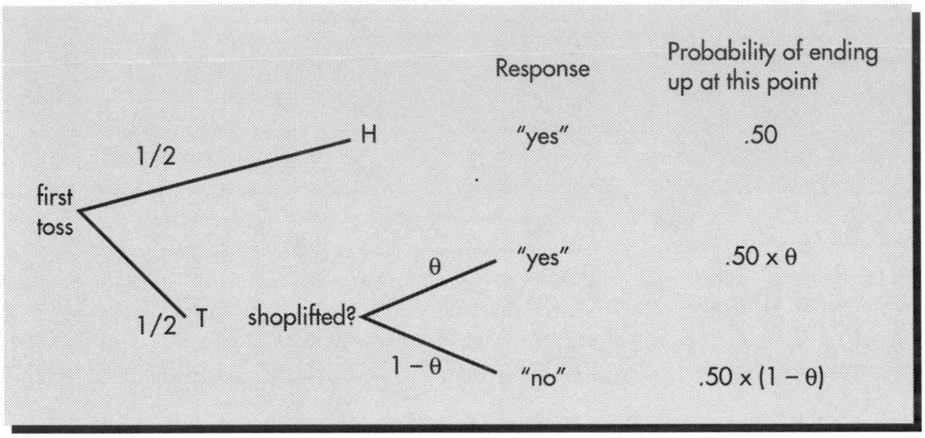

The expected value of the sample proportion of "yes" responses is $.5 + .5 \times \theta$. Setting this equal to the sample proportion of "yes" responses, $\hat{p}$, yields

$$.5 + .5 \times \theta = \hat{p}.$$

Solving for θ, we have

$$\hat{\theta} = 2 \times (\hat{p} - .5).$$

Unlike the case of the first randomized response method described, with this method a "no" response clearly means that the person has not shoplifted.

THE WARNER APPROACH

There are yet other ways to conduct a randomized response survey. For example, have each student choose a random digit. (One way to do this is to have the student take out a dollar and use the last digit of the dollar's serial number.) If the digit is 4, 5, or 6, the student is to give a truthful answer to the sensitive question. If the digit is 0, 1, 2, 3, 7, 8, or 9, the student is to lie—that is, the student is to give the opposite of the truthful answer. So, for example, if my random digit is 7 and I have not shoplifted, I would answer "yes."

In this setting the probability tree is as follows:

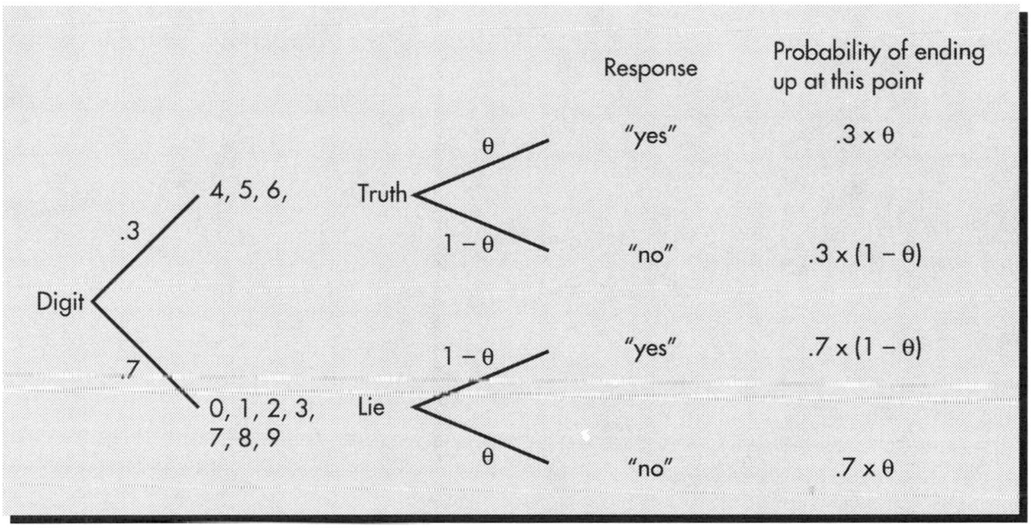

The expected value of the sample proportion of "yes" responses is $.3 \times \theta + .7 \times (1 - \theta)$. Setting this equal to the sample proportion of "yes" responses, $\hat{p}$, yields

$$.3 \times \theta + .7 \times (1 - \theta) = \hat{p}.$$

Solving for θ, we have

$$\hat{\theta} = (.7 - \hat{p})/.4.$$

Note that here the standard error of the estimate of θ is $2.5 \times \sqrt{\hat{p}(1 - \hat{p})/n}$, which differs from the standard error for the first two methods.

Technology Extensions

Rather than rely entirely on the probability trees as a means of comparing the various ways to conduct a randomized response survey, you could use the computer to simulate the randomized response process, produce several estimates of θ using each approach, and compare the variances under the different approaches by graphing the results (e.g., with parallel box plots). In a simulation you (or the student) fix θ and then generate a sample of n Bernoulli variables. For the first approach each Bernoulli should have probability of success $.25 + .5 \times \theta$. For each sample, calculate the estimate of θ and record the result. Repeat this many times (say, 250 times). Then repeat the entire simulation for each of the other approaches you wish to investigate.

For example, Figure 1 contains the results of simulating the 3 above approaches 250 times each, using $n = 100$ in each trial and setting $\theta = .4$. The fourth box plot, labeled "binomial," is the result of simulating a simple random sample, again with

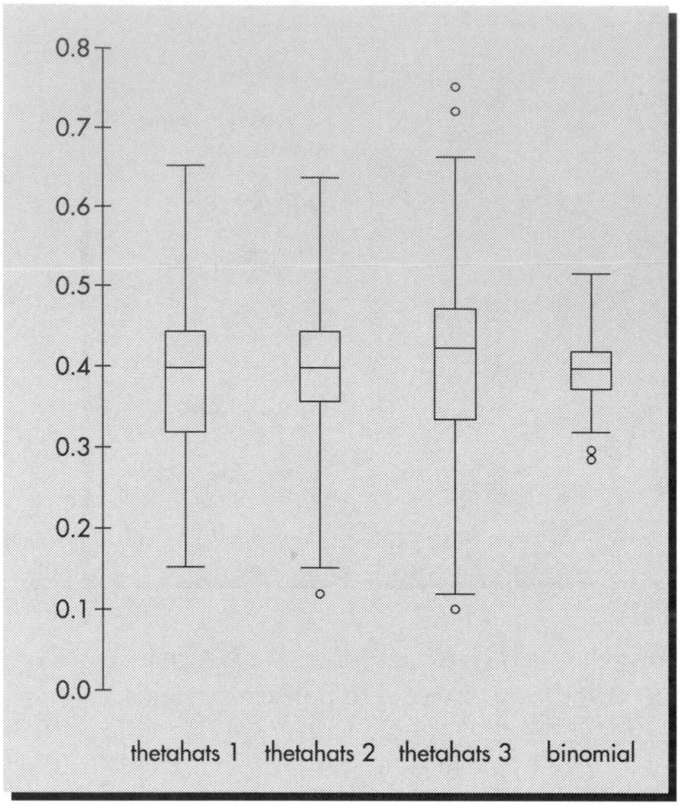

Figure 1

$n = 100$ and $\theta = .4$, in which the randomized response method is not used. Each of the first three box plots summarizes 250 simulated randomized response surveys, while the fourth summarizes 250 simulated simple polls. All four provide estimates of θ that center at about .4, but the third method produces estimates that are more variable than those from the first two methods. Note as well that with $\theta = .4$, the expected value of $\hat{p}$ is .45 for the first method but .7 for the second method, so the results from the second method are a bit less variable than those from the first method (since $.45(1 - .45) > .7(1 - .7)$). The three randomized response simulations yield more variable results than those given by the simple poll simulation. As noted above, using the randomized response method effectively cuts the sample size in half; the higher variability of the estimates is the price one must pay to get truthful answers when asking a sensitive question.

Experimental Design

There are several factors that influence the variability of the estimate of θ when using randomized response sampling. For example, if you use the Warner approach outlined here, then the variability of $\hat{\theta}$ depends on the total sample size n, the true value

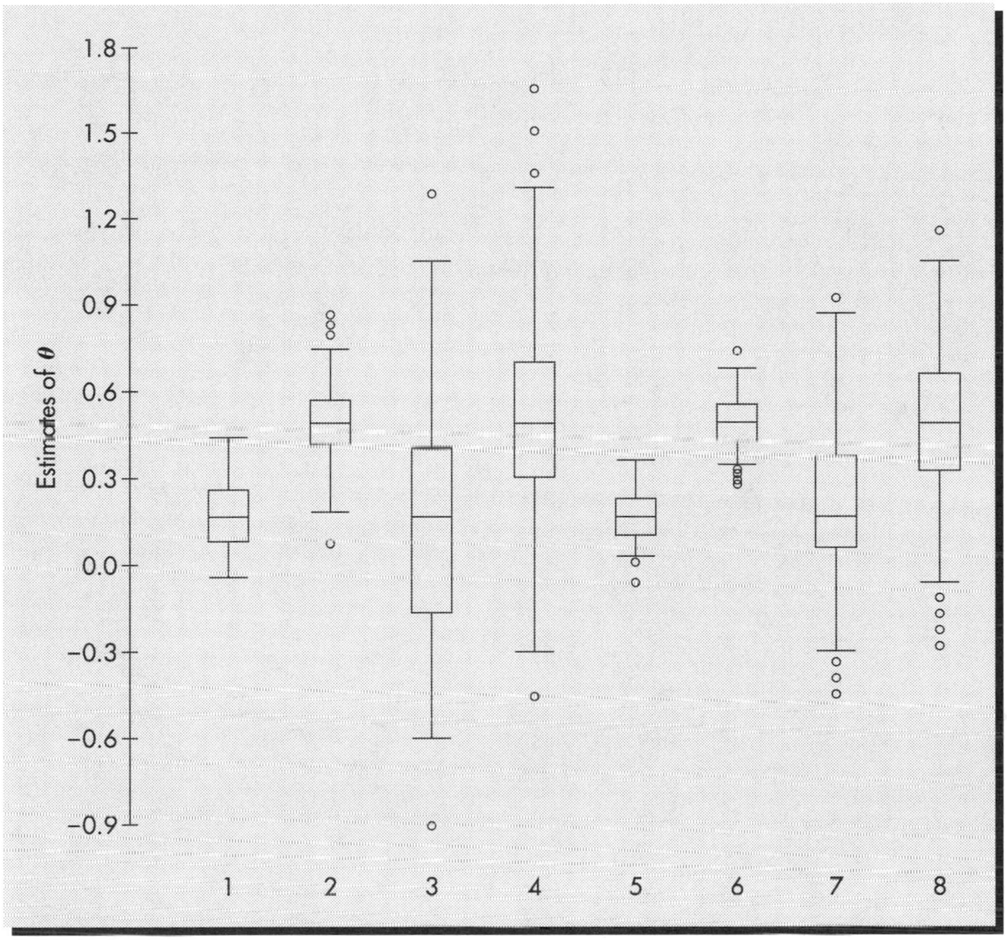

Figure 2

of θ, and the probability (call it π) that a respondent tells the truth ($\pi = .3$ in the example above). Using the simulation ideas just described, students could design and conduct a factorial experiment to study how these three factors influence the variability of $\hat{\theta}$.

For example, Figure 2 shows box plots of 250 values of $\hat{\theta}$ from each of the 8 combinations that arise when n is either 50 (for box plots 1, 2, 3, and 4) or 100, θ is either .2 (for box plots 1, 3, 5, and 7) or .5, and π is either .2 (for box plots 1, 2, 5, and 6) or .4. Note that all eight box plots are centered at θ, which shows that the process is unbiased. The effect of n is modest but noticeable, the effect of θ is small, and the effect of π is considerable. When n is 50, it is quite common to get an estimate of θ that is outside the range (0,1).

Sample Assessment Questions

1. Consider using the first version of the randomized response technique with a group of $n = 60$ persons. Suppose that 36 persons answer "yes" and 24 answer "no." Use these data to estimate θ.

2. Is it possible to get a negative estimate of θ when using the randomized response technique? Can the estimate of θ be greater than 1? If so, how can this happen?

Randomized Response Sampling: How to Ask Sensitive Questions

SCENARIO

Pollsters often want information on issues that are considered sensitive. The IRS cannot estimate the proportion of persons who cheat on their taxes by asking a random sample of persons if they cheat, because everyone (almost) will say "no," including some who are lying when they give that answer. The randomized response technique is a way around the problem of people lying when they are asked a sensitive question.

Question

How many students in the class have shoplifted at some time?

Objectives

The main objective of this activity is to learn how to use the randomized response method. A secondary objective is to review the use of probability trees. After completing this activity, you should be able to plan and conduct a randomized response survey.

Activity

Your instructor will conduct a randomized response sample of the class. In the process you will be asked to answer a question privately. Class data will be tallied and used to estimate the percentage of students in the class who have shoplifted. (Your instructor might choose a different sensitive question for use with your class.)

There are several ways to conduct a randomized response sample. One approach involves having each student toss a coin—privately—to decide whether to answer the "real" question or a "decoy" question. Students are then to tell the truth when answering whichever question is appropriate. The fact that a private coin toss has selected the question to be answered means that no one knows who is answering which question, so there is no reason to be dishonest when responding.

Although the pollster (in this case, the instructor) does not know who answered which question, it is possible to estimate θ, the population percentage who would truthfully answer "yes" to the sensitive question, by considering the following probability tree:

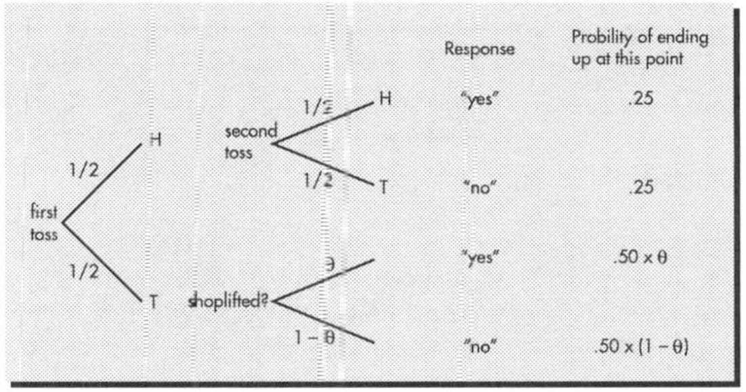

With this probability tree we can construct an unbiased estimate of θ. To see how this is accomplished, let p denote the probability that someone will answer "yes" when a pollster uses the randomized response technique. There are two ways that someone can end up answering "yes": (1) by getting heads twice in a row when tossing a coin or (2) by getting tails on the first toss and honestly answering the sensitive question with a "yes." The probability of getting heads twice in a row is $.5 \times .5$, which is .25. The probability of getting tails on the first coin toss and honestly answering the sensitive question with a "yes" is $.5 \times \theta$. Thus, $p = .25 + .5 \times \theta$.

Let $\hat{p}$ denote the proportion of "yes" responses in the sample. The expected value of the sample proportion of "yes" responses is p, which is $.25 + .5 \times \theta$. Setting this equal to the sample proportion of "yes" responses, $\hat{p}$, yields

$$.25 + .5 \times \theta = \hat{p}.$$

Solving for θ, we have

$$\hat{\theta} = 2 \times (\hat{p} - .25).$$

Wrap-Up

1. Write a brief summary of what you learned in this activity about how randomized response sampling works.

2. Explain how you could use the randomized response technique to estimate the percentage of people who cheat on their taxes.

Technology Extension

Rather than rely entirely on probability trees as a means of studying randomized response surveys, we could use the computer to simulate the randomized response process, produce several estimates of θ, and graph the results in a box plot. In a simulation we fix θ and then generate a sample of n Bernoulli variables. For the approach represented by the probability tree given earlier in this activity, each Bernoulli variable should have probability of success $.25 + .5 \times \theta$. For each sample, we calculate the estimate of θ and record the result. We then repeat this many times (say, 250 times).

For example, here are the results of simulating the above approach 250 times, using $n = 100$ in each trial and setting $\theta = .4$ (so that each Bernoulli variable has probability of success $.25 + .5 \times .4$, or .45). The following box plot summarizes 250 simulated randomized response surveys. Note that the randomized response technique provides estimates of θ that center at approximately the "true" value of .4.

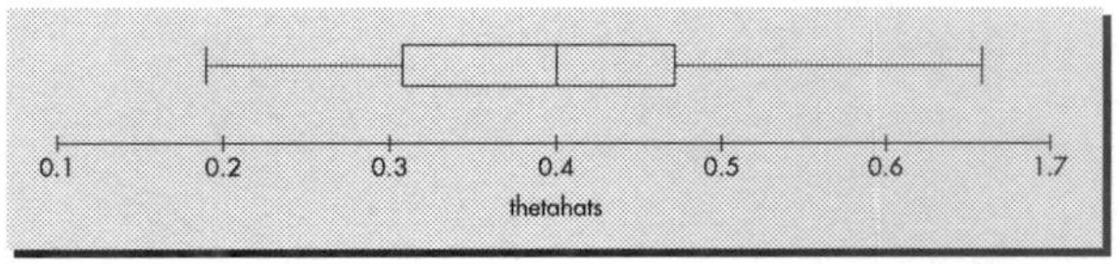

References

1. Sarah Boxer (1987), "AIDS and epidemiology. Women and drugs," *Discover*, **8**(7):12.
2. S.L. Warner (1966), "Randomized response: A survey technique for eliminating evasive answer bias," *J. Amer. Stat. Assoc.*, **60:**63–69.

Estimating the Difference Between Two Proportions

NOTES FOR THE INSTRUCTOR

Statistical Setting

The comparison of proportions by looking at differences between sample proportions occurs very often in the interpretation of survey results, especially in the context of opinion polls. These ideas should be discussed after the single sample case but before the development of confidence intervals for differences.

Prerequisites for Students

Students should understand the notions of sampling distribution for a single proportion and margin of error for a sample proportion, as discussed in the "Estimating Proportions: How Accurate Are the Polls?" activity.

Materials

Teams of students will need to be supplied with a bead box, a box of marbles, a bag of colored candies, or a random number table that will allow sampling of at least three different categories of items.

Procedure

The procedure follows directly the procedure for studying sampling distributions and margin of error for a single proportion, as mentioned under the prerequisites.

It is difficult for students to grasp the difference between independent proportions and dependent proportions of the type studied here (which are actually multinomial proportions). Use other examples to make sure this distinction is clear.

Sample Results from Activity

The plot and numerical summaries for *independent* proportions (see Figure 1) show computer-generated samples of size 20 from a "bead box" with 40% red balls and 60% white balls. Notice that the two component variances are about equal, and the variance of the difference is approximately the sum of these component variances.

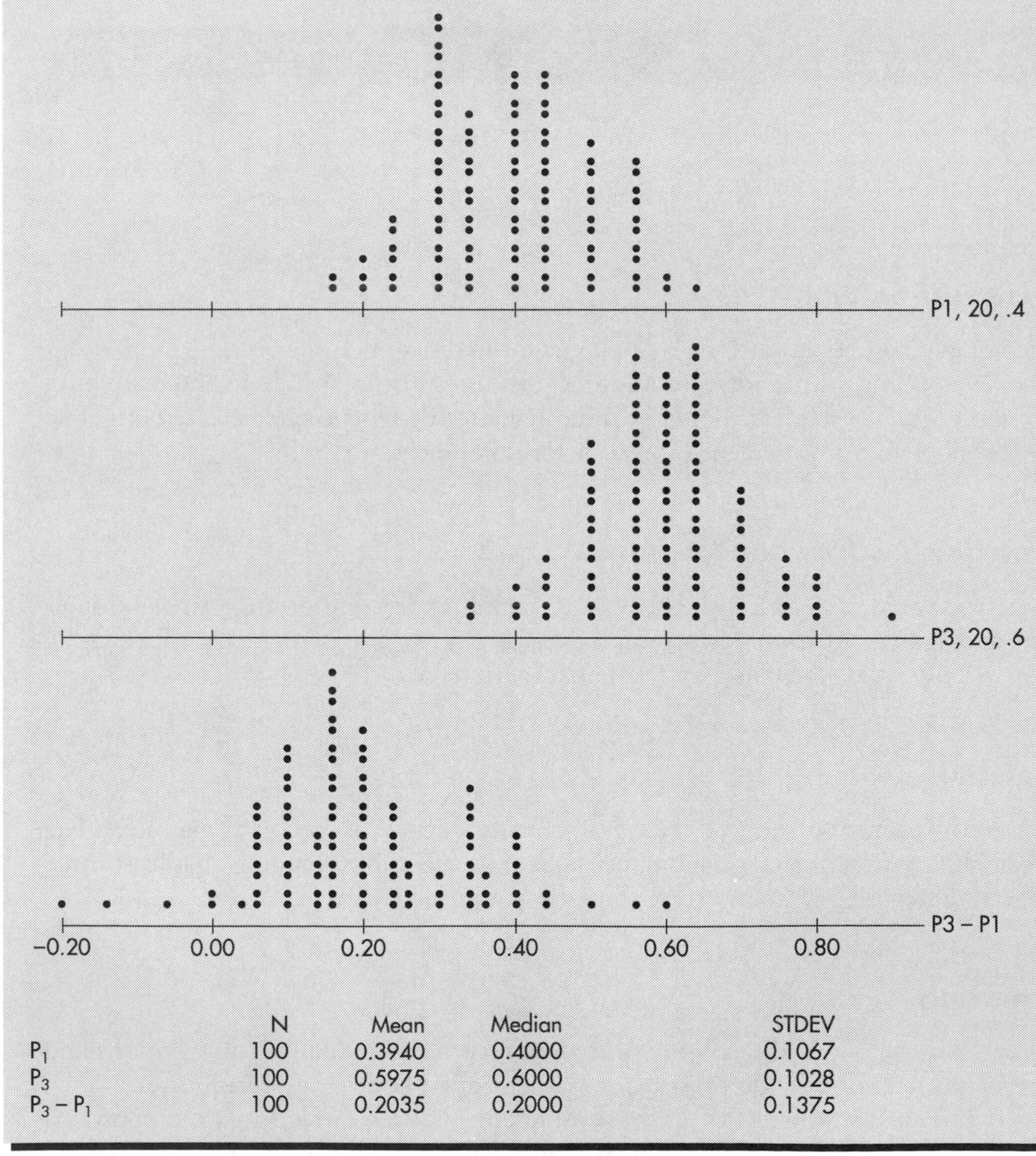

	N	Mean	Median	STDEV
P_1	100	0.3940	0.4000	0.1067
P_3	100	0.5975	0.6000	0.1028
$P_3 - P_1$	100	0.2035	0.2000	0.1375

Figure 1: Differences Between Two *Independent* Proportions

The plot and numerical summaries attached for *dependent* proportions (see Figure 2) show computer-generated samples of size 20 from a "bead box" with 50% red and 40% white beads. Notice that the two component variances are about equal, and the variance of the difference is now approximately four times as large as either component variance.

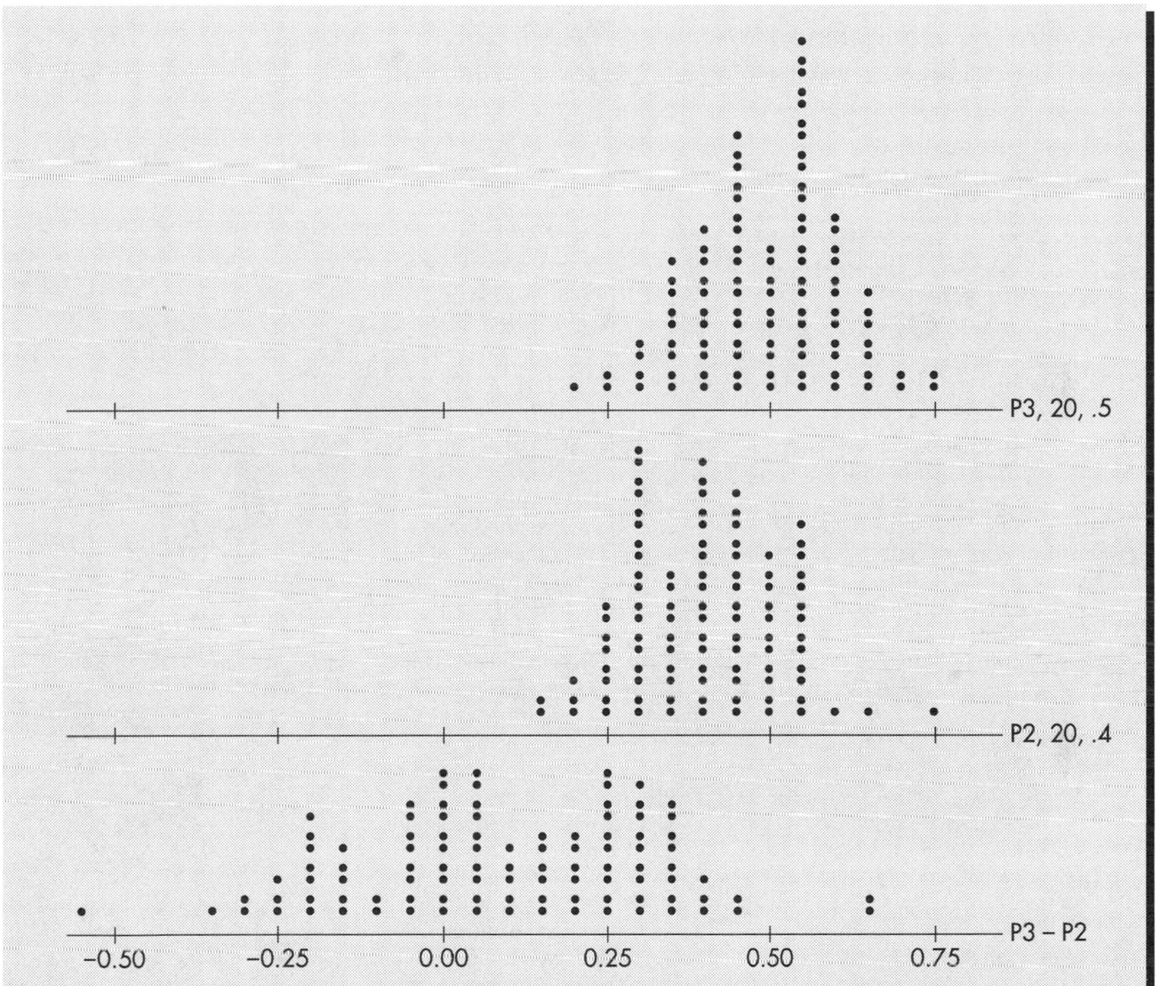

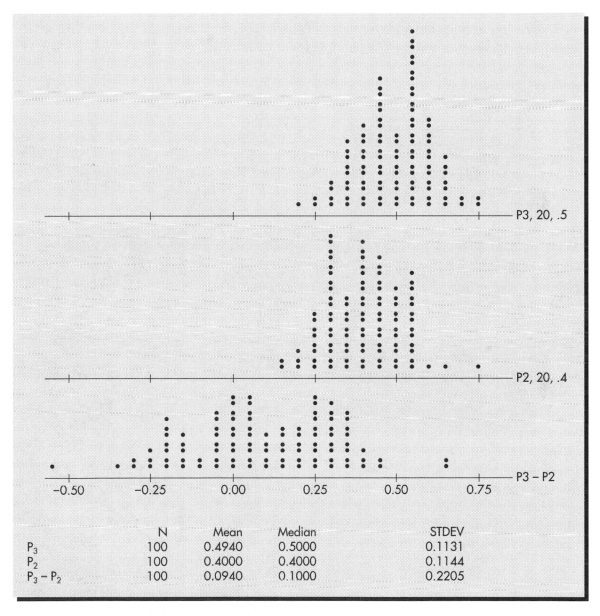

	N	Mean	Median	STDEV
P_3	100	0.4940	0.5000	0.1131
P_2	100	0.4000	0.4000	0.1144
$P_3 - P_2$	100	0.0940	0.1000	0.2205

Figure 2: Differences Between Two *Dependent* Proportions

Sample Assessment Questions

Have students analyze survey results reported in the media, such as those presented at the end of these Notes. Suggested questions include

1. If a margin of error for looking at appropriate differences is reported, is it correct? If a margin of error for differences is not reported, can you find one?

2. Are the conclusions reached in the article correctly stated? If no conclusions are stated, can you state some that are statistically justified?

3. What would you like to know about the design, implementation, and data analysis portions of the study that is not presented in the article?

"If Women Ran America"

Two sexes differ on the issues affecting the nation

NEW YORK— A poll comparing men's and women's attitudes on public issues suggests women want stricter law enforcement against drunken driving, guns and drug dealing.

The poll was commissioned by Life magazine for a story in the June issue headlined "If Women Ran America." Life said its poll found women interested in "safety first. But fairness also, especially fairness for women at work."

- Two in three women polled said consider unequal pay for the same work to be a very serious problem for women in the workplace. Just half the men responded similarly.
- Half the women but only a third of the men think discrimination in promotions is a very serious problem for women at work.
- The poll said 78 percent of women, compared with 64 percent of men think businesses should be required to provide paid maternity leave.
- The poll was taken by the Gallup Organization, which surveyed a national sample of 614 women and 608 men by phone March 30–April 5.
- The margin of error ranges from plus or minus 3 percentage points for the whole sample, up to 6 points for comparisons of results of men and women.
- In other words, the poll indicates a gender gap, rather than chance variation, accounts for differences of opinion such as this: 55 percent of women but only 46 percent of men said the government should make fighting crime and violence an extremely important priority.
- Seventy-six percent of women and only 58 percent of men said the justice system wasn't tough enough on drunken drivers. On drug dealing, 88 percent of women and 77 percent of men wanted the system to be tougher. Seventy percent of women and 63 percent of men wanted to be tougher on illegal gun possession.
- Women were more compassionate than men on some issues: 85 percent would approve of a law requiring businesses to allow employees an unpaid 12–week family medical leave. Women were more likely than men to approve of such a leave for homosexual couples, and to say they would vote for a gay candidate.

(Source: *Gainesville Sun*, May 5, 1992.)

Should smoking be banned from work places, should there be special smoking areas, or should there be no restrictions?

	Non-Smokers	Total
Banned	44%	35%
Special areas	52%	59%
No restrictions	3%	5%

Should the federal tax on cigarettes be raised by $1.25 to pay for health care reform?

Yes	58%	49%
No	36%	46%

Which do you agree with more?

Smoking is a bad habit and society should do everything possible to stamp it out.	31%	25%
It's bad, but everyone should have the right to make his or her own choice whether to smoke	67%	73%

Do people view Congress differently from the way they view their Representative to Congress? A poll of 1161 adults produced the following results, as reported in the New York Times, September 13, 1994. Provide an answer to the question posed above, justifying your answer from the data in the results of the poll

The New York Times/CBS NEWS Poll

How the Public Views Congress As a Whole . . .

Do you approve or disapprove of the way Congress is handling its job?

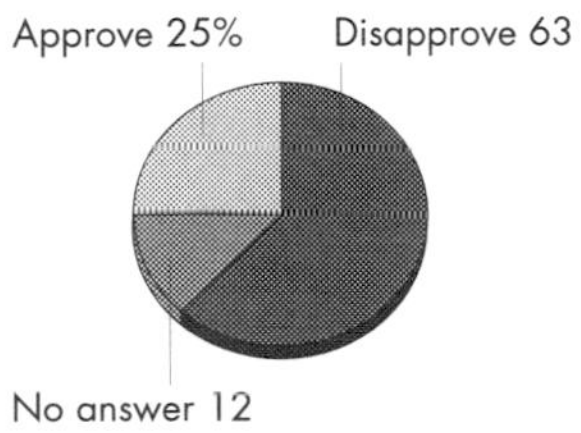

Have most members of Congress done a good enough job to deserve re-election, or is it time to give new people a chance?

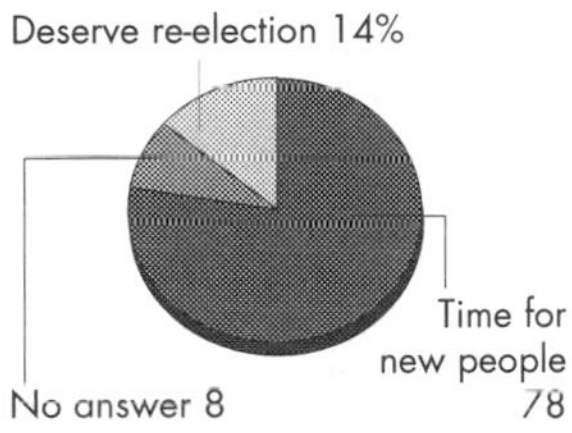

Are most members of Congress more interested in serving the people they represent, or more interested in serving special interest groups?

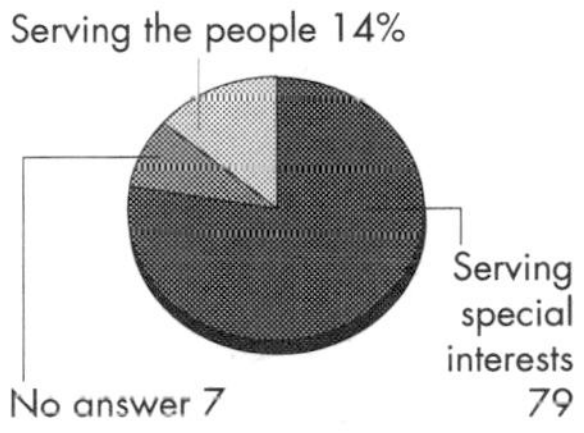

. . . And How They View Their Own Representative

Do you approve or disapprove of the way the representative in Congress from your district is handling his or her job?

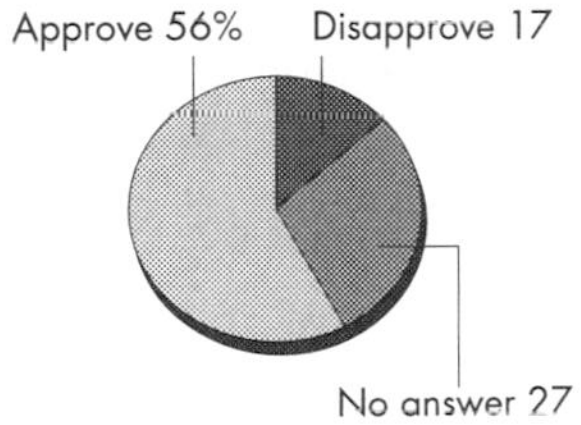

Has the representative from your district done a good enough job to deserve re-election or is it time to give a new person a chance?

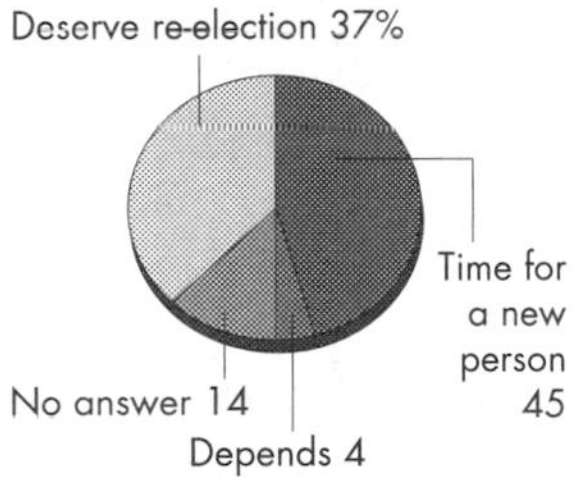

Is the representative from your district more interested in serving the people he or she represents, or more interested in serving special interest groups?

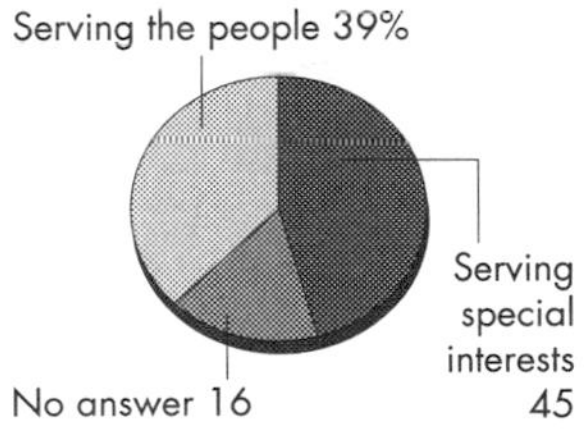

Based on nationwide telephone interviews conducted Sept. 8 to 11 with 1,161 adults.

Estimating the Difference Between Two Proportions

SCENARIO

A Gallup Poll of 1,200 people that stated 60% of the public favor stricter gun control laws also stated that a similar poll five years earlier produced a sample percentage of 65% in favor of stricter gun control. Has the true percentage of people favoring stricter laws actually decreased? In the current poll, the percentage of women favoring stricter laws was 71%, while the percentage of men favoring stricter laws was 48%. Are these significantly different? In the same poll, 42% of those responding agreed that handguns should be banned except for police and other authorized persons, while 50% said they should not be banned, and 8% had no opinion. How can we make a fair comparison between percentages taken in two different polls? How can we make a fair comparison between two different percentages calculated from the same poll? These are the two most common types of comparisons made between sample proportions, and we will study their properties in this activity.

Question

How can two different sample proportions be compared fairly in the light of sampling error?

Objective

In this lesson you will generate sampling distributions for the difference between two sample proportions for the purpose of understanding how to calculate a margin of error for such differences. Independent and dependent sample proportions will both be investigated.

Activity

I. INDEPENDENT SAMPLES

1. Work with a partner throughout the activity. Examine the bead box supplied for this activity. It contains at least three different colors of beads. Choose two colors of beads (say, red and white) that seem to occur quite frequently but for which the proportions in the box do not appear to be the same. The goal of this activity is to compare two ways of estimating the difference between the true proportion of red beads and the true proportion of white beads in the box.
   a. Each member of the team must complete this step separately. Using a sample size of 20 each time, generate 50 samples of beads. Make sure that each sample is returned to the box before the next is selected so that each sample comes from the same population. Record the proportion of red beads and the proportion of white beads for each sample. It is helpful to construct a table for recording the data similar to the one started below.

Sample	Proportion Red	Proportion White
1	.45	.35
2	.40	.40
3	.45	.30
.	.	.
.	.	.
.	.	.

   b. Construct a dot plot of the proportions of red beads. Compare your plot with your partner's plot. (The plot is an approximate sampling distribution for the proportion of red beads in samples of size 20.) Do the same for the proportions of white beads.
   c. Where do the sampling distributions center? How do you think this centering value relates to the true proportions of red beads and white beads in the box?

2. The goal now is to study the behavior of differences between independently selected sample proportions.
   a. Pair your red bead proportions with your partner's white bead proportions on a table similar to the one shown here. Append a column of differences to the table.

Sample	My Proportion Red	Partner's Proportion White	Difference
1	.45	.40	.05
2	.40	.30	.10
3	.45	.50	−.05
.	.	.	.
.	.	.	.
.	.	.	.

b. Plot the differences on a dot plot. How does this plot compare with the plots constructed in step 1b? Do the differences have less variability than the individual sample proportions?
c. From the table in step 2a, calculate the variances of the sample proportions for red beads, the sample proportions for white beads, and the differences. Which is the largest? Is this surprising?
d. Observe that the variance of the differences is approximately the same as the sum of the two component variances.
e. Recall that the margin of error in the case of estimating a single proportion was two times the standard deviation of the sample proportions. Write a formula for the margin of error for the difference between two independent sample proportions, denoting the two population proportions by p_1 and p_2 and the two sample sizes by n_1 and n_2.

II. Dependent Samples

3. Now, let's look at red and white beads in the **same** sample. We have at least three colors of beads, remember, so the proportion of reds and the proportion of whites need not add to one for a single sample. Why do we call these **dependent** proportions?
   a. For 50 samples of 20 beads each from **your own** sampling, as recorded in the Table of step 1a, calculate the difference between the proportions for each sample. Append the column of differences to your table.
   b. Construct a dot plot for the differences. Compare this plot to the plots from steps 1b and 2b. How do the shapes of the four dot plots compare? Do the differences have less variability than the proportions by themselves? Do the dependent differences have less variability than the independent differences?
   c. Calculate the variances for the sample proportions of red beads, the sample proportions of white beads, and the dependent differences. Which is the largest? You should observe that the variance of the differences in this case is about **four times** as large as the variance of either of the component parts.

   d. Write a general, but approximate, rule for calculating the margin of error for the difference between two dependent sample proportions of the type studied here.

III. Conclusions

4. Use the results discovered above to answer the questions posed at the start of this activity.
   a. Was there a significant decrease over the five years in the proportion who favor stricter gun control laws? What margin of error should be used?
   b. Is the true proportion of women who favor stricter gun control laws larger than the proportion of men who favor such laws? What margin of error should be used?
   c. Does the proportion who support a limited ban on handguns differ significantly from the proportion who do not support such a ban? What margin of error should be used?

Wrap-Up

1. Find an article in a recent publication that reports on a poll conducted at different times, with random sampling and provides a margin of error for a change in sample proportions. (News magazines often report such results.) Explain the meaning of the margin of error, and interpret the poll results in the light of this margin of error.

2. A nationwide survey of 1,400 high school students who say they drink alcoholic beverages reported that 14% drink more than once a week, 31% drink once a week, 29% drink once a month, and 26% drink less than once a month. Where are the "real" differences among the percentages of students drinking alcohol?

3. For the 1992 presidential election, the final poll results from three major polls were as follows:

Poll	Size	Clinton	Bush	Perot
CNN	1,562	44	36	14
Gallup	1,579	43	36	15
Harris	1,675	44	39	17

a. Comment on the estimates of the **lead** for Clinton over Bush shown by these three polls. Use the concept of "margin of error" in your discussion.
b. All of these polls declared Clinton the winner. Were they justified, based on the data?

4. On October 20, 1992, the day after the third presidential debate, these same pollsters reported leads for Clinton over Bush as follows:

CNN, 12 points; Gallup, 7 points; Harris, 14 points.

Did the leads as reported by these polls change significantly between October 20 and the day before the election? Are you making any assumptions in your argument?

5. Write a brief summary of what you learned from this lesson.

Extensions

1. To confirm the ideas presented here for approximating the standard deviations of sample proportions, repeat steps 1 through 3 for sample sizes larger than 20. Do the rules for combining standard deviations still hold? Note that the sample standard deviations for individual sample proportions can be approximated by the formula

$$[p(1 - p)/n]^{1/2}$$

and do not have to be calculated from the simulation data.

2. Find examples of polls published in the media for which it is appropriate to compare independent proportions. If the margin of error is given, verify that it is correct. If the margin of error is not given, calculate it to a reasonable degree of approximation.

3. Repeat step 2 for dependent proportions.

References

1. Stephen Ansolabehere and Thomas R. Belin (1993), "Poll faulting," *Chance*, **6**(Winter):22–27.
2. Zbigniew Kmietowicz (1994), "Sampling errors in political polls," *Teaching Statistics*, **16**(Autumn):70–74.
3. A. J. Scott and G. A. F. Seber (1983), "Differences of proportions from the same survey," *The American Statistician*, **37:**319–320.
4. C. J. Wild and G. A. F. Seber (1993), "Comparing two proportions from the same survey," *The American Statistician*, **47:**178–181.

The Bootstrap

NOTES FOR THE INSTRUCTOR

Statistical Setting

The bootstrap is a conceptually simple yet powerful tool that is becoming more and more widely used as computer power increases. This activity gives the student a flavor of how the bootstrap works by considering a fairly simple problem. At the end of the activity you might discuss with the class that the bootstrap can be used to tackle more challenging problems, such as constructing an interval estimate of a correlation coefficient in the absence of normality or estimating a survival curve in the presence of censoring and truncation.

Note that a bootstrap interval estimate of a parameter is not really the same thing as a standard confidence interval. Both types of intervals provide estimates of a parameter, but it would probably confuse students to refer to both as confidence intervals. Hence, we have chosen to use the term "bootstrap interval" for an interval produced using the bootstrap. This terminology is not standard, but it serves to differentiate between two somewhat different ideas.

Prerequisites for Students

Students should be familiar with confidence intervals. The more familiarity and comfort they have with using the computer, the easier it will be for them to complete the activity.

Materials

The only materials needed are copies of the student activity sheet and access to a computer.

Procedure

As with any computer application, you should test the commands given here on your own machine before having your students try this. Some of the calculations take quite a while, up to several minutes or more, depending on the computer used. Thus, you might want to reduce the number of bootstrap samples from 500 to 250.

A Minitab macro for bootstrapping the median is given in the student handout. Other software can be used as well. For example, to use DataDesk, do the following. (You'll need version 4.0 or above of DataDesk, which has a sampling routine built in. If you have version 4.1 or above, then in step 2 you can choose not to generate sample indices. This means that you can eliminate steps 5 and 6.)

1. Store the data in an icon.

2. Go to the **Manip** menu and choose Sample. . . . Choose random sample, click on the "Sample with replacement" box, and set the percent of cases to be sampled to 100. (If the program won't let you specify 100%, then use 99.9%.) Ask for 500 samples. This will produce a window called "Random samples" with 500 sample icons and 500 icons of sample indices.

3. Go to the **Calc** menu, choose Calculation options, then choose Select Summary Statistics, and check the box for medians only (i.e., un-select the other summary statistics).

4. Select all 1,000 icons in the window from step 2 by clicking on the small rectangle near the top right-hand side of the window. Then go to the **Calc** menu, choose Summaries, then As Variables. This will produce a new icon, named Medians, that holds the sample medians for the bootstrap samples—along with the medians of the sample indices, which we'll want to dispose of.

5. Go to the **Manip** menu and choose Generate Patterned Data. . . . Generate numbers from 1 to 2 in steps of 1, repeat each number once, and repeat the entire sequence 500 times. This will generate an icon, named Pattern1, with ones corresponding to the bootstrap medians and twos corresponding to the medians of the sample indices.

6. Select the Medians icon, then the Pattern1 icon. Go to the **Manip** menu and choose Split into Variables by Group. This will create two new icons. The one named 1 will hold the 500 bootstrap medians. Rename this icon "boot medians." (The icon named 2 will hold the medians of the sample indices, which we don't want, so we'll ignore it.)

7. Select the "boot medians" icon, go to the **Plot** menu, and create a histogram.

8. Sort the cases in the "boot medians" icon (the sort command is under the **Manip** menu). Then find the 13th largest value and the 488th largest value. These are the endpoints of a 95.2% bootstrap interval for the population median.

To bootstrap a different statistic, such as the mean, using DataDesk just replace the median by the mean in step 3.

Note: The data used here form a time series, which may raise concerns about the usual assumption that the observations are independent of one another. However, these data exhibit no serial correlation.

Sample Results from Activity

Figure 1 is a histogram of bootstrapped medians generated using DataDesk.

A typical 95.2% confidence interval for the rainfall median is (10.81,15.37). Of course, repeating the bootstrap process will yield a slightly different confidence interval each time.

Sample Results from Extensions

You might want to verify that the bootstrap gives sensible results by using it in a situation in which normal theory works. That is, you could bootstrap the mean with data that came from a normal distribution and compare the bootstrap interval with the usual CI for a population mean.

As was mentioned here in the student notes, the rainfall data do not appear to come from a normal distribution, so a bootstrap confidence interval will differ a bit

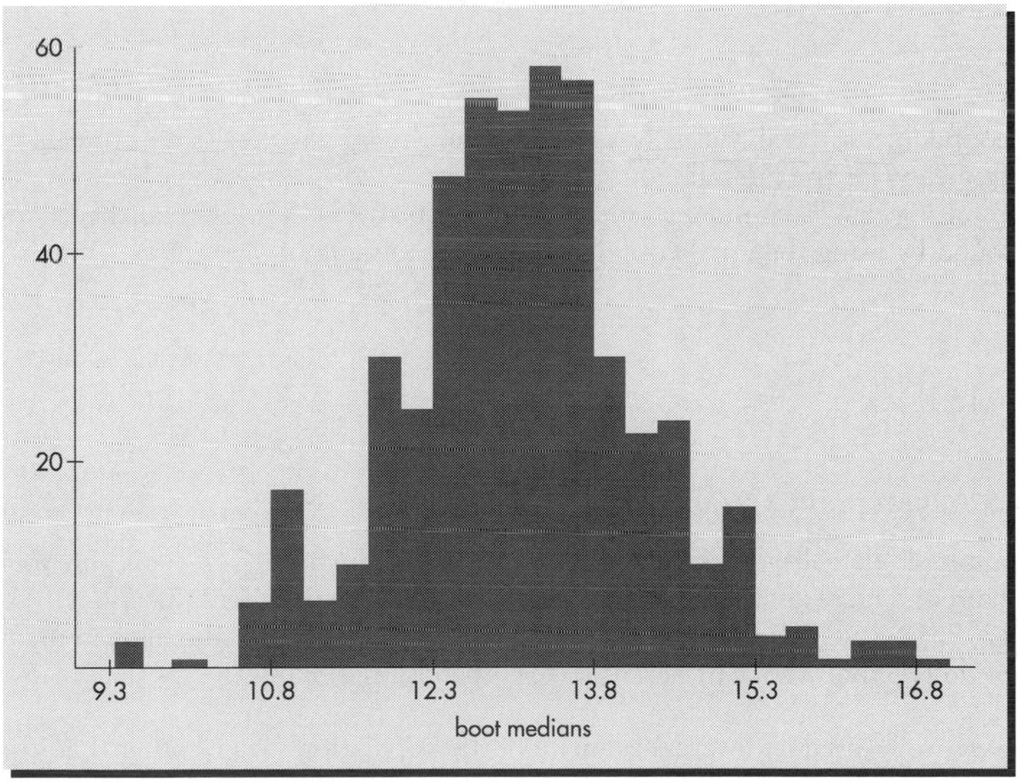

Figure 1

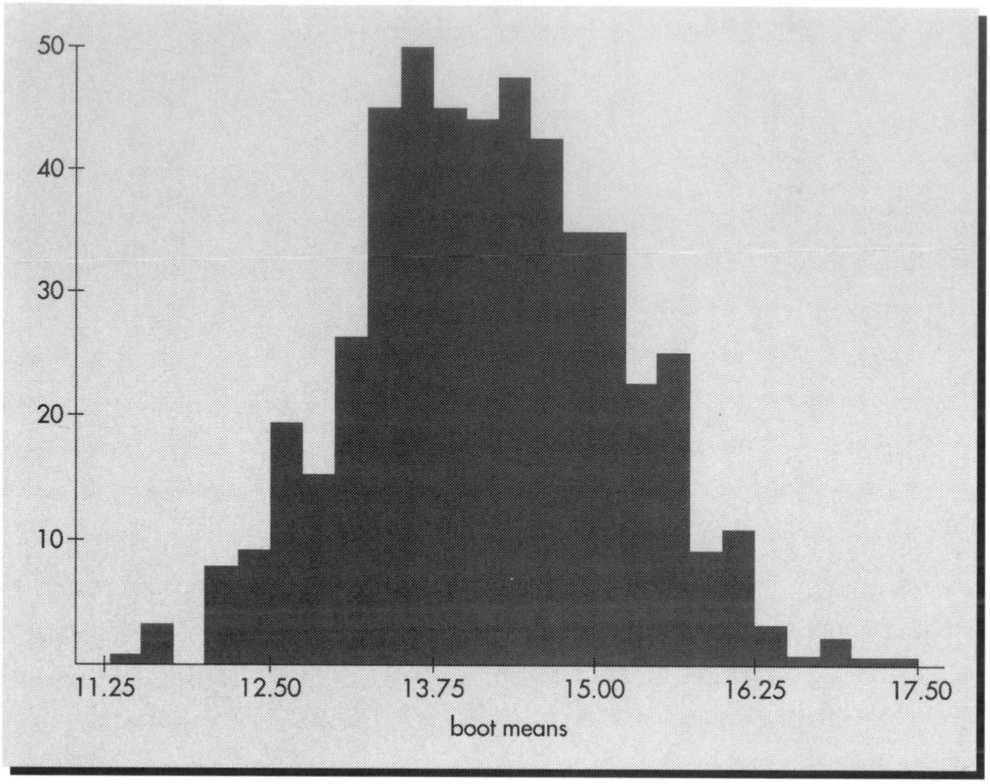

Figure 2

from a confidence interval that is based on normal theory. Figure 2 is a histogram of bootstrap means for the rainfall data.

The resulting 95.2% confidence interval is (12.26,16.15). The normal theory interval, found by using the usual confidence interval formula of

$$\bar{x} \pm t*\frac{s}{\sqrt{n}},$$

is (12.16,16.18).

Sample Assessment Questions

1. A sample of 13 college students were asked "How much did you spend on your last haircut?" Here are their responses (in dollars): 0, 10, 10, 0, 12, 23, 77, 35, 17, 14, 12, 7, 12. Use this data to construct a bootstrap interval for the median number of dollars that a student at this college spends on a haircut.

2. Suppose we wanted an interval estimate of the population mean, rather than the population median, in question 1. Why might we prefer using the bootstrap technique to using the standard formula $\bar{x} \pm t*(s/\sqrt{n})$?

The Bootstrap

SCENARIO

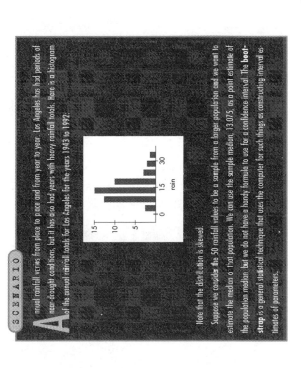

Annual rainfall varies from place to place and from year to year. Los Angeles has had periods of near-drought conditions, but it has also had years with heavy rainfall totals. Here is a histogram of the annual rainfall totals for Los Angeles for the years 1943 to 1992:

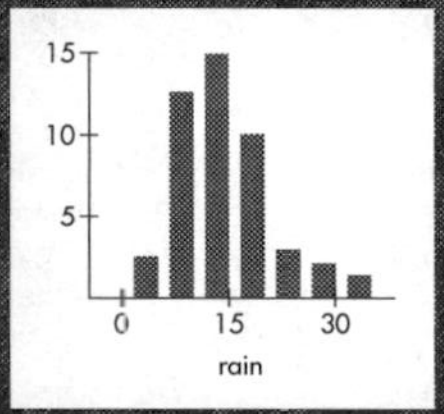

Note that the distribution is skewed.

Suppose we consider the 50 rainfall values to be a sample from a larger population and we want to estimate the median of that population. We can use the sample median, 13.075, as a point estimate of the population median, but we do not have a handy formula to use for a confidence interval. The **bootstrap** is a general statistical technique that uses the computer for such things as constructing interval estimates of parameters.

Question

How much rain can Los Angeles expect to get in a typical year?

Objectives

The goal of this lesson is to develop an understanding of the bootstrap technique. After completing this lesson, you should understand how the bootstrap works and be able to use the bootstrap method to construct confidence intervals for parameters such as the median of a distribution.

Activity

Here are the rainfall totals (in inches) for the years 1943 to 1992:

22.57, 17.45, 12.78, 16.22, 4.13, 7.59, 10.63, 7.38, 14.33, 24.95, 4.08, 13.69, 11.89, 13.62, 13.24, 17.49, 6.23, 9.57, 5.83, 15.37, 12.31, 7.98, 26.81, 12.91, 23.66, 7.58, 26.32, 16.54, 9.26, 6.54, 17.45, 16.69, 10.70, 11.01, 14.97, 30.57, 17.00, 26.33, 10.92, 14.41, 34.04, 8.90, 8.92, 18.00, 9.11, 11.57, 4.56, 6.49, 15.07, 22.56

We wish to use this data to make an inference about Los Angeles rainfall.

The sample median of 13.075 describes the data we have. If we got a new set of 50 years of rainfall data, we would expect the new sample median to differ from 13.075. The question is, "How does the sample median behave in repeated samples of size 50?"

Ideally, we would study this by taking many samples of size 50 from the population and computing the sample median for each one. This would require having the entire population before us, so that we could sample from it! However, all we have are the 50 values listed above. The key idea behind the bootstrap technique is to *use the distribution of the 50 sample values in place of the true population distribution*. That is, we will think of the distribution we have, as represented by the histogram above, as being an estimate of the population distribution of rainfall values. We will use a model that says that in some years Los Angeles will get 22.57 inches of rain, in other years it will get 17.45 inches, etc., and the only rainfall totals possible are the 50 values listed above. (Of course, we would be better off if we had a sample of size 100, rather than of size 50—the larger the sample size, the better we expect the sample histogram to represent the true population distribution.)

In order to see how medians of repeated samples of size 50 behave, we will sample *with replacement* from the set of 50 values listed earlier. Thus, in a sample of 50 years, we might get one year in which 22.57 inches of rain fall, two years in which 17.45 inches of rain fall, no years in which 12.78 inches of rain fall, etc.

It is a good idea to simulate this process by hand before moving to the computer. Write the numbers 22.57, 17.45, etc., on slips of paper and put the slips in a bag. Then draw out a slip, record the value, and replace the slip in the bag. Do this 49 more times, so that you have a total of 50 observations. This collection of 50 draws is called a **bootstrap sample**. (It is possible that you will draw each of the 50 slips exactly once, but this is highly unlikely. It is more likely that you will draw some slips several times and others not at all.) Now compute the median of your bootstrap sample. This is a bootstrap estimate of the population median.

Using the Computer

We would like to see how the sample median behaves in repeated samples of size 50. Drawing a sample by hand takes a lot of time. We'll use the computer to repeat the process very quickly, drawing 500 bootstrap samples and computing 500 bootstrap estimates of the population median.

For example, here is a Minitab macro for bootstrapping the median:

```
noecho
# note k1 = sample size
# note k3 should be set to 1 at the start
sample k1 c1 c2;
replace.
let k2 = median(c2)
let c3(k3) = k2
let k3 = k3 + 1
end
```

To use this macro with the rainfall data, you must set k1 = 50 and set k3 = 1. The 50 rainfall values should be stored in column c1. The macro tells Minitab to take a sample of size k1 (50) with replacement from c1 and put the results in c2. The median of this bootstrap sample is calculated and stored in c3. If you run this macro 500 times, you end up with 500 bootstrap estimates of the median stored in column c3.

Once you have 500 bootstrap estimates, you can use them to construct an interval estimate of the population median. (You can use the 500 values to study other aspects of the median as well.) For example, here is a histogram of 500 bootstrap medians from the rainfall data:

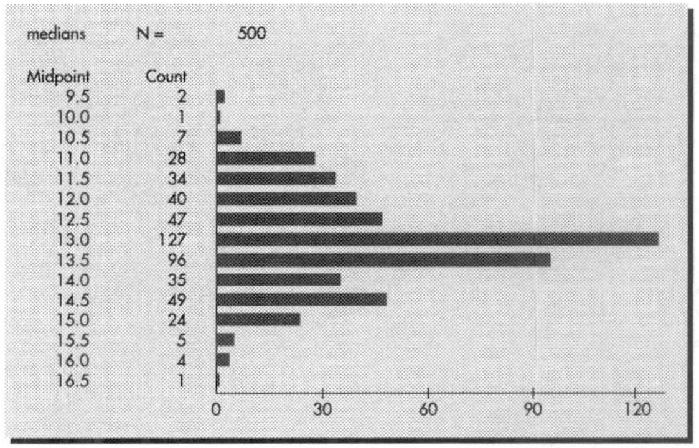

To get a 95% "bootstrap interval," we take the middle 95% of this distribution. We throw away the top 2.5% and the bottom 2.5% of the distribution. That is, we sort the 500 values and throw out the first 12 (2.5% of 500 is 12.5, but we'll round down) and the last 12. This leaves the middle 476 values, which gives us a 95% bootstrap interval for the population median. (Actually, this gives us a 95.2% interval, since 476/500 = .952.) In this example the bootstrap interval is (10.81,15.17). You might get a slightly different interval if you repeat the bootstrap simulation.

By taking 500 bootstrap samples we get a pretty good idea of what the distribution of all possible bootstrap samples looks like. However, this does not necessarily mean that our interval will be good. The accuracy of the bootstrap interval depends on how well the original 50 observations represent the true population—the key feature is the quality of original sample. If the original sample is biased in some way or if the sample size is simply too small for the sample histogram to give a good representation of the population, then taking 5,000 or even 50,000 bootstrap samples (rather than 500) won't produce a good result.

Wrap-Up

1. Write a brief summary of what you learned in this activity about the bootstrap technique and how it works.

2. Explain how the bootstrap technique could be used to construct an interval estimate of the 90th percentile of a distribution.

Extension

The bootstrap allows us to make inferences without making the usual assumption that the data came from a normal population. However, the bootstrap works with data from normal populations, and it can be used in situations in which a formula is available, such as for constructing a confidence interval for a population mean.

Use the bootstrap technique to construct an interval estimate of the population mean. To do this with Minitab, you must modify the macro by replacing the line "let k2 = median(c2)" with the line "let k2 = mean(c2)". The rainfall data do not appear to come from a normal population, so the usual confidence interval formula of

$$\bar{x} \pm t^* \frac{s}{\sqrt{n}}$$

will differ from the bootstrap interval

References

1. B. Efron (1982), *The Jackknife, the Bootstrap, and Other Resampling Plans*, Philadelphia: Society for Industrial and Applied Mathematics.
2. B. Efron and P. Diaconis (1983), "Computer-intensive methods in statistics," *Scientific American*, **248**(5):116–130.
3. B. Efron and J. Tibshirani (1993), *An Introduction to the Bootstrap*, New York: Chapman & Hall.
4. P. Hall (1992), *The Bootstrap and Edgeworth expansion*, New York: Springer-Verlag.

Estimating the Total of a Restaurant Bill: Bias in Estimation

NOTES FOR THE INSTRUCTOR

Statistical Setting

One way to estimate the total cost of the items on a fast food restaurant bill is to round each of the prices to the nearest multiple of 10¢ before adding them. There are two problems with this method. First, not all last digits are equally likely. If prices tend to end in 9's, then the estimated total will tend to be too high. Second, the usual rule of rounding a last digit of 5 up to the next multiple of 10 also results in an estimated total that tends to be too high. This setting is used to introduce students to the idea of bias in estimation. This activity also develops the idea of a probability distribution and its mean and the idea that the expected value of a sum of individual errors is equal to the sum of the expected values of the individual errors.

Prerequisites for Students

Students should be able to compute the mean of a probability distribution, or they can learn this concurrently. The standard deviation is needed in Extension 2.

Materials

You will need copies or an overhead transparency of a current menu from a fast food restaurant, or of the McDonald's menu included with this activity.

For Extension 3, at least one grocery bill per student is needed. Students should be asked to bring one in to work with—it does not have to be their own.

Sample Results from Activity

1. Answers will vary.

2. The histogram will be centered above zero.

3. There are three possible reasons why the rounding scheme is biased. The first is that the dinners aren't a random sample from the items available. The second, and more important, reason is that the last digits are not equally likely. Prices tend to end in a 9. Thus, most rounding will be "up" rather than "down," and the estimated total will tend to be too high. A third reason is the use of the rule that if the last digit is a 5, it is rounded up. This means again that the estimated total will tend to be too high:

Last Digit	0	1	2	3	4	5	6	7	8	9
Error	0	−1	−2	−3	−4	5	4	3	2	1

If all last digits are equally likely, the expected error is

$$0.1[0 + (-1) + (-2) + (-3) + (-4) + 5 + 4 + 3 + 2 + 1] = 0.5¢ \text{ per item.}$$

Last Digit	Frequency	Error When Rounded	Probability
0	1	0	1/40
1	0	−1	0
2	0	−2	0
3	0	−3	0
4	0	−4	0
5	4	5	4/40
6	0	4	0
7	0	3	0
8	1	2	1/40
9	34	1	34/40

5. a. For the McDonald's menu, the expected error is $0(1/40) + 5(4/40) + 2(1/40) + 1(34/40) = 56/40 = 1.4¢$ per item.
   b. 0.

6. The hard way is to find the error for each item on the menu and then add these errors. The sum is 56¢. You may have to suggest the easier way:

$$(40 \text{ items})(\text{expected error of } 56/40\ ¢ \text{ each}) = 56¢.$$

7. Students may suggest rounding down in all cases except when the price ends in a 9. If the items are randomly selected, this method gives an expected error of

$$0(1/40) + (-5)(4/40) + (-8)(1/40) + 1(34/40) = 6/40 = 0.15¢ \text{ per item,}$$

which is close to unbiased.

Sample Results from Wrap-Up

1. 10(1.4¢) = 14¢.

2. All of the prices end in a 0, 5, 8, or 9, with a 9 by far the most likely. If the price ends in a 0, there is no error in the estimate. If the price ends in a 5, 8, or 9, the price will be rounded up and so the estimate will be too large. Since there will be no underestimate of the price of any item, the total can't be underestimated.

Sample Results from Extensions

1. The absolute average error tends to increase with the number of items purchased. This is the case even if the estimation procedure is unbiased.

2. For this question, students must know how to compute the standard deviation of a probability distribution:

$$\sigma_x = \sqrt{\Sigma(x - \mu_x)^2 \, P(x)}.$$

In this case, $\sigma_x \approx 1.22$. If the total error is 60¢ for 40 items, the average error is 1.5¢. The z-score for an average error of 1.5¢ for 40 items is

$$z = \frac{1.5 - 1.4}{1.22/\sqrt{40}} = 0.52.$$

Using the central limit theorem, the probability of a total error larger than 60¢ is 0.3015.

3. Grocery store prices also tend to end in a 9. Consequently, an unbiased method of rounding will have to round down more digits than 0, 1, 2, 3, and 4.

Sample Assessment Questions

1. Here is the frequency distribution of the last digits of the prices of items in a grocery store. You are using the rule of rounding prices that end in 0, 1, 2, 3, or 4 down and the rest up to estimate the total cost of the items you are purchasing.

Last Digit in Price of Item	Frequency (Number of Items)	Error When Rounding	Probability
0	42		
1	13		
2	26		
3	11		
4	6		
5	50		
6	32		
7	56		
8	15		
9	28		

a. Complete the remaining two columns of the table.
b. What is the average error per item with this rounding rule?
c. Suppose you buy 50 items. On the average, what would be the difference between your estimated grocery bill and the estimated bill? Be sure to specify if the estimated total tends to be too high or too low.

2. Explain what the term "biased" means.

Estimating the Total of a Restaurant Bill: Bias in Estimation

SCENARIO

One way to estimate the total cost of the items on a restaurant bill is to round each of the prices to the nearest multiple of 10 cents before adding them. If you use this method, sometimes your estimate will be too high and sometimes it will be too low. But you would hope that the overestimates and the underestimates would just about balance out. That is, such a method of estimation should be *unbiased*.

Questions

If we estimate the total cost of the items on a restaurant bill by rounding the prices to the nearest multiple of 10 cents before adding them, will we get an unbiased estimate? If not, can we devise a method of rounding so that the resulting estimate of the total is unbiased?

Objective

In this activity you will learn the meaning of *bias* in estimation.

Activity

1. a. From the McDonald's menu shown here, order dinner for yourself and a friend. Your friend is paying.
   b. Find the total cost of the two dinners, ignoring taxes.

   c. Estimate the total by rounding each item to the nearest multiple of 10¢.
   d. Find the error in the estimate by subtracting the total cost from the estimate:

estimated total − actual total.

If the answer is negative, report it as such.

McDonald's Dinner Menu

McLean	1.89	Chunky Chicken Salad	2.99
McLean with Cheese	1.99	Chef Salad	2.99
Big Mac	1.99	Garden Salad	2.09
Quarter Pounder	1.99	Side Salad	1.39
Filet of Fish	1.49	Carrot Sticks	.69
Double Cheeseburger	1.29	Soft Drinks	.89
Hamburger	.59		.99
Cheeseburger	.69		1.09
Chicken Fajita	.99	Child-Sized Drink	.69
McChicken Sandwich	2.09	Milk Shake	1.09
Chicken McNuggrts			1.35
6 piece	1.79	Coffee	.49
9 piece	2.49		.69
20 piece	4.69		.89
McGrilled Chicken Classic	2.69	Hot Tea	.49
Fries	.79	Milk	.70
	1.19	Orange Juice	.75
	1.48		.95

Sundae	.99
Cone	.39
Apple Pie	.89
Cookies	.65
Chocolate Chip Cookies	.69

2. Make a histogram of the errors for all of the students in your class.
3. Compute the mean error for the dinners in your class. Is the rounding procedure biased? If so, in what direction? Can you suggest a reason for this?

4. Using the McDonald's menu, complete this probability distribution of the errors when prices are rounded to the nearest 10¢.

Last Digit in Price of Item	Frequency (Number of Items)	Rounding Error	Probability
0			
1			
2	0	−2	0
3			
4			
5			
6			
7			
8	1	2	1/40
9			

5. a. Find the mean of this probability distribution using the formula

$$\mu_x = \sum x \cdot P(x).$$

b. What would the mean be if the method of estimation was unbiased?

6. If you were to buy one of each item on the McDonald's menu and estimate the total by rounding each price to the nearest 10¢, how far off would your estimate be? Do this problem two different ways.

7. Devise a method of rounding prices at McDonald's so that the estimate of the total bill is unbiased, or as close to it as possible.

Wrap-Up

1. Suppose you buy 10 items selected at random from the McDonald's menu. If you estimate the total cost by rounding each price to the nearest 10¢, how far off from the actual total would you expect to be?

2. Explain why it is impossible to underestimate the total of a McDonald's bill by rounding the price of each item to the nearest 10¢.

Extensions

1. Design an experiment to determine if the absolute value of the error in estimating the total of a bill (not necessarily McDonald's) tends to increase, decrease, or remain the same as the number of items purchased increases. Does it matter if the rounding procedure is biased or not?

2. Use the central limit theorem to estimate the probability that the error will be less than 60¢ on a purchase of 40 items selected randomly at McDonald's. (See Samuel Zahl (1973), "Grocery shopping and the central limit theorem," in Frederick Mosteller et al. (eds.), *Statistics by Example: Detecting Patterns*, Reading, MA: Addison-Wesley.)

3. Collect some register tapes from grocery store purchases. Make a probability distribution of the last digits of the prices. Can you devise a method of rounding grocery store prices so that the estimate of the total bill is unbiased?

How Many Tanks?

NOTES FOR THE INSTRUCTOR

Statistical Setting

This activity can be used either as an introduction to the topic of estimation or as a summary of properties of estimators. In the former case, it is important to allow an open-ended discussion of what constitutes an estimator and what properties make one estimator better than another. Students have difficulty separating the estimator (or rule) from an estimate (the rule applied one time to a set of data). Being lucky one times does not mean that they have a good general rule.

In the latter case, this activity allows students to see that not all good estimators depend on the mean or median. The largest and smallest observations (extreme order statistics) are very informative in many cases.

Prerequisites for Students

Students should be familiar with plots of univariate data and with summary statistics like the mean, median, interquartile range, and standard deviation. They need not have previous experience in estimation, but they should be familiar with the importance of random sampling.

Materials

The class will need a set of chips numbered consecutively from 1 to N, with N usually being in the 30s or 40s.

Procedure

Students should not be able to view the chips from which they are selecting the sample of size 3. The extra information obtained from seeing the entire collection of chips can alter the results.

Encourage students, working in small groups, to come up with as many different estimators as possible. When using their rule on different data sets, make sure that there are at least 20 different samples of size 3 available for study. Each rule should be computed on at least 20 different data sets, each of size 3. If you have time, you can investigate what happens for larger data sets.

Sample Results from Activity

Students will usually come up with a rule of the form 2(center) or 2(center) − 1 with either the mean or the median as the measure of center. Rules involving the order statistics are more difficult to see. If no one seems to be moving in that direction, plot a sample of three points on a real number line and encourage students to look at the pattern. Adding the average gap to the largest observed integer then begins to make sense.

For a sample of n integers from N, the best estimator of N is

$$[(n + 1)/n](\text{largest sample integer}) - 1,$$

which is nearly the same as the simpler

$$[(n + 1)/n](\text{largest sample integer}).$$

Some of the actual results for the Mark V tank are shown in the following table. The actual data were obtained from German records after the war ended.

Month	Actual Number of Tanks Produced	Serial Number Estimate	Estimate by Intelligence Agencies
June 1940	122	169	1,000
June 1941	271	244	1,550
September 1942	342	327	1,550

Sample Assessment Questions

1. In the context of the activity on estimating N,
   a. explain the difference between an estimator and an estimate.
   b. discuss at least two properties that a good estimator should possess.

2. You are called upon to design a plan for estimating the number of people at a shopping mall on a Friday evening. Discuss how you might design this plan.

How Many Tanks?

SCENARIO

During World War II, Allied intelligence reports on German production of tanks and other war materials varied widely and were somewhat contradictory. Statisticians set to work on improving the estimates. In 1943 they developed a method that used the information contained in the serial numbers stamped on captured equipment. One particularly successful venture was the estimation of the number of Mark V tanks, whose serial numbers, they discovered, were consecutive. That is, the tanks were numbered in a manner equivalent to 1, 2, 3, . . . , *N*. Capturing a tank was like randomly drawing a random integer from this sequence.

Question

How can a random sample of integers between 1 and *N*, with *N* unknown to the sampler, be used to estimate *N*?

Objectives

The goal of this lesson is to see how estimators are developed and what properties of estimators might be studied to determine a good estimator. Generally, there is no one answer to the question of how to estimate an unknown quantity, but some answers are better than others.

Activity

1. Collecting the data
   The chips in the bowl in front of you are numbered consecutively from 1 to *N*, one number per chip. Working with your group, randomly select three chips, without replacement, from the bowl. Write down the numbers on the chips, and return the chips to the bowl. Each group in the class should repeat this process of drawing three integers from the same population of integers.

2. Producing an estimate and an estimator
   Think about how you would use the data to estimate *N*. Come to a consensus within the group as to how this should be done.

   Our estimate of *N* is ______.
   Our rule or formula for the estimator of *N* based on a sample of *n* integers is:

3. Discovering the properties of the estimators
   a. The instructor will collect the data, the rules (estimators), and the estimates of *N* from each group and write them on a chart. Use each of the data sets (all of size 3) with your rule to produce an estimate of *N*. Construct a dot plot of these estimates.
   b. Calculate the mean and the standard deviation of the estimates you produced in step 2. The instructor will place the means and standard deviations for each estimator on the chart.
   c. Collect copies of the dot plots of estimates for each of the estimators used in the class.
   d. Study the dot plots, means, and standard deviations from all estimators produced by the class. Reach a consensus on what appears to be the best estimator.
   e. The instructor will now give you the correct value of *N*. Did you make a good choice in step 3a? Why or why not?

Wrap-Up

1. Explain how you could use the technique developed here to estimate the number of taxis in a city, the number of tickets sold to a concert, or the number of accounts in a bank.

Extensions

1. On examining the tires of the Mark V tanks, it was discovered that each tire was stamped with the number of the mold in which it was made. A sample of 20 mold markings from one particular manufacturer had a maximum mold number of 77.

Estimate the number of molds this manufacturer had used. Does this help you estimate the number of tires produced by this manufacturer? Explain.

2. Suppose the sequence of serial numbers does not begin at zero. That is, you have a sample of integers between L and N ($L < N$) and want to estimate how many integers are in the list. How would you modify the rule used above to take this nonzero starting point into account?

References

1. Roger W. Johnson (1994), "Estimating the size of a population," *Teaching Statistics*, **16**(2).
2. James M. Landwehr, Jim Swift, and Ann Watkins (1987), *Exploring Surveys and Information from Samples*, Palo Alto, CA: Dale Seymour Publications, pp. 75–83.
3. R. Ruggles and H. Brodie (1947), "An empirical approach to economic intelligence in World War II," *J. American Statistical Assoc.*, **42**:72–91.

Estimating a Total

NOTES FOR THE INSTRUCTOR

Statistical Setting

This lesson provides another activity that analyzes data by looking at means, but it extends that idea to produce estimates of totals. Thus, students see the mean used in a new context. The size of the sampling unit is a critical factor in deciding what multiplier to use to expand the estimate of a mean to the estimate of a total. In addition, there is a trade-off here between the number of sample plots and the size of a sample plot. This forces careful thought about the sample size question.

The activity can be used anywhere in the course after some basic ideas of using the sample mean as an estimator have been covered. It works well as an illustration of the behavior of standard errors when an estimator is multiplied by a constant. It also works well as an introduction to systematic sampling as an alternative to simple random sampling.

Prerequisites for Students

Students should have some familiarity with the use of a sample mean to estimate a population mean, the notion of random sampling, and the concept of standard error of the mean.

Materials

The only materials needed are the rectangular arrays of dots provided in Figures 1 and 2, a quarter, and a dime.

Procedure

Make sure the students understand the nature of random sampling in this context. If a quarter is randomly dropped onto the rectangle and straddles the edge, instruct the students to simply slide it onto the figure so that the quarter is entirely within the rectangle. More complicated methods for randomly locating plots can be used, but that is not the intent of this lesson. Similarly, systematic sampling can be understood as sampling with a regular pattern so as to get sample plots from across the rectangle. A technical definition of systematic sampling need not be given.

Students can effectively work in groups of two for this activity.

With the small samples being used here, there will be much variation among the estimates of the total, within all methods. Thus, it is important to collect the estimates from the class, plot them, and discuss the center and variability of the resulting distribution. The goal is to compare the quarter-sized plots with the dime-sized plots, within the figures, to see if systematic sampling has any serious advantages or disadvantages.

Sample Results from Activity

The data on the following table show the results from one sample conducted under each of the eight plans. The sample plot size is Q (quarter) or D (dime), the figure number is 1 or 2, and the sampling scheme is R (random) or S (systematic). The multiplier for the quarter-sized samples is (180/4.9) = 36.73, and the multiplier for the dime-sized samples is (180/2.5) = 72. Figure 1 contains a random array of 300 dots; all of the methods should be approximately equivalent as estimators of the total. Notice that the systematic arrangement of the quarters does a little better than the others here, but that is the result of one lucky sample.

	N	Mean	Median	STDev	SEMean	Total	SETotal
Q1R	6	9.83	11.00	3.06	1.25	361.06	45.91
Q1S	6	8.000	7.500	1.26	0.516	293.84	18.95
D1R	12	4.167	4.000	1.95	0.562	300.24	40.46
D1S	12	4.917	5.000	1.98	0.570	354.24	41.04
Q2R	6	3.17	2.50	3.82	1.56	116.43	57.30
Q2S	6	12.83	11.50	12.72	5.19	471.24	190.63
D2R	12	3.58	2.50	3.73	1.08	257.76	77.76
D2S	12	6.58	5.50	5.87	1.69	473.76	121.68

Figure 2 also contains 300 dots, but there is some clustering of the dots toward the bottom of the rectangle. Small random samples do not do well here because many of the sample plots will miss the high-density areas. The random arrangement of quarters looks like it is doing well in terms of standard error, but it is actually underestimating the total by a large amount. (The true total is not even within two standard errors of the estimate.) Systematic sampling will tend to produce estimates of standard error that are very large, but they may be more accurate indications of the variability inherent in this estimation problem.

Sample Assessment Questions

1. You are to estimate the number of cars per hour that flow into a mall parking lot over a weekend. You have 4 hours of time in which to observe the lot and collect the data. Should you observe the lot for many small time periods over the weekend or for a few large ones? Explain.

2. You are to estimate the total number of employees in local businesses by taking a sample of businesses from the yellow pages of the local telephone directory.
   a. Explain how you would select the businesses that comprise the sample.
   b. Assuming you can contact the sampled businesses and determine how many employees they have, explain how you would construct an estimate of the total number of employees.
   c. Do you think estimates conducted in this way might tend to be too low? Explain.
   d. Explain how you would attach a measure of error to the estimate found in part b.

3. You are to estimate the total number of trees growing on public property in a moderately large city. You are provided with a map of the city which highlights the public property. You must decide whether to sample a few large parcels of property or a much larger number of small parcels of property, given that the total acreage covered would be the same in both cases. What factors would influence your decision, and why?

Estimating a Total

SCENARIO

Pine forests are periodically infected with beetles, and the infestations often seem to jump from tree to tree in a random manner. Diseases that spread through the soil, however, tend to infect large clumps of trees in a few locations. How can a forester choose a good sampling method to use to count diseased trees? Should the sampling method change for the two types of disease described above?

Question

How can we construct a good estimate of the number of diseased trees in a forest by counting diseased trees in sampled plots? How should we select the size and number of the sampled plots?

Objectives

In this activity you will learn one way to estimate a total and how systematic sampling may be used to good advantage in certain situations.

Activity

Figure 1 shows a number of dots scattered within a rectangle that measures approximately 18 centimeters by 10 centimeters. The goal is to estimate the number of dots in the rectangle. (The dots may represent diseased trees in a forest.)

1. Collecting the sample data randomly
   a. To select a sample plot, randomly drop a quarter inside the rectangle. To obtain the data on a sampled plot, draw a circle around the quarter and count the dots inside or touching the circle. Repeat this procedure until a sample of six plots has been selected and the data collected. (The sampled plots may overlap. Do not use information in the pattern when selecting sample plots; if this were a real forest, you would not be able to see the pattern of diseased trees.)
   b. Calculate the mean and the standard deviation for these six counts.
   c. A quarter is approximately 4.9 square centimeters in area. Use this information, along with the data from the random sample, to estimate the total number of dots inside the rectangle of Figure 1.

2. Collecting the sample data systematically
   a. Another way to select the sample plots is to arrange the circles systematically across the rectangle of Figure 1. For example, you could have two rows of three evenly spaced circles each. (A chess board is a systematic arrangement of black and red squares. Student desks in a classroom are usually arranged systematically rather than randomly.) Sampling with a quarter, select six sample plots systematically and count the number of dots in each.
   b. Calculate the mean and the standard deviation for these six counts.
   c. Use the data from the systematic sample to estimate the total number of dots inside the rectangle of Figure 1. Compare this estimate with the one from random sampling.

3. Changing the size of the sample plot
   The size of each sample plot will now be changed to a dime instead of a quarter. A dime is approximately 2.5 square centimeters in area, about half the size of a quarter. Thus, to keep the sampling effort equivalent to that used above, 12 sample plots should be used when sampling with a dime.
   a. Follow the instructions of part 1 with 12 dime-sized sample plots.
   b. Follow the instructions of part 2 with 12 dime-sized sample plots.
   c. Do you see any advantages of using 12 dime-sized plots rather than 6 quarter-sized plots for estimating the total number of dots?

4. Combining the data for the class
   The instructor will now provide you with the estimates of the totals from each member (or team) in the class. Keep in mind that we are using four different estimators: quarter-random, quarter-systematic, dime-random, and dime-systematic.
   a. Construct four plots (dot plots or stem plots work well), one for each of the four sets of estimates.
   b. Compare the four plots. How do the four methods compare? Does one method look like it is clearly better than the others?

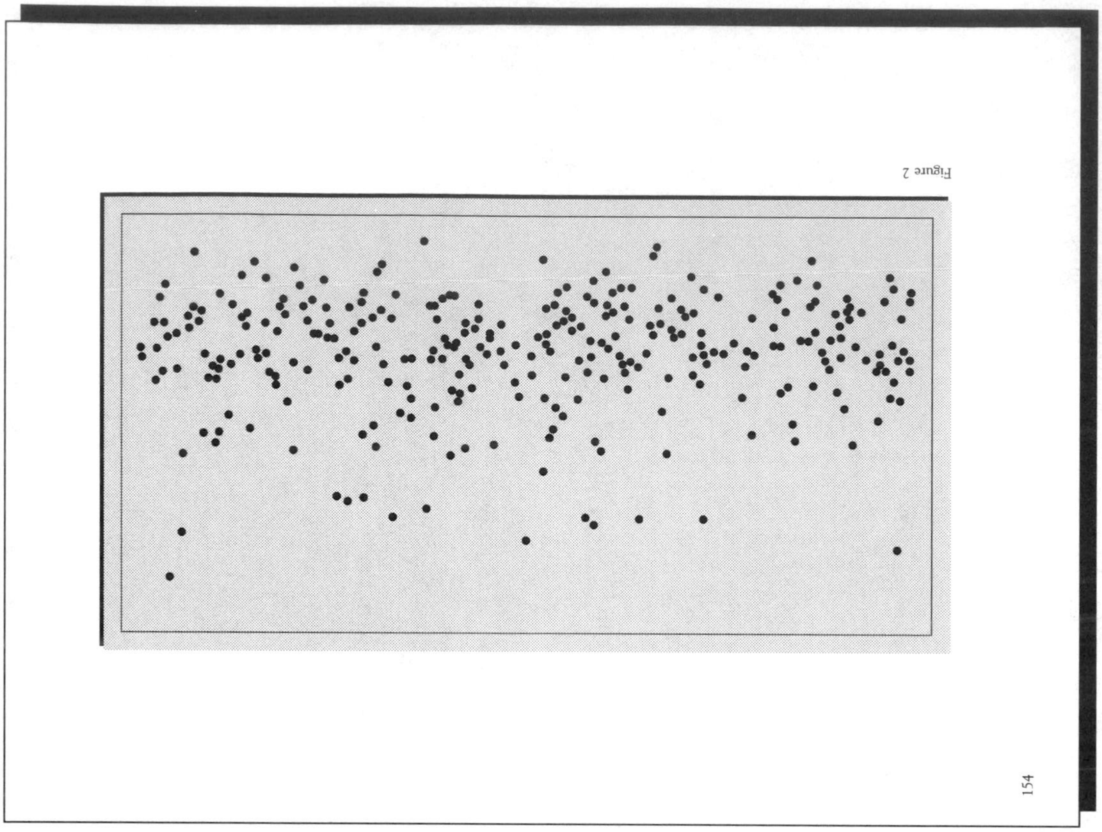
Figure 2
154

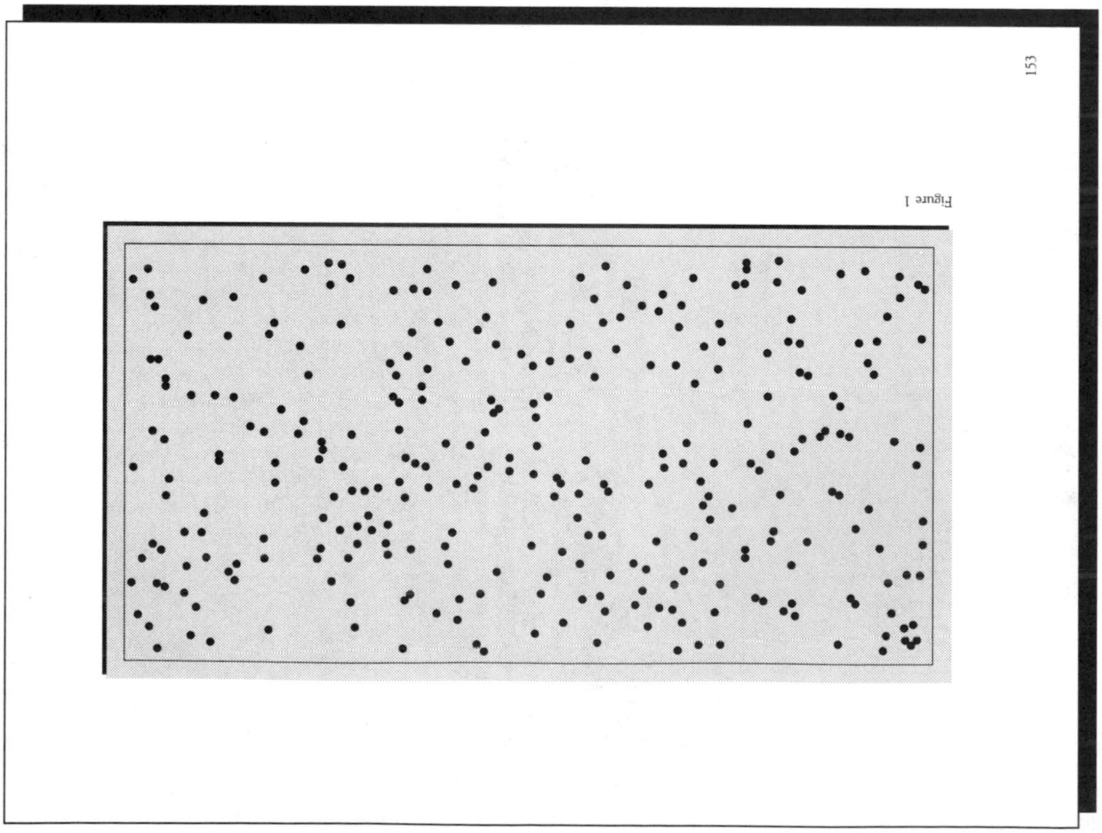
153
Figure 1

c. Where do the plots appear to center? How many dots do you think are in Figure 1?

5. Figure 2 contains another array of dots. Repeat steps 1 through 4 above to compare estimates of the total number of dots in Figure 2. Is there a preferred method, among the four, for estimating the total in this case? (The rectangle of Figure 2 is approximately 18 × 10 centimeters.)

Wrap-Up

1. You are to estimate the total number of TV sets in your city or county.
   a. Explain how you would design the sampling plan to collect the data.
   b. Explain how you would estimate the total number of TV sets from the data you would collect.

Extensions

1. You may recall from previous studies that the standard deviation of a sample mean, sometimes called the standard error of the mean, is estimated by dividing the standard deviation of the sample data by the square root of the sample size. This estimate of error can be used to form an estimate of error appropriate for the estimation of a total.
   a. The estimates of the total number of dots were of the form $K \times$ (sample mean), where K is some numerical constant. Produce a formula for the standard error of an estimate of this form.
   b. Using the appropriate K, calculate an approximate standard error for the estimates of the totals constructed earlier in this activity. Use only your own sample data in this calculation. (Recall that there were eight estimates.)
   c. Do these standard errors help in deciding how the methods compare? If so, do there appear to be major differences among the four methods for Figure 1? How about for Figure 2?

HYPOTHESIS TESTING EXPERIMENTS

Introduction to Hypothesis Testing

NOTES FOR THE INSTRUCTOR

Note: This activity is not in Student Guide.

Question

Are pennies really fair coins?

Statistical Setting

This activity is used to introduce the logic of hypothesis testing. The goal of the activity is to show students how it is that people reject a hypothesis upon observing data that are unlikely to arise if the hypothesis is true. The activity should be used in the first class period in which hypothesis testing is discussed.

Prerequisites for Students

There are no true prerequisites for this activity; it *could* be used on the first day of class with students who have never studied statistics before (although most instructors will want to postpone using the activity until later in the course, when discussing inference). However, it will help if the students have enough familiarity with probability that they can find, for example, the probability of getting an odd number four times in a row when rolling a fair die.

Materials

This activity requires one or more pennies.

Procedure

Several variations of this activity are possible (see the Options ahead); some involve deception. We describe the version of the activity that involves standing pennies on edge. (Warning: If you plan to use the "Pennies on Edge" activity elsewhere in the course, then you need to choose a version of this activity that does not involve standing coins on edge.)

Take a penny and stand it on edge on a table (or another hard, flat surface). (This may take some time and effort. Try placing the penny so that Lincoln is upside down; be patient.) Once the penny is standing on edge, announce that you will play a game with the class. (It helps if you have played games of some type earlier in the course, lest they immediately become suspicious.) Tell them that you want to get to know some of them better by having lunch with them and that you are going to let the penny determine who pays for lunch. You will tap the table to make the penny fall. If it lands showing heads, then you win and the student pays for lunch; if it lands showing tails, then the student wins and you pay for lunch. Try to convince the class that this makes the game fair, since heads and tails should be equally likely.

Tap the table to make the penny fall. You are almost certain to get heads with this game. Pretend to be pleasantly surprised at having won. Now play the game a second time. After winning a second time, act quite surprised. If someone doubts your fairness is setting up the game, act hurt at the suggestion that you would cheat.

You can let a student volunteer tap the table for each of the next several rounds. It does not matter where the table is tapped (i.e., to the right of the penny, to the left, in front of the penny, etc.); heads will show almost every time. Things go wrong only if someone pounds on the table with such force that the penny bounces before falling. This makes heads the probability of getting heads close to 50% and is thus to be avoided.

Continue playing (one penny at a time) until the class stops believing that the game is fair. Eventually the class will become incredulous upon seeing your winning streak. Ask them, "Isn't it possible that I have just been very lucky?" Someone will say, "Yes, but it isn't likely if the game is fair." At this point tell them that they have just demonstrated the logic of hypothesis testing—they have rejected the hypothesis that heads and tails are equally likely not because the data are *impossible* under this hypothesis, but because the data are not *likely* under the hypothesis. (You should probably agree to buy lunch for the students who lost the game.)

Options

USING MARBLES

Before coming to class, put four blue marbles in a box. Bring the box with you but don't let the students see the contents of the box. Begin the activity by announcing that you are going to play a game with the class (as above). Tell them that you are going to determine who pays for lunch by drawing a marble out of a box. If a blue marble is drawn, then the student pays for lunch; if a green marble is drawn, then you will pay for lunch.

Announce that you have put three green marbles and one blue marble in your box, so that the student has a 3/4 chance of winning. (Remember—the box actually has four blue marbles in it, so the student has no chance of winning.) Ask for three volunteers from the class to play the game. Have a different student blindly reach into the box and draw a marble. Act pleasantly surprised when the marble is blue, meaning that the first student volunteer has to pay.

Now replace the first marble and play the game again, drawing to see whether the second volunteer will have to pay. After winning a second time, act quite surprised—and don't let anyone look inside the box during the game. Keep playing the game until the class is convinced that something is wrong. At this point you can show them the contents of the box. You might point out that they became suspicious after only two draws, which corresponds to a *p*-value of 1/16.

You might want to enlist a student to help with the deception. Meet with the helper in advance and explain how the game is rigged. The helper will then be the one to blindly draw marbles from the box during the game, after examining the box and declaring to the class that it has three green marbles and one blue marble.

USING DICE

Get a die that has only odd numbers on it. Some stores sell such dice. You can make your own by buying a blank die and using a permanent pen to put 1 on two (opposite) sides, 3 on two sides, and 5 on two sides. When looking at the die from any angle, one can only see three sides. These will have a 1, a 3, and a 5, so that the die will appear to be a regular, fair die. In fact, you can casually hold the die in front of the class without generating suspicion.

Play the game by rolling the die. You win on odd numbers; students win on even numbers. You can roll the die on the floor in front of the front row of students and have a student announce the result to the class—just don't let students actually handle the die to examine whether or not it is fair. Keep playing until people are getting suspicious. (Winning four times corresponds to a *p*-value of 1/16.)

TOSSING COINS

If possible, get a two-headed coin and proceed as with the die above. If you don't have a two-headed coin, you can use an ordinary coin, flip it, and announce "heads" each time. One approach is to toss the coin and catch it with your right hand. If it shows heads, then stop. If it shows tails, then quickly flip it onto the back of your left hand. This way you can show everyone the result of the first toss. After that you will have to call out the results without showing the coin to the students. Of course, if you are lucky and the coin lands showing heads, then you can show the result to the class. You have a pretty good chance of getting heads several times without needing to lie.

SPINNING A 1962 PENNY

Most people expect that if you *spin* a penny, rather than toss it, it will fall showing heads half of the time. In fact, for pennies minted around 1962, the probability of getting heads is only around 10%. Thus, you could let the students win on heads, with you winning on tails. With a spinning penny you have quite a good chance of winning several times in a row without resorting to lying. (This will not work if you have used the "Spinning Pennies" activity earlier in the course.)

Sample Assessment Questions

1. Write a brief summary of what you learned in this activity about hypothesis testing.

2. Does rejecting a hypothesis mean that the hypothesis has been proven to be false? If not, then what can we conclude when we are told that a hypothesis has been rejected?

Dueling Dice

NOTES FOR THE INSTRUCTOR

Note: This activity is not part of Student Guide.

Question

Are all dice created equal?

Statistical Setting

This activity can be used to show how statistics can help us discover things we otherwise would not learn. The activity can also be used show students how a chi-square test will reject a false hypothesis (if enough data are collected).

Prerequisites for Students

Students should have some familiarity with probability and with sampling distributions, although this is not strictly necessary. They should know how to find the standard error of a sample percentage. If you wish to conduct a hypothesis test, then students must be familiar with a z-test for a proportion or with the chi-square test.

Materials

You will need one fair die and one "altered" die; these should be distinguishable by casual observations; for instance, they could be of different colors. To make the altered die, take an ordinary die and change the 2 to a 5 by adding 3 dots. An alternative is to start with a blank die and put the numbers 1, 3, 4, 5, 5, and 6 on the six sides. (Note: On a regular die the 2 and 5 are on opposite sides. Thus, even after the 2 is changed to a 5, one never sees both 5's at the same time—this is important.)

Procedure

This activity involves deception. For the rest of this description we will assume that the fair die is green and the altered die is red. Tell the class that you want to study an example of a sampling distribution. The particular example involves a comparison of your two dice. The variable of interest is X, the number of times out of 10 trials that the green die wins a roll-off with the red die. Do *not* tell the class that the red die has been altered.

Roll the dice and note which die has a larger number; that is, does the green die "win" or does the red die win? If the dice tie, then discard the trial and roll the dice again. Repeat this until you have competed 10 trials, keeping track of how often the green die wins.

Now tell the class that you want each of them to conduct 10 roll-offs (not counting any that end in ties) and record how often the green die wins. Pass the dice around the room, giving each student a chance to do 10 roll-offs. This will take about 40 seconds per student, on average, so 30 students can do this in 20 minutes. (During that time you can discuss other course material.) Have them keep track of how many of the 10 roll-offs are won by the green die. Remind them that if there is a tie, then they are to repeat the trial; they need a total of 10 winners.

You will want a total of 200 or more trials (i.e., 20 or more students) in order to have power of at least 80% in the hypothesis test at the end of the activity. If you have a large class, then consider having only the first 30 students conduct 10 roll-offs each.

The students will not realize that the fair die has only a 40% chance of winning a roll-off. They might examine the dice carefully, but it is unlikely that they will notice that the red die has been altered. (If someone does notice this, you can hope they will not announce it to the class. If they do, then the activity loses much of its punch, but you can continue anyway.)

After the students have completed the roll-offs, collect the data and construct the sampling distribution for X. That is, ask the class how many got $x = 0$, how many got $x = 1$, etc. Tally the results and graph the distribution. Tell them that this representation of the sampling distribution of X shows how a statistic varies from one experiment to the next.

At about this time someone should notice that the distribution is centered below .5, whereas one would expect it to be symmetric about $x = 5$. If the class does not notice, then tell them. Next, calculate the total number of trials won by the green die, and find the percentage of green wins. Point out that this is quite a bit less than the "expected" value of 50%.

If you have previously discussed hypothesis tests, then you can do a z-test or a chi-square test of the null hypothesis that $p = .5$. Since p is actually .4, a two-sided test at the .05 level has power .81 if $n = 200$ and power .94 if $n = 300$. Thus, you are almost certain to reject the null hypothesis, provided you have enough sample data.

If the class has not studied hypothesis testing, you can still calculate the standard error of the percentage of green wins and point out that the observed percentage (which should be around .4) is several standard errors from .5.

Thus, the statistical analysis will have shown that $p \neq .5$. At this point, give the green die to one of the students and ask the student to examine it. Likewise, have someone examine the red die. If all goes well—and it usually does—this will be the first time that anyone will have noticed that the red die has two 5's on it. You can then announce to the class that "the statistics noticed something none of you have noticed."

Sample Assessment Question

Write a brief-summary of what you learned in this activity about the power of statistical analysis.

Statistical Evidence of Discrimination: The Randomization Test

NOTES FOR THE INSTRUCTOR

Statistical Setting

This activity uses the randomization test to introduce students informally to the idea of testing the hypothesis that two population proportions are equal. The initial setting is statistical evidence of discrimination in hiring and employment tests. Other settings are explored in the Extension exercises.

This activity can be used at any stage of the introductory statistics class, but it fits naturally right before students begin hypothesis testing. It has been used successfully at that stage as an introduction to hypothesis testing.

Prerequisites for Students

Students should have had some previous experience with simulation. Otherwise they may get bogged down in that idea.

Materials

Each group of students will need one deck of playing cards. Students can usually supply their own decks. Each group will also need a set of 90 blank 3″ by 5″ cards.

Procedure

Students should work on this activity in small groups. The simulations will go more quickly if the groups can share results.

Each of the simulations should be repeated as many times as possible without students getting bored. This generally means no more than 5 trials per student or 10 trials per group.

If you have a large class, you can assemble more than 100 trials. If you assign the simulations for homework, cut the number of trials from 100 to 5 per student. During the next class session, combine the results and send the students home to write their conclusions.

You may want to suggest that students use 2 × 2 tables to organize the given data and their data from the simulation.

Sample Results from Activity

1. a. Answers will vary.
   b. Answers will vary.
   c. 35/48; 35/48; no discrimination
   d. Answers will vary. The theoretical results, using the hypergeometric distribution, appear in the following table:

Number of Males Promoted	Probability
11	0.000
12	0.000
13	0.004
14	0.021
15	0.072
16	0.162
17	0.241
18	0.241
19	0.162
20	0.072
21	0.021
22	0.004
23	0.000
24	0.000

   e. Each simulation will vary. Theoretically, the probability that 21 or more males will be promoted if there was no discrimination on the basis of gender is 0.021 + 0.004 = 0.025.
   f. This situation often leads students to a discussion of the idea of a one-tailed test versus a two-tailed test. A student will frequently remark that promoting 21 or more women is equally unusual under the hypothesis that there is no discrimination. The students then tend to decide that a two-tailed test is more "fair." The *p*-value for a two-tailed test is 0.05. The students now have a close call to make. They typically conclude that, in spite of their initial impression and their belief that the bank supervisors discriminated, the hard evidence here for discrimination is not compelling. There is reasonable doubt.

Students who have studied the z-test for the difference of two proportions may notice that this situation falls into that category but does not meet the assumptions for that test. Here one of the expected counts is $24(1 - 0.875) = 3 < 5$. The randomization test requires no similar assumption.

Students may be interested to know that there was a second half of this experiment. This time the branch manager's job was described as "complex." In this case 25 (completely different) male bank supervisors got the file of the female candidate and 5 of them recommended promotion, and 20 (completely different) male bank supervisors got the file of the male candidate and 11 of them recommended promotion. This difference cannot reasonably be attributed to chance.

Here is a different way to simulate the problem in the Scenario. Remove 4 black cards from a deck of cards, leaving 48 cards. The 13 hearts will represent the 13 files not recommended for promotion. The remaining 35 clubs, diamonds, and spades will represent the files recommended for promotion. Shuffle the cards thoroughly (at least seven times) and deal them one at a time into two piles, the "male" pile and the "female" pile. After you have 2 piles of 24 cards each, count the number of hearts in the "male" pile. Repeat this simulation two more times.

2. a. Here is one way the simulation could be designed. On 26 cards, write "Chicano." On 64 cards, write "white." Shuffle the 90 cards thoroughly and count out 17 cards to represent the people who pass the exam. Count the number of "Chicano" cards. Repeat this procedure many times, keeping track of the number of Chicanos who pass. Finally, estimate the probability that three or fewer of the cards that were counted out will say "Chicano."

b. Results from the simulations will vary. From the hypergeometric distribution, the one-sided p-value for this experiment is 0.2038.

c. Paragraphs will vary, but students should realize that it is very likely to get the difference that occurred in *Stover* by chance. There is no convincing evidence of disparate impact here.

3. a. Here is one way the simulation could be designed. On 26 cards, write "white." On 9 cards, write "African-American." Shuffle the 35 cards thoroughly, and count out 30 cards to represent the nurses who pass the exam. Count the number of "white" cards. Repeat this procedure many times, keeping track of the numbers of whites who pass. Finally, estimate the probability that 26 of the cards that were counted out will say "white." Students may realize that it is easier to count out five cards to represent the nurses who fail the examination.

b. Results from the simulations will vary. From the hypergeometric distribution, the one-sided p-value for this experiment is 0.0004.

c. Conclusions will vary, but students should realize that it is *very* unlikely to get the difference that occurred in *Dendy* by chance. The results cannot reasonably be attributed to chance variation. There is compelling evidence of disparate impact in this case.

Sample Results from Extensions

1. These results are difficult to get by chance if carbolic acid makes no difference. The *p*-value is less than 0.01.

2. These results are difficult to get by chance if there was no difference in the proportion of people who have been promised painful shocks and the proportion of people who have been promised painless shocks who choose to wait together. The *p*-value from Fisher's exact test is 0.01.

Sample Assessment Questions

1. Suppose that on the 1994 test for the fire department, 5 out of 10 women pass and 9 out of 10 men pass. Assume for a moment that the chance a man passes this exam is the same as the chance a woman passes this exam. Describe a way to simulate with cards the probability of getting a difference between the proportion of men who pass and the proportion of women who pass as large or larger than the situation in 1994.

2. Suppose that on a hiring examination, 35 out of 40 women pass and 20 out of 40 men pass. Under the assumption that men and women are equally likely to pass, a simulation was performed 500 times. The following results show the number of men who passed. Write a paragraph explaining what conclusion should be drawn from this simulation.

Number of Males	Frequency	Number of Males	Frequency
19	3	28	76
20	4	29	62
21	6	30	59
22	6	31	36
23	18	32	15
24	29	33	11
25	47	34	7
26	49	35	2
27	69	36	1

References

Some very readable references on the randomization test (sometimes called the permutation test) include these three:

1. Peter Barbella, Lorraine J. Denby, and James M. Landwehr (1990), "Beyond exploratory data analysis: The randomization test," *The Mathematics Teacher*, **83** (February):144–149.
2. Frederick Mosteller and Robert E. K. Rourke (1973), *Sturdy Statistics: Nonparametric and Order Statistics*, Reading, MA: Addison-Wesley, pp. 12–23.
3. Sandy L. Zabell (1989), "Statistical proof of employment discrimination," in Judith M. Tanur et al. (eds.), *Statistics: A Guide to the Unknown*, third ed., Pacific Grove, CA: Wadsworth, pp. 79–86.

The following references are also of use:

1. Eugene S. Edgington, "Randomization Tests," UMAP Module 487, COMAP, Lexington, MA 02173 (Phone 617/862-7878).
2. Eugene S. Edgington (1987), *Randomization Tests*, second ed., New York: Marcel Dekker.
3. Richard J. Larsen and Donna Fox Stroup (1976), *Statistics in the Real World: A Book of Examples*, New York: Macmillan, pp. 205–207.
4. Paul Meier, Jerome Sacks, and Sandy L. Zabell (1984), "What happened in Hazelwood: Statistics, employment discrimination, and the 80% rule," *American Bar Foundation Research Journal* (Winter):139–186.
5. John A. Rice (1988), *Mathematical Statistics and Data Analysis*, Pacific Grove, CA: Wadsworth, pp. 434–436 and 452.
6. Sandy L. Zabell, "Statistical proof of employment discrimination," in Judith M. Tanur et al., eds. (1989), *Statistics: A Guide to the Unknown*, third ed., Pacific Grove, CA: Wadsworth, pp. 79–86.

Statistical Evidence of Discrimination: The Randomization Test

SCENARIO

In 1972, 48 male bank supervisors were each given the same personnel file and asked to judge whether the person should be promoted to a branch manager job that was described as "routine" or the person's file held and other applicants interviewed. The files were identical except that half of them showed that the file was that of a female and half showed that the file was that of a male. Of the 24 "male" files, 21 were recommended for promotion. Of the 24 "female" files, 14 were recommended for promotion. (B. Rosen and T. Jerdee (1974), "Influence of sex role stereotypes on personnel decisions," *J. Applied Psychology*, 59:9–14.)

Questions

Is this convincing evidence that the bank supervisors discriminated against female applicants? Or could the difference in the numbers recommended for promotion reasonably be attributed to chance? That is, there was no discrimination on the basis of gender: It just happened that 21 out of the 35 bank managers who recommended promotion got files marked "male."

Objectives

In this activity you will learn how to use the randomization test to tell if the difference between two proportions is statistically significant. The randomization test will be a simulation of the situation under the hypothesis that there was no discrimina-

tion on the basis of gender. "Statistically significant" means that a difference as big as or bigger than that which occurred is unlikely to have happened without a cause other than random variation. When you have finished this activity, you should be able to use the randomization test to decide whether you can reasonably attribute the difference between two proportions to chance or whether you should search for some other explanation.

Activity

1. Remove all four aces from a deck of playing cards. There will now be 24 red cards in the deck, which will represent "male" files, and 24 black cards in the deck, which will represent "female" files. Shuffle the cards at least seven times, and then cut them.
   a. Count out the first 35 cards to represent the files recommended for promotion. How many males were in this pile?
   b. On a number line like the following, place an X above the number of males you got in step a.

0 5 10 15 20 25 30 35

   c. In this simulation what is the probability that a male will be promoted? a female? Does this simulation model the case of discrimination on the basis of gender or the case of no discrimination?
   d. Repeat the simulation with the cards until you have a total of 100 X's on the number line. You may want to combine results with other groups. (If you have 10 groups, each group would do only 10 simulations.)
   e. Use the results on your number line to estimate the probability that 21 or more of the 35 recommended for promotion will be male if there was no discrimination on the basis of gender.
   f. Do you believe your simulation provides evidence that the bank supervisors discriminated against females?

Few situations in real life are as clear cut as that of the bank supervisors. In real life the files are never identical. Education, experience, character, recommendations, and test scores vary. Statistics can tell us only whether the difference between two groups can reasonably be attributed to chance. If it cannot reasonably be attributed to chance, then further digging is required to determine whether the explanation is discrimination or difference in qualifications.

In *Griggs* v. *Duke Power Company* (1971), the Supreme Court established the idea of "disparate impact." Disparate impact occurs, for example, when the pass rate of one group on an employment test is substantially lower than that of another. Such an employment test is illegal unless the employer can prove that the use of the test is a business necessity. For steps 2 and 3,

a. Design a simulation with cards to help you determine if the difference in pass rates can reasonably be attributed to chance or whether there is clearly disparate impact on this examination.
b. Repeat your simulation 100 times, sharing the work with other groups. Place your results above a number line.
c. Write a paragraph explaining your conclusions.

2. In the 1975 court case *Chicano Police Officers Association* v. *Stover*, 3 of 26 Chicanos passed an examination and 14 of 64 whites pass the same examination.

3. In the 1977 *Dendy* v. *Washington Hospital Center* case, 26 out of 26 white nurses passed an examination and 4 out of 9 African-American nurses passed the same examination.

Wrap-Up

1. Discuss why the procedure you have been using may have been named the "randomization" test.

2. For a classmate who is absent today, write an explanation of how to use the randomization test to determine if there is discrimination in hiring.

Extensions

1. Joseph Lister (1827–1912), surgeon at the Glasgow Royal Infirmary, was one of the first to believe in Pasteur's germ theory of infection. He experimented with carbolic acid to disinfect operating rooms during amputations. Of 40 patients where carbolic acid was used, 34 lived. Of 35 patients where carbolic acid was not used, 19 lived.

a. Design a simulation with cards to help you determine if the difference can reasonably be attributed to chance or whether there is evidence that carbolic acid saves lives.
b. Repeat your simulation 100 times, sharing the work with other groups. Place your results above a number line.
c. Write a paragraph explaining your conclusions.

2. In a psychology experiment (A. Schachter (1959), *The Psychology of Affiliation*, Stanford, CA: Stanford University Press), a group of 17 people were told that they would be subjected to some painful electric shocks and a group of 13 people were

told they would be subjected to some painless electric shocks. The subjects were given the choice of waiting with others or waiting alone. Here are the results:

	Wait Together	Wait Alone
Painful Shocks	12	5
Painless Shocks	4	9

a. Design a simulation with cards to help you determine if the difference in the proportions that chose to wait together can reasonably be attributed to chance.
b. Repeat your simulation 100 times, sharing the work with other groups. Place your results above a number line.
c. Write a paragraph explaining your conclusions.

References

Some very readable references on the randomization test (sometimes called the permutation test) include these three:

1. Peter Barbella, Lorraine J. Denby, and James M. Landwehr (1990), "Beyond exploratory data analysis: The randomization test," *The Mathematics Teacher*, **83** (February):144–149.
2. Frederick Mosteller and Robert E. K. Rourke (1973), *Sturdy Statistics: Nonparametric and Order Statistics*, Reading, MA: Addison-Wesley, pp. 12–23.
3. Sandy L. Zabell (1989), "Statistical proof of employment discrimination," in Judith M. Tanur et al. (eds.), *Statistics: A Guide to the Unknown*, third ed., Pacific Grove, CA: Wadsworth, pp. 79–86.

The following references are also of use:

1. Eugene S. Edgington, "Randomization Tests," UMAP Module 487, COMAP, Lexington, MA 02173 (Phone 617/862-7878).
2. Eugene S. Edgington (1987), *Randomization Tests*, second ed., New York: Marcel Dekker.
3. Richard J. Larsen and Donna Fox Stroup (1976), *Statistics in the Real World: A Book of Examples*, New York: Macmillan, pp. 205–207.
4. Paul Meier, Jerome Sacks, and Sandy L. Zabell (1984), "What happened in Hazelwood: Statistics, employment discrimination, and the 80% rule," *American Bar Foundation Research Journal* (Winter):139–186.
5. John A. Rice (1988), *Mathematical Statistics and Data Analysis*, Pacific Grove, CA: Wadsworth, pp. 434–436 and 452.
6. Sandy L. Zabell, "Statistical proof of employment discrimination," in Judith M. Tanur et al., eds. (1989), *Statistics: A Guide to the Unknown*, third ed., Pacific Grove, CA: Wadsworth, pp. 79–86.

Is Your Class Differently Aged? The Chi-Square Test

NOTES FOR THE INSTRUCTOR

Statistical Setting

This activity can be used as an introduction to the chi-square distribution and the chi-square test of goodness of fit. Students construct the sampling distribution for the chi-square statistic for a table with three rows and see where a value of chi-square computed from the ages of the students in the class lies in this distribution.

Prerequisites for Students

Students should know how to display a frequency distribution on a dot plot.

Although this is reviewed on the students' pages, it is helpful if students know how to use a random digit table to simulate drawing samples of size 40 from a population divided into three categories of 16%, 23%, and 61%.

Materials

This activity requires the ages of 40 statistics students, a table of random digits for each student, Minitab statistical software (optional, for the two Extension questions), and a bag of 120 plain M&M's (for Assessment question 2).

Procedure

This activity can be done individually, in small groups, or by the entire class. However, the entire class will need to share the results of Activity questions 2 and 4.

Sample Results from Activity

1. Answers will vary.

2. The observed column will vary. The rest of the table should be as follows:

Age	Proportion in U.S. Colleges	Observed Number in Your Class (O)	Expected Number in Your Class (E)
Under 25	0.61		24.4
25–34	0.23		9.2
35 and older	0.16		6.4
Total	1.00	40	40

3. Chi-square values will vary.

4. a. 61–83
   b. 84–99
   c. Tables will vary.
   d. Answers will vary.
   e. Relatively far.
   f. Answers will vary.
   g. If your students have only 100 simulations, the sampling distributions can vary a bit from the theoretical one. For example, here are two simulated sampling distributions from Minitab for 100 trials of the chi-square statistic for a table with three rows (df = 2):

```
MTB > random 100 c1;
SUBC > chisquare 2.
MTB > dotplot c1
```

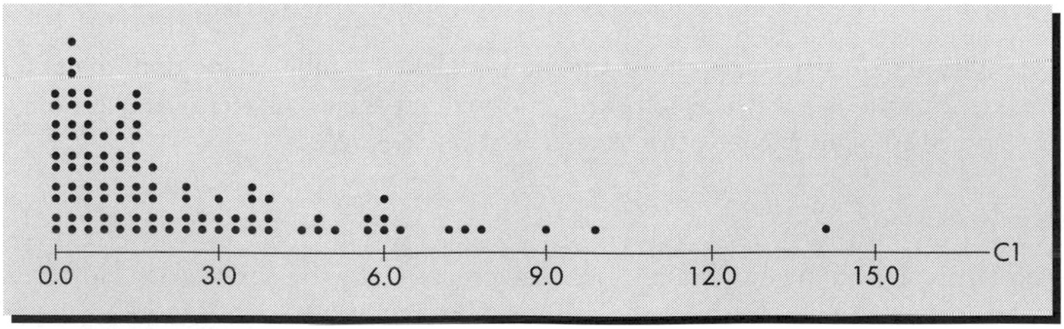


```
MTB > random 100 c2;
SUBC > chisquare 2.
MTB > dotplot c2
```

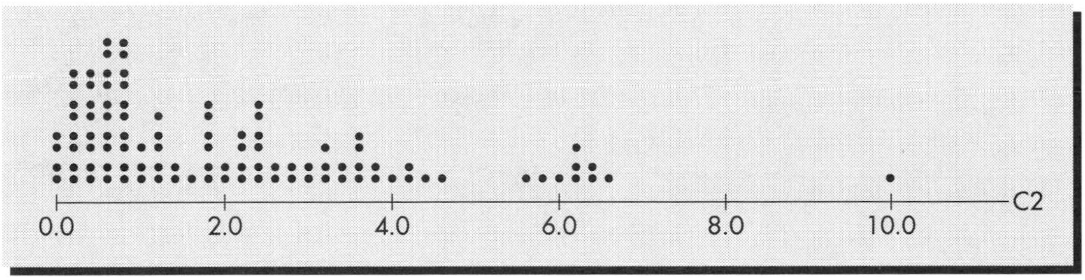

Here is a simulated sampling distribution of the chi-square distribution for 2500 trials with df = 2.

```
MTB > random 2500 c1;
SUBC > chisquare 2.
MTB > dotplot c1
```

Each dot represents 24 points

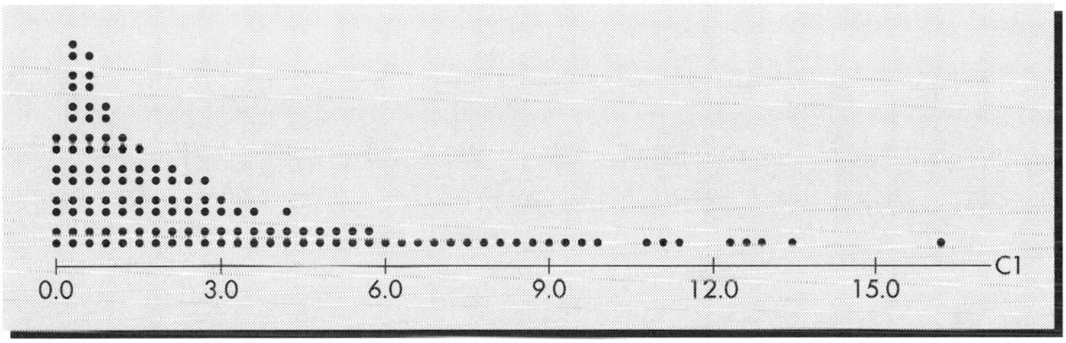

5. Answers will vary.

Sample Results from Wrap-Up

1. a.

Answer	Proportion if Equally Likely	Observed	Expected
a	0.2	12	17
b	0.2	22	17
c	0.2	19	17
d	0.2	17	17
e	0.2	15	17
Totals	1.0	85	85

The value of chi-square is 3.41.

b. We must be able to assume that the answers given are a random sample from the distribution of all answers to the verbal section of the SATs. Although "b" and "c" do appear more frequently on this test, we cannot reject the hypothesis that there are equal numbers of the five answers in the population of all answers. If we are still suspicious, we should get a bigger sample of answers. (See Table 1 here for a sample SAT answer form.)

Table 1 CORRECT ANSWERS FOR SCHOLASTIC APTITUDE TEST FORM CODE 6G

VERBAL		MATHEMATICAL	
Section 2	Section 6	Section 5	Section 1
1. D	1. E	1. D	1. A
2. D	2. A	2. D	2. B
3. B	3. A	3. C	3. D
4. B	4. A	4. D	4. D
5. C	5. B	5. E	5. B
6. E	6. E	6. C	6. D
7. B	7. D	7. A	7. C
8. A	8. B	8. B	*8. B
9. A	9. D	9. C	*9. C
10. B	10. B	10. E	*10. D
11. C	11. A	11. D	*11. B
12. B	12. A	12. A	*12. A
13. E	13. C	13. E	*13. C
14. E	14. C	14. A	*14. B
15. C	15. A	15. A	*15. D
16. D	16. C	16. C	*16. B
17. A	17. E	17. E	*17. D
18. C	18. D	18. C	*18. C
19. C	19. C	19. D	*19. B
20. B	20. B	20. C	*20. B
21. B	21. B	21. B	*21. A
22. B	22. D	22. C	*22. A
23. C	23. E	23. E	*23. C

(Continued)

Table 1 CORRECT ANSWERS FOR SCHOLASTIC APTITUDE TEST FORM CODE 6G

VERBAL		MATHEMATICAL	
Section 2	Section 6	Section 5	Section 1
24. B	24. B	24. D	*24. D
25. C	25. D	25. B	*25. A
26. A	26. E		*26. C
27. C	27. B		*27. B
28. E	28. D		28. D
29. E	29. C		29. D
30. B	30. B		30. C
31. D	31. A		31. E
32. D	32. C		32. D
33. B	33. B		33. C
34. E	34. D		34. A
35. E	35. E		35. B
36. C	36. A		
37. E	37. C		
38. C	38. C		
39. D	39. D		
40. E	40. D		
	41. B		
	42. D		
	43. D		
	44. C		
	45. B		

*Indicates four-choice questions. (All of the other questions are five-choice.)

2. Answers will vary. Students should include instructions on how to set up the table and how to construct the sampling distribution of chi-square.

Sample Results from Extensions

1. As the number of rows increases, the distribution spreads out, becomes generally more symmetrical, and moves to the right. For example, here is the distribution for 2 rows followed by one for 11 rows.

```
MTB > random 1000 c1;
SUBC > chisquare 1.
MTB > dotplot c1
```

Each dot represents 14 points

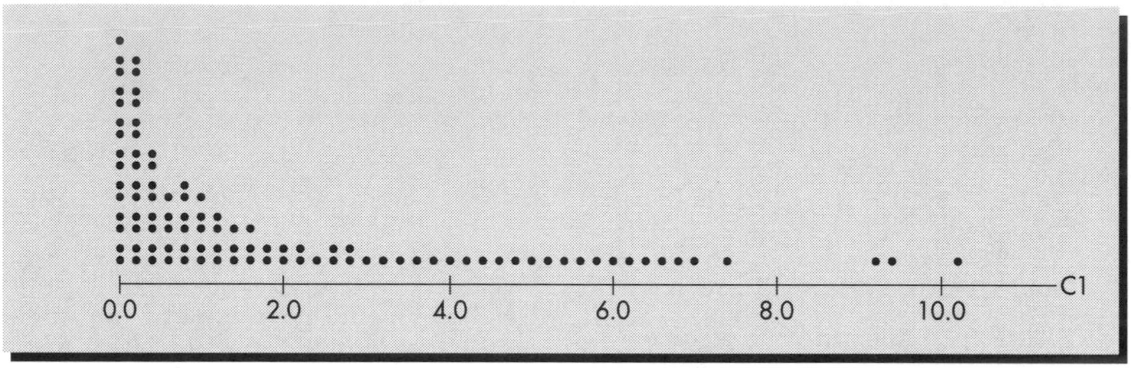


```
MTB > random 1000 c2;
SUBC > chisquare 10.
MTB > dotplot c2
```

Each dot represents 4 points

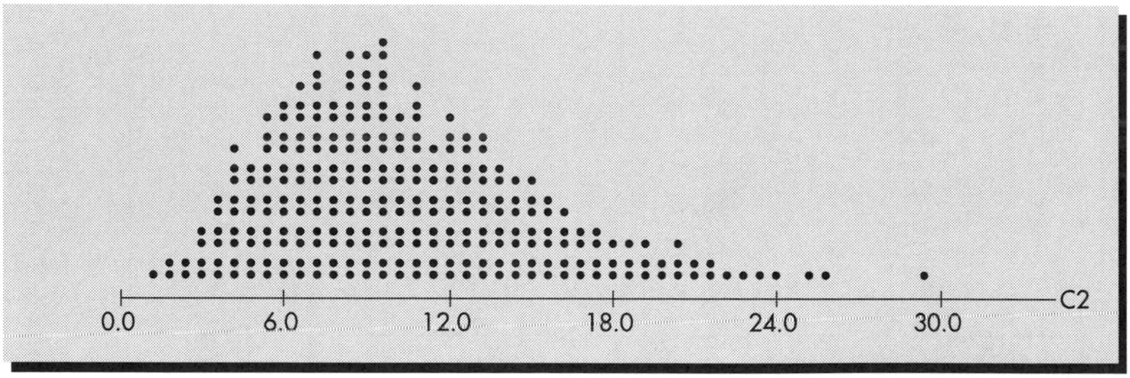

2. a. The expected values are 313, 104, 104, and 35, respectively. The value of chi-square is 0.5.

b.

```
MTB > erase c1
MTB > random 2500 c1;
SUBC > chisquare 3.
MTB > dotplot c1
```

Each dot represents 21 points

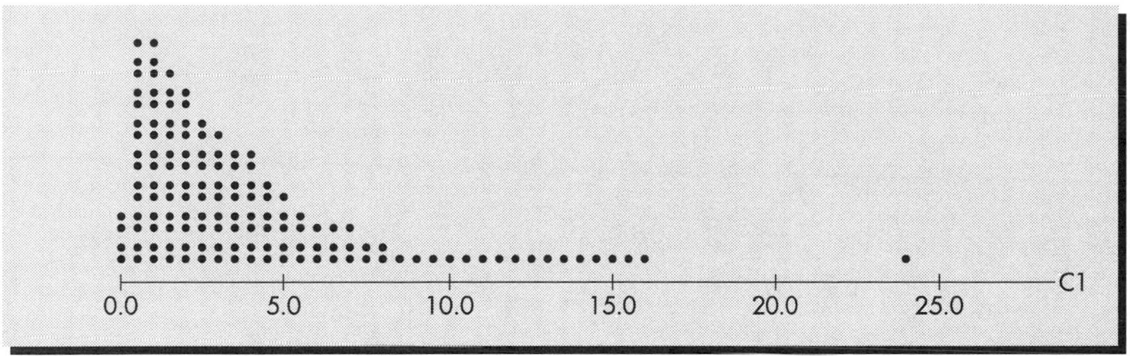

c. There is only about an 8% chance of getting a chi-square value of 0.5 or smaller if one were selecting peas randomly from Mendel's theoretical distribution. The result Mendel claims from his experiment is suspicious, but not conclusive.

Sample Assessment Questions

1. What would be the chi-square value if the proportions in the sample were exactly equal to the proportions in the population? Explain.

2. Several years ago the M&M company said that a bag of plain M&M's contained, on the average, 30% browns, 20% each of yellows and reds, and 10% each of oranges, greens, and tans. Your instructor has a bag of plain M&M's. Describe how you would decide whether or not you think the distribution of colors is still the same as it was then. If you have access to Minitab, construct the appropriate distribution and state your conclusion.

Is Your Class Differently Aged? The Chi-Square Test

SCENARIO

Not so many years ago, almost all college students were of "traditional" college age. Today, about 39% of students enrolled in colleges in the United States are 25 years old or older. About 16% are 35 or older.

Question

Is the distribution of the ages of students in your class typical of that of all college students in the United States?

Objective

In this activity you will learn to construct a chi-square distribution that can be used to answer questions such as that proposed here.

Activity

1. Would you expect the age distribution of students in your class to be typical of the age distribution of all U.S. college students? Why or why not?

2. If your class has more than 40 students, select 40 of them at random and complete this table. If your class has fewer than 40 students, get enough additional ages from students in a similar class to bring the total up to 40.

Age	Proportion in U.S. Colleges	Observed Number in Your Class (O)	Expected Number in Your Class (E)
Under 25			
25–34			
35 and older			
Total	1.00	40	40

You would not expect the numbers in the last two columns to match exactly, even if the students in your class were drawn randomly from the U.S. college population. But are the numbers observed in your class far enough away from those expected to convince you that the students who enroll in your class are differently aged?

3. The chi-square statistic is a measure of how different the observed column is from the expected column. For your table, compute the chi-square statistic,

$$\chi^2 = \sum \frac{(O - E)^2}{E}$$

Is the chi-square statistic that you computed in step 3 large enough to convince you that the ages of the students in your class aren't similar to those of a typical random sample of U.S. college students? To answer this question, you will simulate drawing random samples of ages from the population of U.S. college students and see how often you get a value of chi-square that is as large as or larger than the one from your class.

4. Use a random digit table to take a random sample of the ages of 40 U.S. college students. Since 61% of the students are under the age of 25, let the 61 pairs of digits 00, 01, 02, . . . , 60 represent students whose age is under 25.
   a. What pairs of digits will represent students aged 25 to 34?
   b. What pairs of digits will represent students aged 35 and over?

c. Using the first 40 pairs of digits in the lines of the table assigned to you, take a random sample of the ages of 40 U.S. college students. Place your results in the following table.

Age	Random Digits	Observed Number in Your Sample (O)	Expected Number in Your Sample (E)
Under 25	00–60		
25–34			
35 and older			
Total		40	40

d. Compute the value of chi-square for your sample.
e. Does a large value of chi-square mean that the proportions in the sample are relatively close or relatively far from the proportions in the population?
f. Place an X representing your value of chi-square above a number line that extends from 0 to 15.
g. Repeat this simulation with other members of your class until you have at least 100 values above the number line.

5. What is your estimate of the probability that a random sample of 40 U.S. college students have a chi-square value as large as or larger than that from your class? Is the distribution of ages of the students in your class similar to those of a typical random sample of U.S. college students in the United States? What do you conclude?

Wrap-Up

1. People sometimes say that "b" and "c" answers occur most frequently on multiple-choice tests. In this activity you will decide if there is any evidence of this on the basis of the answer form from the verbal section of a real SAT. (This form was selected randomly from The College Board, 10 SATs, New York: College Entrance Examination Board, 1988.) The correct answers are given here.

1.	d	15.	c	29.	e	43.	a	57.	e	71.	a
2.	d	16.	d	30.	b	44.	a	58.	d	72.	c
3.	b	17.	a	31.	d	45.	b	59.	c	73.	b
4.	b	18.	c	32.	d	46.	e	60.	b	74.	d
5.	c	19.	c	33.	b	47.	d	61.	b	75.	e
6.	e	20.	b	34.	e	48.	b	62.	d	76.	a
7.	b	21.	b	35.	e	49.	d	63.	e	77.	c
8.	a	22.	b	36.	c	50.	b	64.	b	78.	c
9.	a	23.	c	37.	e	51.	a	65.	d	79.	d
10.	b	24.	b	38.	c	52.	a	66.	e	80.	d
11.	c	25.	c	39.	d	53.	c	67.	b	81.	b
12.	b	26.	a	40.	e	54.	c	68.	d	82.	d
13.	e	27.	c	41.	e	55.	a	69.	c	83.	d
14.	e	28.	e	42.	a	56.	c	70.	b	84.	c
										85.	b

a. Make a table showing the observed number of answers for each of "a," "b," "c," "d," and "e," and the expected number of answers, assuming they are equally likely to occur. Compute the chi-square statistic for this table.
b. The following distribution is the result of computing 1,000 values of chi-square for samples taken randomly from a population with equal numbers of answers "a," "b," "c," "d," and "e."

Each dot represents 5 points

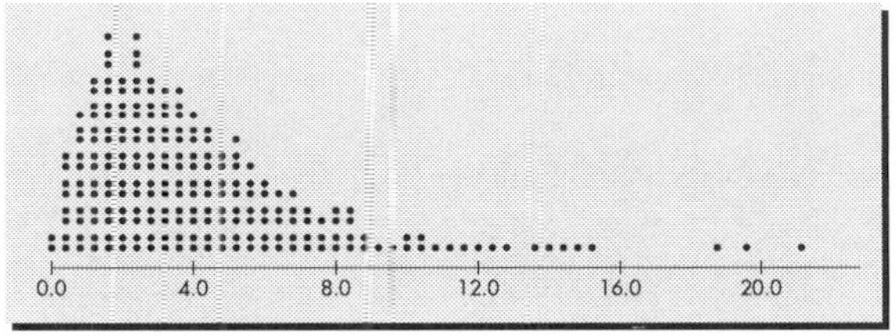

What do you conclude about whether the answers "a," "b," "c," "d," and "e" occur with the same frequency on this type of test? What assumption are you making?

2. Write instructions for students who were absent today, telling how they could test to see if a die is fair.

Extensions

You can use the statistical software Minitab to construct the distribution of the chi-square values for a table with *r* rows. The set of commands shown here will clear the first column, compute 1,000 values of a chi-square statistic for a table with 3 rows, place the values in the first column, and display them on a dot plot.

```
MTB > erase c1
MTB > random 1000 c1;
SUBC > chisquare 2.
MTB > dotplot c1
```

Each dot represents 9 points

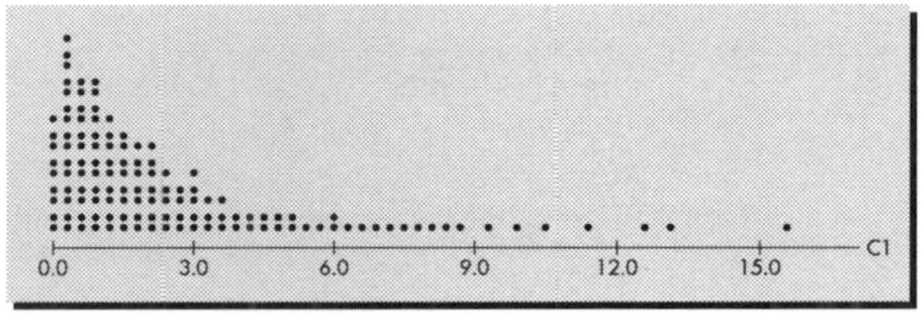

The number in the SUBC > line is always one less than the number of rows in the table.

1. Use Minitab to construct a distribution of chi-square values for tables with various numbers of rows. Describe how these distributions change as the number of rows increases.

2. Gregor Mendel performed experiments with peas to test his genetic theory. It has been said that the experimental results were *too* close to his theory. In one ex-

periment Mendel predicted that he should get 9/16 round yellow peas, 3/16 round green peas, 3/16 wrinkled yellow peas, and 1/16 wrinkled green peas. He writes that he found 315 round yellow peas, 108 round green peas, 101 wrinkled yellow peas, and 32 wrinkled green peas in a sample.
   a. What is the chi-square value for this experiment?
   b. Use Minitab to construct an appropriate distribution of chi-square values for this experiment.
   c. What do you conclude?

Reference

David Freedman et al. (1991), *Statistics*, second ed., New York: Norton.

Coins on Edge

NOTES FOR THE INSTRUCTOR

Statistical Setting

This activity shows how the results of hypothesis tests depend on sample size. That is, it shows how power increases as n increases. The "Coins on Edge" activity can be used as a continuation of the "Introduction to Hypothesis Testing" activity.

Prerequisites for Students

Students should be familiar with the idea of a hypothesis test. It is possible to do the activity without having the students perform the statistical analysis (of finding the *p*-values, etc.). However, if you want to use the activity as it is written, then the students need to know what a *p*-value is, how to find binomial probabilities, and how to use the normal approximation to the binomial. One option is to use the activity after students have worked with hypothesis tests for a while. Another option is to use the activity as soon as the idea of a hypothesis test has been introduced and to lead the class to the ideas of type I error, type II error, *p*-value, and power.

Materials

Each students (or team of students) needs 10 pennies, a copy of the data recording sheet, and a hard, *flat* surface on which to work.

Procedure

The goal of the activity is to let students see how power increases as the sample size increases. You want to start by having them believe that "tipping" a penny is a fair process that gives heads with probability .5. The more you can get them to believe that this is true, the more effective the activity will be, since they will want to retain H_0 until the evidence is overwhelming. For example, if you have used the "Spinning Pennies" activity or the coin-flipping version of the "Introduction to Hypothesis

Testing" activity, you might start by telling the class that spinning or tossing a penny might not be fair and that you will use "tipping" as a way to get heads with probability .5. Try to get them to agree at the onset that this is sensible.

This activity works well in groups of 2 or 3 students. The probability of getting heads when "tipping" a penny is much larger than .5. However, it is important that the table be tapped gently. If you strike the table violently, then the pennies bounce before falling and the probability of getting heads is close to .5 (which is not what we want).

It will take the students about 30 minutes to collect the data and get started with the hypothesis-testing calculations. They may well need some help setting up the appropriate binomial calculations when finding *p*-values, particularly if the students have limited experience with hypothesis tests in general and *p*-values in particular. You might want to do the first *p*-value calculation with the entire class. For example, suppose the first 2 pennies give 1 head and 1 tail. Then the *p*-value is *P*(sample % heads would differ from .5 by at least as wide a margin as obtained in this sample, under the assumption that heads and tails are equally likely) if we were to repeat the experiment. Thus, the *p*-value is 1. If the first 2 pennies both give heads, then the *p*-value is $P(0 \text{ or } 2 \text{ heads} \mid \text{probability of heads is } .5)$, which is .5.

Sample Results from Activity

Here are the data from one class:

Number of Pennies	Number of Heads	Cumulative Heads	Cumulative Pennies	p-value (rounded)
2	1	1	2	1.00
2	2	3	4	0.625
2	2	5	6	0.219
2	2	7	8	0.070
2	1	8	10	0.109
5	4	12	15	0.035
5	4	16	20	0.012
5	4	20	25	0.004
5	3	23	30	0.006
10	9	32	40	0.0004
10	8	40	50	0.00006

Suggested Extension

Consider the data for 50 trials from just one student or group. As a class you could construct a confidence interval for the true probability, *p*, of getting heads. One way to approach this is to ask the class, "Could *p* be .5?" This question will already have been answered (as "no") at the end of the activity. Then ask, "Could *p* be .6?" You can do a quick z-test of this hypothesis by finding the z-score and using the rule "reject H_0 if the z-score is outside the range $\pm$ 2."

Now consider other possible values of p. If you have nine groups of students, for example, then ask one student or group to check whether p might be .55, another group to check whether p might be .60, and so on up to .95. The values for which $-2 < z\text{-score} < 2$ (i.e., the answers to the hypothesis-testing question is "yes, this might be the true value of p") are inside the confidence interval, and other values are outside the CI.

You can refine the CI by checking other values of p. For example, using the data in the preceding table (40 out of 50 heads), we get $|z\text{-score}| < 2$ for p = .70, .75, .80, and .85. Checking values of p between .65 and .70, we get $|z\text{-score}| < 2$ for p = .67, .68, and .69. Checking values of p between .85 and .90, we get $|z\text{-score}| < 2$ for p = .86, .87, and .88. Thus, the refined CI for p is (.67,.88).

Sample Assessment Question

Suppose a researcher wants to take a sample of 15 voters in order to test, using α = .05, the hypothesis that half of the population of voters approve of the job performance of a certain politician. Suppose that in fact 70% of the population approve of this person's job performance. Discuss the researcher's plan and what the researcher might conclude.

Coins on Edge

SCENARIO

People often flip a coin to make a "random" selection between two options, for example, to choose which team goes first in a competition. This depends on the assumption that the probability of getting heads is .5, that is, that the coin is "fair."

Question

Suppose you stand a penny on edge and then make it fall. What is the probability that you will get heads?

Objective

The purpose of this activity is to develop an understanding of the concept of *power* in a statistical test. [Power is the ability of a test to reject a null hypothesis.]

Activity

You need 10 pennies and a flat table. Stand two of the pennies on edge on the table. This may take a while—be patient. (*Hint:* You might find it easiest to make a penny stand on edge if you place it so that Lincoln is upside-down.) Then, with a gentle downward stroke (not sideways), strike the table, so that the coins fall over. Don't pound on the table; just strike it hard enough so that the pennies fall over. If they don't fall, then hit the table again—with a downward stroke. Count the number of

heads facing up; record your results in the first row of Table 1. Then record your feelings about the question "Is the probability of getting heads .5?" Write down how sure you feel about your answer to the question.

Now take two more pennies, stand them on edge, and strike the table to make them fall. Record your results in the second row of the table. Also record your feelings about the question "Is the probability of getting heads .5?" Do you feel that you are now sure of the answer to the question?

Repeat the above experiment for each row of the table, using the number of pennies indicated in the first column. Be sure to record your reaction after each repetition. When you have completed the activity, you will have a total of 50 individual results. After each repetition, record not only the number of heads and the cumulative number of heads, but what conclusions you feel you can draw regarding the question "Is the probability of getting heads .5?" and how sure you feel about your answer.

STATISTICAL ANALYSIS

Use the data from the first row of the data sheet to perform a test of the null hypothesis that the probability of getting heads is .5; use a general (two-sided) alternative hypothesis. You only have $n = 2$ observations, so you will want to use the binomial distribution to find the p-value. Note that the p-value is the probability of getting data at least as extreme as the data you obtained, under the assumption that probability of getting heads is indeed .5. That is, the p-value is P(sample % heads would differ from .5 by at least as wide a margin as obtained in this sample, if we were to repeat the experiment). If $\alpha = .05$, do you reject H_0?

Repeat the hypothesis test, but now use the data from the second row of the data sheet. That is, do a test based on $n = 4$ observations.

Repeat the hypothesis test for each row of the data sheet. When n is large (say, starting with $n = 25$), you might want to use the normal approximation to the binomial distribution when finding the p-value. At what point do you reject H_0? That is, how large a sample do you need before you can reject H_0?

Wrap-Up

1. Summarize how the p-value is related to the sample size. Give an explanation of this phenomenon.

2. Write a brief summary of what you learned in this activity about how power—the ability to reject H_0—is related to sample size.

COINS ON EDGE 171

Table 1 DATA SHEET FOR COIN EXPERIMENT

Number of Pennies	Number of Heads	Cumulative Heads	Cumulative Pennies	Thoughts and Remarks to This Point
2			2	
2			4	
2			6	
2			8	
2			10	
5			15	
5			20	
5			25	
5			30	
10			40	
10			50	

EXPERIMENTS

Jumping Frogs

NOTES FOR THE INSTRUCTOR

Statistical Setting

This activity seems to liven up a class on analyzing data from an experiment. It is also a good introduction to designed experiments and to concepts such as main effects and interaction.

Prerequisites for Students

Students should be familiar with stem and leaf plots and with the use of sample means as estimators.

Materials

Students will need the instructions for making the frogs, square sheets of paper of two different sizes and weight (notebook paper and heavier paper such as copier paper), and a measuring tape showing measurements in centimeters.

Procedure

The design of the experiment calls for assigning an equal number of students to each of the four possible combinations of size and weight. To assign the frogs randomly to the students, you may wish to assign a number to each student, perhaps alphabetically, and then use random number tables to select the students. The first one-fourth of the students selected could be assigned the first combination of size and weight, the next one-fourth the second combination, and so on. Unless the class can be broken into four equal groups, there will be some students left who will not make a frog.

The sizes of the paper were selected mainly for convenience. Using either very small sizes of paper or very heavy paper will result in difficulties in folding the paper.

Sample Results from Activity

The following data show results from a class. We tried this experiment using regular notebook paper and paper used for a laser printer, which was heavier. The heavier paper in the small size produced the best frogs. Obviously, your results will depend on the paper used.

We made the frogs jump on a table and measured the distance jumped in centimeters, giving us a larger range of values.

		Size	
		Large	Small
Weight	Heavy	39, 45, 60, 111, 76 76, 64, 52, 96, 63	69, 136, 146, 151, 31 72, 128, 64, 100, 98
	Light	22, 35, 33, 34, 29 19, 54, 13, 67, 15	41, 46, 53, 75, 59 29, 95, 42, 36, 30

Some students may have trouble folding the paper at the first try. In that case you could let them try with a new square since crumpled paper would be like a beaten-up frog.

The means used for this analysis of effects are given in the following table

	Large (+)	Small (−)	Overall
Heavy (+)	68.2	99.5	83.85
Light (−)	32.1	50.6	41.35
	50.15	75.05	62.60

From this summary,

$$A = (1/2)(83.85 - 41.35) = 21.25$$

and

$$B = (1/2)(50.15 - 75.05) = -12.45$$

It looks like weight has a larger effect than size on the distance the frogs jump.

To look at interaction, we compare

$$D1 = (1/2)(68.2 - 32.1) = 18.05$$

with

$$D2 = (1/2)(99.5 - 50.6) = 24.45$$

Since D1 and D2 are not far apart $[(1/2)(D1 - D2) = -3.2]$ on the scale of these measurements, we would say that there is not much of an interaction effect. That is, the effect of the weight is about the same for large frogs as it is for small frogs.

The plot of the so-called "main effect" means is provided below, and again shows that weight has more of an effect on mean distance than does size.

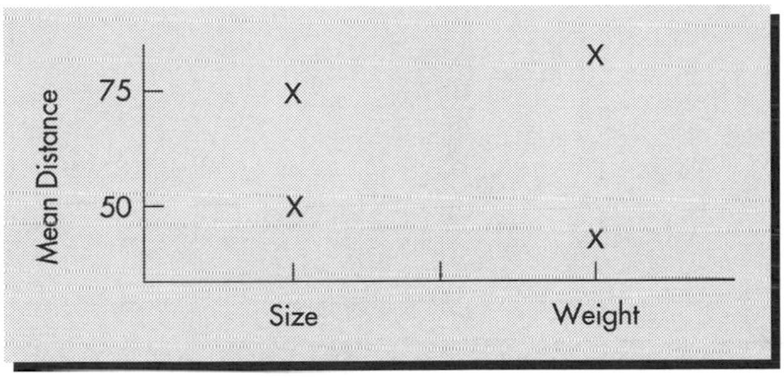

The "interaction" plot is shown below. Since the lines are nearly parallel, we conclude that there is little interaction. The change in mean distance as we move from large to small frogs is about the same for heavy paper as it is for light.

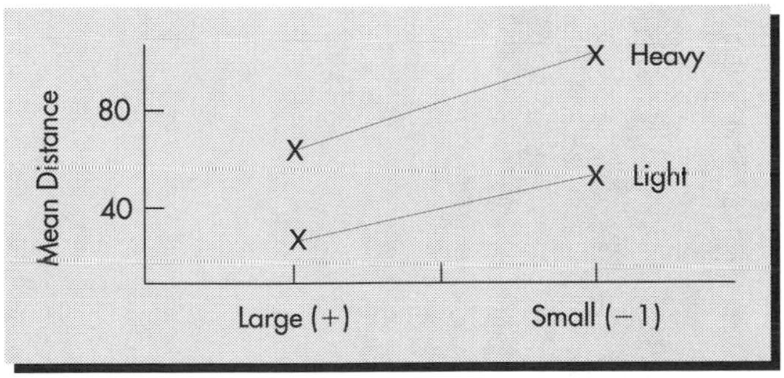

Jumping Frogs

SCENARIO

We often notice that our favorite athletes do not perform exactly the same every time. We reason that their performance is affected by several factors, such as the strength of their opponents, their own schedules, or the state of their health. Statistics has ways of estimating the effects of such factors on the outcomes. Our sports personality for this activity is going to be a paper frog, and we are interested in the distance it can jump. In particular, we want to find out how the weight and the size of the frog affect the jump of a paper frog?

We can carry out an experiment to measure the effects of these two factors. The directions for making the frogs follow this activity. We have a choice of the type and the size of the paper used. We will use two sizes of the paper (squares less than 8 inches) and two different weights of paper. Getting the ultimate frog is far more complicated and is thus beyond the scope of this activity.

Question

How can we measure the effect of a factor on the outcome when we have two factors that can influence the outcome of an experiment?

Objective

This activity will illustrate the analysis of an experiment where the outcomes are affected by two factors, each factor taking on two values. It will discuss how to estimate the effect of each factor and the effect of the interaction of the factors on the outcome, without using a formal design and analysis of experiments approach.

Activity

1. Our objective here is to see how the size and the weight of the frog influence the jump of a paper frog. Since we will vary them, the size and the weight of the frog are the two factors that will affect the outcome. In this activity we will use only two sizes and two weights (i.e., we will use two levels of each factor). The factors that we will work with here are

   S: Size of the frog (2 levels = 2 sizes),
   W: Weight of the frog (2 levels = 2 weights).

   We can then use the data on the distances the frogs jump to find out how size and weight affect the frog's jump and use these results to guide you in coming up with the champion frog.

2. To construct a frog you will need square sheets of paper. Your instructor will have 7.5″ and 6″ squares in two weights of paper. Each of you will be randomly assigned one of the squares. Construct the frog with your sheet of paper. Practice with your frog five times, and report to the instructor the length of the last jump. (Measure the jump in centimeters.)
   Enter the data for the class in Table 1.

Table 1 SIZE OF FROG

		Large	Small
Weight	Heavy		
	Light		

3. We will first look at the results of our experiments graphically.
   a. Construct a back-to-back stem and leaf plot with all measurements of the large paper frog jumps on one side and the small paper frog jumps on the other side. Note that you are using both heavy and light paper measurements on each side. You may wish to show them with different symbols.

b. Repeat the back-to-back stem and leaf plot with heavy paper frog jumps on one side and light paper frog jumps on the other side. Again you could indicate the large paper and small paper frog jumps with different symbols.

c. What do the stem and leaf plots tell us about the relative distances jumped by large and small frogs? Are there differences in the patterns of jumps of the heavy and light frogs in the first back-to-back stem and leaf plot?

d. Now turn your attention to the second back-to-back stem and leaf plot. Compare the distances jumped by the heavy and light frogs in the second stem and leaf plot. Does the size of the frog seem to affect the jump?

e. Combine your analyses to reach a conclusion on which factor has more of an effect on the jumps, size or weight.

4. We will now construct arithmetic summaries for the data. We will try to get an estimate of the effect of size and weight on a frog's jump. Let's look at the following table of means, Table 2.

Table 2 SIZE OF FROG

	Large (+)	Small (−)
Heavy +	Mean of heavy/large frogs = $W+S+$	Mean of heavy/small frogs = $W+S-$
Light −	Mean of light/large frogs = $W-S+$	Mean of light/small frogs = $W-S-$

You now need to get the mean of all the heavy frogs and that of all the light frogs by getting the two row averages. So, for example, the mean jump for the heavy frogs is $1/2((W+S+ + (W+S-))$. Based on these averages, which weight gives better jumping frogs?

In order to estimate the difference in the length of the jumps due to the different weights, calculate

$$A = 1/2(\text{mean jump of the heavy frogs} - \text{mean jump of the light frogs}).$$

We will use this as an estimate of the effect of the weight of the frog. Now calculate the mean of all the large frogs and all the small frogs by getting the column averages. Which size frog jumps better? The effect of the paper size is given by

$$B = 1/2(\text{mean jump on the large frogs} - \text{mean jump of the small frogs}).$$

In the previous section you reached a conclusion on the relative effects of weight and size. Do the values of *A* and *B* confirm it? Using the values of *A* and *B*, complete the following statements:

Changing the weight of the paper (frogs) from light to heavy (increases/decreases) the average distance jumped by ______.
Changing the size of the paper (frogs) from small to large (increases/decreases) the average distance jumped by ______.

5. Do the two factors interact with each other? In other words, is the difference in the mean jumps of the large and small frogs different for the heavy paper and the light paper? We can estimate the interaction term as follows. First, for the large paper frogs we calculate the average difference due to weight as

$$D1 = 1/2(\text{mean of the large/heavy} - \text{mean of the large/light}),$$

and for the small frogs we get

$$D2 = 1/2(\text{mean of small/heavy} - \text{mean of the small/light}).$$

The quantity $1/2(D1 - D2)$ gives us a measure of the interaction term or the effect of changing the levels of both factors. Looking at the table of means and the estimate of the interaction between the two factors, are the differences in the average jumps between large and small frogs the same for both weights of paper?

Wrap-Up

1. We were able to estimate the effect of the two factors using means. Do the following graphical analysis. Find the overall mean. Complete the following graph using the row and column means from Table 2. Above the label "Size" you will plot the two column means, and plot the two row means above the label "Weight." Use the appropriate vertical scale.

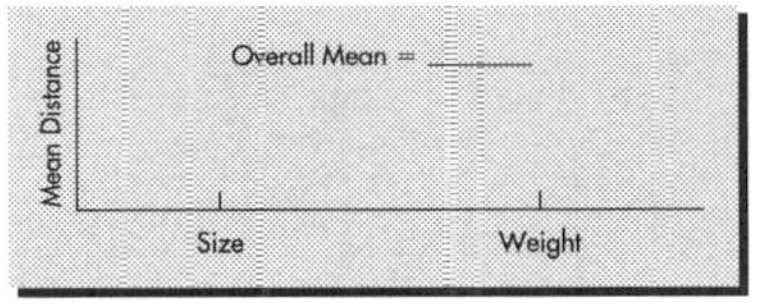

Look at the distances of the two means for size from the overall mean. Are the two means equidistant from the overall mean? Compare the magnitude of this distance to the value of B that you computed earlier. Do a similar analysis on the distances of the two means for weight from the overall mean, comparing them to the value of A. Does it seem reasonable to use A and B as measures of the effects of the two factors? Comment.

2. Complete the following graph using the means $W{+}S{+}$, $W{-}S{+}$, $W{+}S{-}$, and $W{-}S{-}$ from Table 2. You will plot $W{+}S{+}$ and $W{-}S{+}$ above $S{+}$, and $W{+}S{-}$ and $W{-}S{-}$ above $S{-}$.

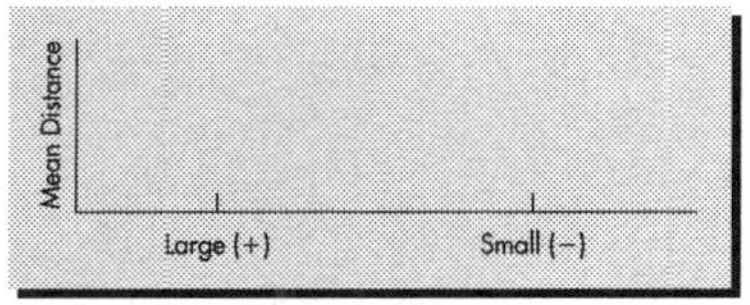
Mean Distance
Large (+)
Small (−)

Now join the points $W{+}S{+}$ and $W{+}S{-}$. Also join the points $W{-}S{+}$ and $W{-}S{-}$. You now have two lines. Compare the slopes of the two lines. Are the lines parallel? If we replace $S{+}$ and $S{-}$ on the horizontal axis by $+1$ and -1, respectively, we can get the slopes of the two lines. Calculate the slopes, and compare them to $D1$ and $D2$ that you calculated earlier. Explain how the measure for interaction of the two factors is related to the slopes of the two lines.

Extensions

This activity is an analysis of data from a designed experiment. The data values were affected by two factors, the size and the weight of paper, each factor taking on two values. We estimated the effects of the size and the weight of paper used on the jumps of the frogs as well as the interaction of these two factors.

1. Your instructor randomly assigned a frog to you. Why did we need to do that? Give examples of how nonrandom assignments could affect the analysis.

2. Write a brief report describing the design and analysis of an experiment to find out how test scores are affected by the time the test is administered and the level of difficulty of the test.

Reference

Richard L. Scheaffer (1989), *Planning and Analyzing Experiments*, unpublished manuscript.

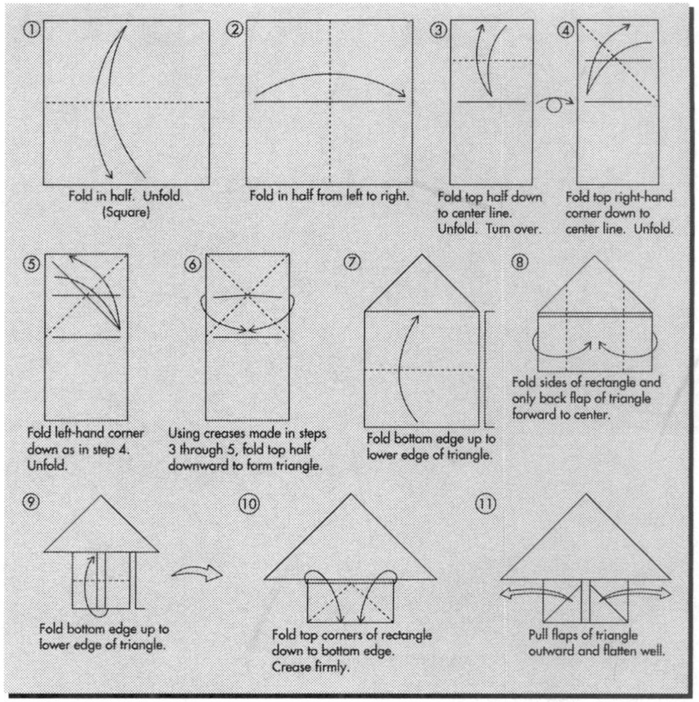
①
Fold in half. Unfold. (Square)
②
Fold in half from left to right.
③
Fold top half down to center line. Unfold. Turn over.
④
Fold top right-hand corner down to center line. Unfold.
⑤
Fold left-hand corner down as in step 4. Unfold.
⑥
Using creases made in steps 3 through 5, fold top half downward to form triangle.
⑦
Fold bottom edge up to lower edge of triangle.
⑧
Fold sides of rectangle and only back flap of triangle forward to center.
⑨
Fold bottom edge up to lower edge of triangle.
⑩
Fold top corners of rectangle down to bottom edge. Crease firmly.
⑪
Pull flaps of triangle outward and flatten well.

Constructing a Paper Jumping Frog

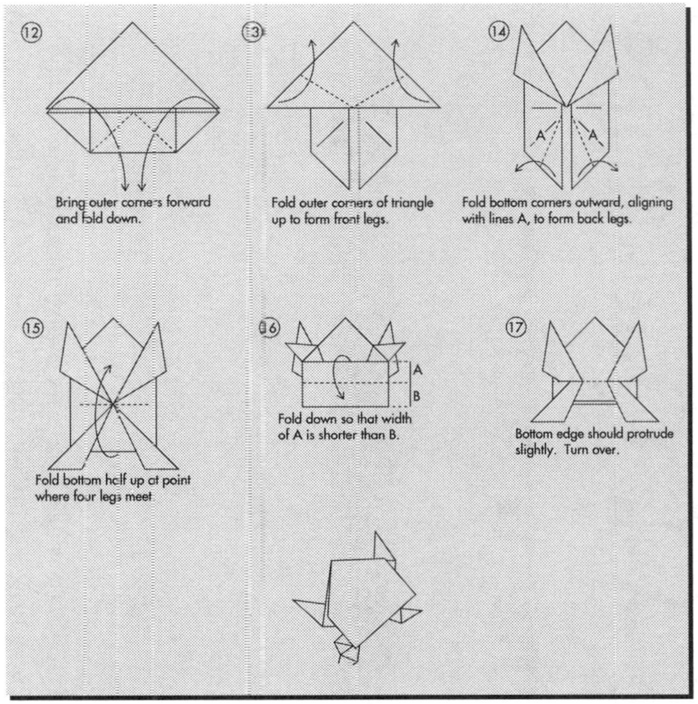

12
Bring outer corners forward and fold down.
13
Fold outer corners of triangle up to form front legs.
14
A
A
Fold bottom corners outward, aligning with lines A, to form back legs.
15
Fold bottom half up at point where four legs meet.
16
A
B
Fold down so that width of A is shorter than B.
17
Bottom edge should protrude slightly. Turn over.

Constructing a Paper Jumping Frog (cont.)

How to Ask Questions: Designing a Survey

NOTES FOR THE INSTRUCTOR

Statistical Objectives

The purpose of this activity is to make students aware that factors in the design of a survey such as question wording are an important source of variation in the results. Students will also get practice in conducting a hypothesis test that two population proportions are equal and in determining a reasonable sample size.

This activity is a good way to introduce students to projects. It has been successful with students who have never worked on a group project in statistics before. The project they are to do is relatively well defined, could be completed in one week, and requires cooperation to design the survey, collect the data, and present the results.

Prerequisite for Students

Students should be familiar with the hypothesis test for the equality of two population proportions.

Materials

Two versions of a questionnaire appear just after these Notes. Make enough copies of each for half of your class.

Procedure

Randomly mix the two versions of the questionnaire, and pass them out to your students. Half of the students should get each version. Do not reveal that there are two versions. Students should be told they don't need to put their names on the questionnaires, that you are collecting some data for the next activity. Collect the questionnaires and tabulate the responses. Compare the proportion of students who would allow such speeches to the proportion of students who would not forbid such speeches using a t-test for the difference of two proportions or a randomization test. Students usually tend to be more reluctant to "forbid" than they are willing to "allow."

If your class is too small to expect significant results, you might give the questionnaire to your other classes beforehand and combine the results.

Encourage the students to discuss whether the wording of these two questions is logically equivalent, whether it makes a difference which wording is used in a campus survey, and, if so, which one is most likely to give a "fair" response.

Divide students into groups of three or four, and let the groups pick one of the questions on the student activity sheet to work on for their project. If you wish, you might encourage groups to think of issues different from those given here to investigate. Remind the students how to compute the sample size needed to detect an effect of the size they suspect will exist.

After about a week, students should be asked to report on their results to the rest of the class.

References

Most college libraries carry books on the design of surveys. Three of the many good ones are these:

1. C. A. Moser, and G. Kalton (1972), *Survey Methods in Social Investigation*, second ed., New York: Basic Books.
2. A. N. Oppenheim (1966), *Questionnaire Design and Attitude Measurement*, New York: Basic Books.
3. H. Schuman and S. Presser (1981), *Questions and Answers in Attitude Surveys*, New York: Academic Press.

Mark an X next to your answer.
Should this college forbid public speeches that might incite violence?
____ YES
____ NO

Mark an X next to your answer.
Should this college allow public speeches that might incite violence?
____ YES
____ NO

How to Ask Questions: Designing a Survey

SCENARIO

Political polls do a remarkably good job of predicting the winner of national elections. However, we occasionally hear of a case where a poll has gone badly wrong. A recent example was the 1992 election in Colorado, where a measure was on the ballot to prohibit the state legislature and cities from passing anti-discrimination laws concerning homosexuals. Polls showed that the Colorado measure would be defeated. The measure passed. (See the *New York Times*, November 8, 1992.)

Question

What may have accounted for the difference between the result in the polls and the result in the election?

Objective

In this activity you will examine how the design of a survey can affect the answers that are given.

Activity

With your group, select one of the following questions and design and carry out a simple experiment to answer it. Use the students at your college as the population of interest.

Before doing the survey, your group will have to decide on a sample size that will be large enough to establish the statistical significance of any difference you feel is practically significant.

1. Does the order in which two candidates appear on a ballot make a difference in the percentage of votes they receive?
2. Is it possible to word a question in two different ways that are logically equivalent but that have a different percentage of students agree with them?
3. Does the order in which two statements appear in a survey make a difference in the percentage of students who agree with them?
4. Can the percentage who agree with a statement be changed by having respondents read some introductory material?
5. If a statement is rewritten to be logically equivalent but to have a more complicated sentence structure and bigger words, will it affect the percentage of students who agree with it?
6. Does the appearance of the interviewer make a difference in how students will respond to a question? For example, do students tend to respond the same way about a controversial issue when the interviewer is female as when the interviewer is male?
7. If the interviewer does not know how a student responds (as on a secret ballot), does it make a difference in the percentage of students who agree with a controversial statement?
8. If a student knows absolutely nothing about an issue, will he or she give an opinion anyway? Will students admit it if they don't know the answer to a question?
9. Do students report events (such as how many days last week had rain or the description of a person who just walked by) as accurately as they think they do?
10. If you let students volunteer to be in your poll, do you get a different result than if you approach the students?

Wrap-Up

1. Your group should prepare a report about what you have learned and present it to the rest of the class.

2. Write a set of guidelines that the school paper might use to conduct surveys of student opinion.

References

1. American Statistical Association, *What Is a Survey?*, in Jonathan D. Cryer and Robert B. Miller (1994), *Statistics for Business: Data Analysis and Modeling*, second ed., Belmont, CA: Duxbury, pp. 388–399.
2. David S. Moore (1991), *Statistics: Concepts and Controversies*, Third ed., San Francisco Freeman.

Gummy Bears in Space: Factorial Designs and Interaction

NOTES FOR THE INSTRUCTOR

Statistical Setting

This activity should come after lessons on the importance of randomizing, such as the "Random Rectangles" activity. There are several aspects of the concept of interaction to emphasize in the context of this activity. The most important of these are the following:

- The structure of (a two-factor) interaction has three components: a first factor, a second factor, and a response.
- Arithmetically, interaction is a difference of differences: The effect of the first factor, as measured by the response, is different for different levels of the second factor.
- Geometrically, the difference of differences corresponds to interaction graphs with line segments of different slopes.
- Abstractly, the structure of an interaction is symmetric, in that you can interchange the first and second factors. (In particular, there are two interaction graphs for every two-factor experiment.)

Although it is not explicitly a part of this activity, the data can serve as an example in which the standard deviation is roughly proportional to the mean. Transforming such data sets to logarithms or taking square roots stabilizes the variances (i.e., makes them essentially constant, as required by the assumptions of standard statistical tests such as the F-tests of analysis of variance).

Prerequisites for Students

The only true prerequisite is that students should know how to compute and interpret averages. Beyond this, a knowledge of either parallel dot plots or scatter plots will be helpful, but it is not essential. (Because this activity does not depend on measuring variability, it can be done even before lessons on measures of spread.)

Materials

The following items are needed for this activity:

Popsicle or craft sticks (or tongue depressors, or other small flat sticks)
Rubber bands
Short length of thin wooden dowel (3/16″ or 1/4″ diameter), or thin pencil
Pennies
Narrow boards (very roughly 20″ × 2′ × 3/4″; rigid wooden yardsticks will work)
Several identical books (or blocks of wood about 1.5″ thick, like pieces of 2 × 4)
Tape measures or yardsticks
Items to launch: small, uniform in size and weight, ideally a bit soft and sticky. We recommend Gummy Bears, a candy made of gelatin.

Procedure

1. Launch equipment
   Prepare launchers as shown in Figures 1 and 2 here. Either prepare several complete sets, one for each team, or else let the launch teams build their own.
   Wrap a rubber band tightly around one of the craft sticks about 2 cm from the end. Use a second rubber band to fasten the other end of this stick to a second as shown in the figure. Insert the pencil or dowel as a fulcrum.
   Note that you want the largest launch angle to be around 60 degrees to make the interaction pronounced. Thus, the number of books to use (we suggested nine) depends somewhat on how thick the books are.

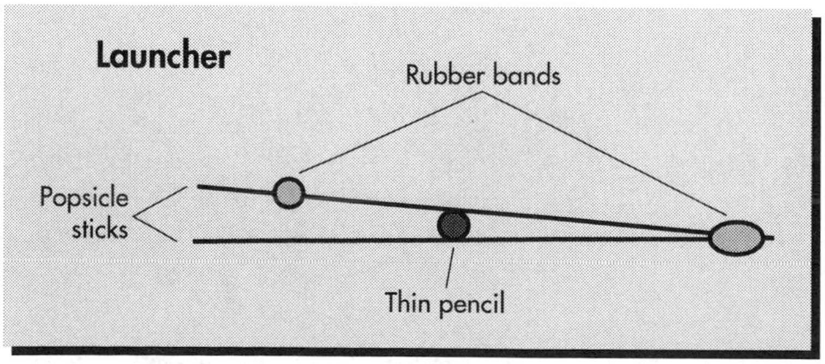

Figure 1: Launcher

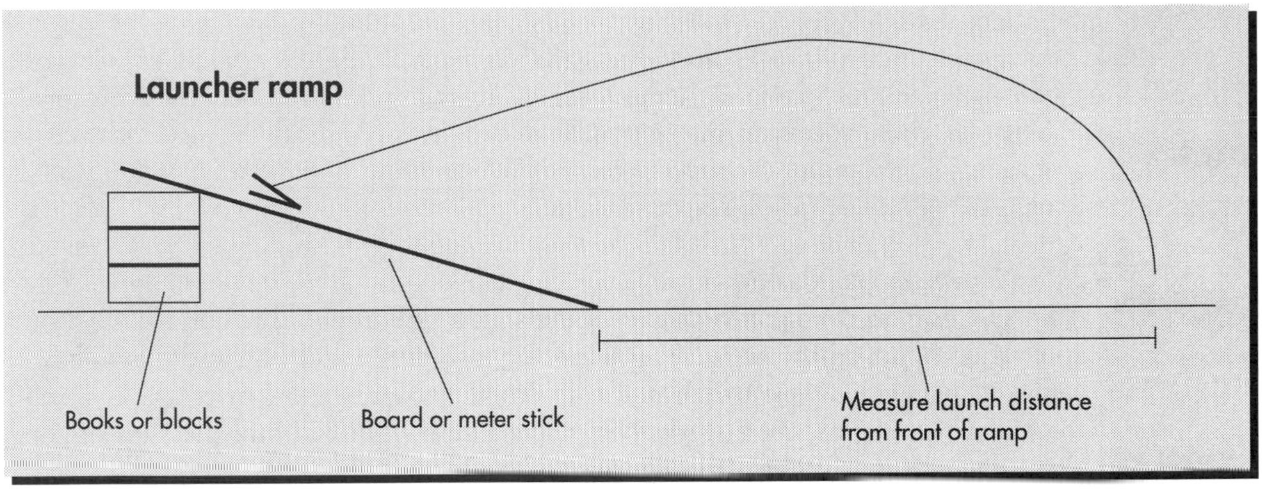

Figure 2: Launch ramp

2. Form teams
   Although data can be generated by one person, a team of people works much better. As few as two and as many as eight on a team will work. Here's a sketch of the various roles, but the main thing to keep in mind is that there's lots of room for flexibility, and you can leave it to the teams to divide up the tasks themselves. Ideally, two people are responsible for setting up each launch: adding or removing books under the ramp, positioning the fulcrum, and positioning the launcher on the ramp, etc. (If the order of the conditions is properly randomized rather than systematic, it requires some attention to get this right.) These two people can also be responsible for holding the ramp and launcher in place. One person should be designated to load projectiles onto the launcher, and another to do the actual launching. To avoid sore fingers, students should use pennies instead of their fingers to release the spring. Measuring is best done by two people, at either ends of the tape, and it doesn't hurt to have recording done by two people, as protection against mistakes.

3. Anticipate the results
   You might ask the students to anticipate the results before conducting the experiment. That is, you could ask them, "Which works better, launching from the front of the ramp, or launching from the back? Notice that if you launch from the back of the ramp, you start higher up, and so you would expect the Gummy Bear to travel farther before it lands. On the other hand, the rules say you have to measure distances from the front of the ramp, so there is a cost to moving back in order to gain height. Try to guess how much distance you'll gain or lose from moving the launcher to the back of the ramp if, under the ramp, you have one book, five books, or nine books. Do you expect to gain more distance when the angle is steep or flat? What about the middle angle (five books)? Do you expect it to be more like one book, or more like nine books?"

3. Collect data
   Because launch distances are quite variable, with an SD of around 15″, it works best to make multiple launches and report averages. Thus the instructions to students ask them to make sets of four launches and report an average for each set of four. (The interaction effect is roughly 30, or about 3 standard errors if you use averages of four for each response value.)

4. Randomize order of conditions
   This activity assumes, rather than explains, that the order of the conditions should be randomized. In the activity randomizing is done by shuffling and drawing from slips of paper, but it is worth pointing out to students that, in practice, using a table of random numbers works better, because it's hard to mix the slips of paper well enough to give truly random results.

Sample Results from Activity

Here are the raw data, along with averages and standard deviations, from one team.

Ramp	Blocks	Distances				Avg.	SD
f	1	22	28	4	32	21.5	12.4
b	1	27	25	8	24	21.0	8.8
f	5	39	66	60	25	47.5	19.0
b	5	30	53	46	28	39.3	12.2
f	9	75	71	77	73	74.0	2.6
b	9	106	65	134	93	99.5	28.7

These data lead to a summary table that corresponds to steps 3 and 4 of the activity:

	Number of Blocks		
Position	1	5	9
Front	21.5	47.5	74.0
Back	21.0	39.3	99.5
Diff	0.5	8.2	−25.5

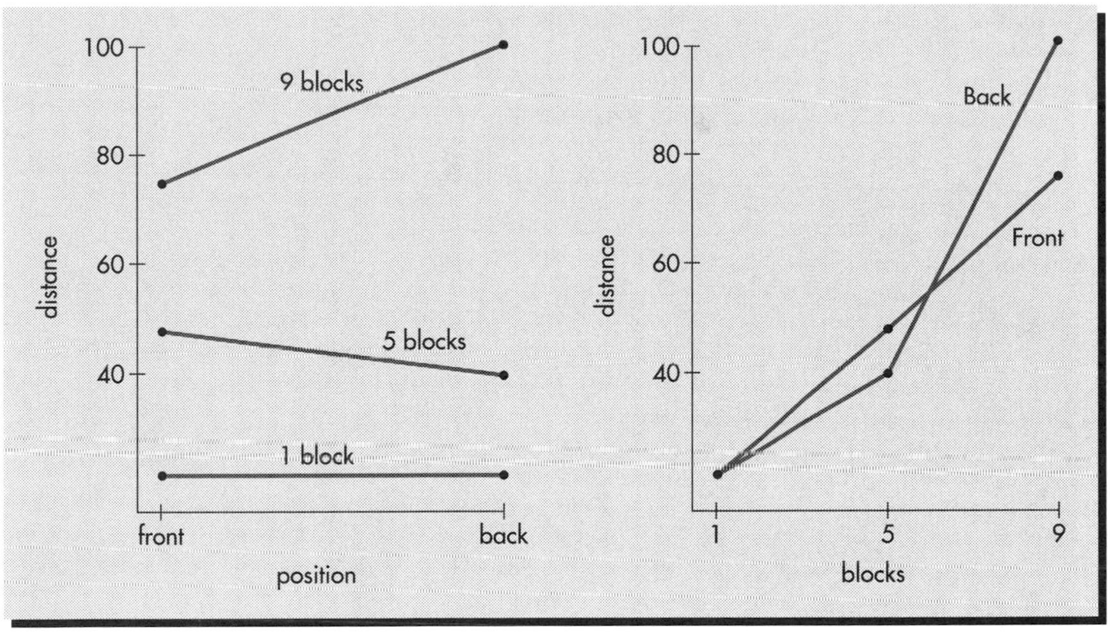

Figure 3

The two formats for the interaction graph are shown in Figure 3. Note that if you restrict the data to one and five books, there is no interaction (line segments are parallel). For five and nine books, there is a very pronounced interaction of launch angle with position.

Extensions

The design used here is a standard factorial design. One extension of this idea is to use a randomized complete block design with student teams as the blocks. For example, pick a launch position, say "back," and a position for the fulcrum. Then the only factor of interest is the number of books. Have each team of students (block) conduct four launches under each of the three levels of the books factor. Now make a scatter plot of the average distance obtained by a team versus number of books, with one point on the plot for each team at each of the three levels (one, five, or nine books).

Next, have each team compute a team average (the average of all 12 launches) and subtract it from each condition average, to get the condition effects (for 1, 5, or 9 books) for their team. Construct another scatter plot, this time of condition average versus the number of books. If there is a lot of variability between teams, then this second scatter plot should show less variability than the first plot, since the condition averages plotted show differences between factor levels after removing the effects of the teams.

Sample Assessment Questions

1. Rectangles

   Congratulations! You have just been appointed Director of Research in the Department of Rectangle Science. For your first project, you have designed a two-way factorial experiment:

   Response: Area in square inches
   Factor 1: Length—4″ or 10″
   Factor 2: Width—1″ or 4″

   a. Make a table summarizing the results you would get from this study. Then draw and label a graph showing the interaction pattern.
   b. Clever scientist that you are, you decide to use the same design over again, this time to study the perimeter of the rectangles. Make a table summarizing the results you would get using perimeter as your response. Then draw and label an interaction graph for these results.
   c. Discussion: Write a paragraph comparing your interaction graphs from steps a and b. What is the major difference in the patterns of the two graphs? Why do area and perimeter behave differently?

2. Growth and age

   Imagine a two-way factorial design to study the following scientific hypothesis: "Toddlers get taller; adults don't." Here's a quick summary:

   Response: Height in inches
   Factor 1: Age groups—2-year-olds and adults
   Factor 2: Time—at the start of the study, and three years later

   Make a two-way table summarizing the results you would expect to get from this study. Then draw and label a graph showing the interaction pattern.

3. Athletic training and heart rate

   If you compare the heart rates of athletes and nonathletes at rest, just after they climb three flights of stairs, and then again after two minutes of rest, you will find that on the average the athletes have a lower resting pulse and that their heart rate increases less during moderate exercise and returns to resting rate more quickly than for nonathletes.

   a. Summarize the factorial structure that a study would need to confirm this statement. (List the response, and each of the two factors, as in #1 above.)
   b. Make a two-way table summarizing the results you would expect to get, and show these results in an interaction graph.

4. Warm blooded, cold blooded

   a. Complete the following sentence by describing the interaction in words: "If you compare the body temperatures of humans and turtles in the summer when the surrounding temperature is 80° and in the winter when the surrounding temperature is 40°, you find that. . . ."

b. Summarize the factorial structure that a study would need to verify the interaction. (List the response and each of the two factors.)
c. Draw and label an interaction graph showing the kind of results you would expect to get from the study.

5. Thread color and background color
Design an experiment to study possible interaction between the effects of thread color and background color on the speed of needle threading. Tell how you would measure the response (speed of needle threading) and how you would choose the combinations of conditions to compare.

Gummy Bears in Space: Factorial Designs and Interaction

SCENARIO

The experimental designs used in this activity are called **factorial designs.** They have two main advantages: They allow you to study the effects of two or more influences (called **factors**) in a single experiment, and they allow you to measure the **interaction** between the two (or more) factors.

Here is an example of interaction from the field of cognitive psychology. How do you answer the question "Does a cat have claws?" Do you imagine a picture of a cat, and check to see whether it has claws, or can you answer without a mental image, just because you associate claws with cats? The psychologist Stuart Kosslyn figured that young children would have to rely on a picture but that older children and adults would have learned the kind of word associations that allow them to answer more quickly, without a picture. To study this theory about language development, Kosslyn used a factorial design.

Factor #1: Age groups: (a) 1st grade, (b) 4th grade, and (c) adult
Factor #2: instructions: (A) imagery instructions—inspect a mental image in order to answer the question, or (B) no imagery instructions—just answer as quickly as you can
Response: The time it took a subject to answer the question

For his experiment, Kosslyn used all six possible combinations of age group and instructions. The results fit his prediction of an interaction between age and instructions: Adults were slower when they had to call up a mental image than with the other instructions. First graders were equally slow both ways, suggesting that they used a picture regardless of the instructions.

Question

Your instructor will give you a catapult. Do the angle of the launcher and its position on the launch ramp interact in their effects on how far you can shoot a gummy bear?

Objectives

The goals of this lesson are to see how to use a two-way factorial design to study the effects of two factors at once, as well as to see how to summarize the results in an interaction graph.

Activity

Your instructor will divide your class into teams and provide you with the equipment you need to launch gummy bears. Imagine that your team is a scientific research group that is trying to study the effects of two factors on launch distance:

Factor #1: Launch angle (the number of books under the end of the launch ramp)
Factor #2: Launch position (the launcher is either at the front of the ramp or at the back of the ramp)

1. Plan the experiment.
   a. Each team should assign jobs: (1) loader/launcher. (2) holder (keeps the launch equipment steady), (3) measurer, and (4) data recorder and supervisor.
   b. List the conditions.
      Factor #1. Use three different launch angles: (a) 1 book under the back end of the launch ramp, (b) 5 books, and (c) 9 books. (The angles, or numbers of books, are called **levels** of the factor.
      Factor #2. Use two different positions for the launcher on the ramp: (A) front of the ramp, and (B) back of the ramp positions. (This factor has two levels, front and back.)
      For a **two-way factorial** experiment, you **cross** the two factors, that is, you use *all possible combinations* of the levels of the two factors as conditions for your experiment. In this case, there are six possible combinations of launch angle and launch position.

c. Randomize the order. Use the six conditions to label six slips of paper, and fold them so you can't see the labels. Then mix them thoroughly and open them one at a time. The first one you draw tells you which condition to use first, and so on. Write the order next to the list of conditions from step 1b.
d. Keep other conditions fixed. Decide where to position the fulcrum (pencil or dowel) along the bottom popsicle stick. Mark the place on the popsicle stick, and be sure to check before each launch to make sure you keep the position the same for all your launches.

2. Produce the data. You will record six sets of four launches each, one set for each of the six conditions.

a. Set up your equipment for the first combination of launch angle and launch position on the ramp. Launch a gummy bear, and measure how far it went. Measure the distance from the *front* of the ramp, and measure only in the direction parallel to the ramp, as shown in this illustration.

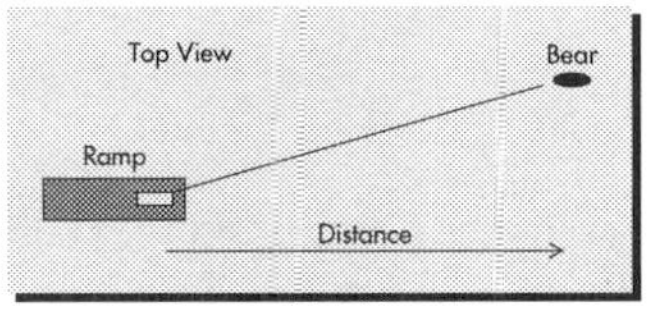

Do three more launches (remember to check the position of the fulcrum!), and record the distance each time. Then compute the average of the first four launches.

b. Now reset your launch equipment for the second combination of angle and position on the ramp. Then do a second set of four launches, and compute their average.
c.–f. Repeat for the third combination of angle and position, and then the fourth, fifth, and sixth.

3. Tabulate the data. Construct a table with two rows and three columns for summarizing your data. Label the rows to correspond to the position of the launcher (front or back) on the ramp. Label the columns to correspond to the number of books (1, 5, or 9) under the back of the ramp. Put your average distances in the body of the table.

Here, as a guide, is a similar table showing actual results from a version of Kosslyn's imagery experiment described in the Scenario section. The response times are in milliseconds.

	First Grade	Fourth Grade	Adult
Imagery instructions	2,000	1,650	1,600
No imagery instructions	1,850	1,400	800

4. Interpret interaction as a "difference of differences." How much distance do you gain or lose by launching from the back of the ramp instead of the front? Use your data to answer this question by computing the difference "back average minus front average." Compute this difference first for the two sets of launches with 1 book under the ramp. Compute a second difference for the two sets of launches with 5 books under the ramp, and finally compute a third difference for 9 books. Show the differences in your table by adding a third row called "Difference." If the two factors (launch angle and position on the ramp) interact, then the three differences will be different. Is interaction present in your launch data?

5. Graph your results. The following graphs show two ways of representing the data from Kosslyn's imagery experiment. Using them as a guide, construct similar graphs to present the data from your table in step 3.

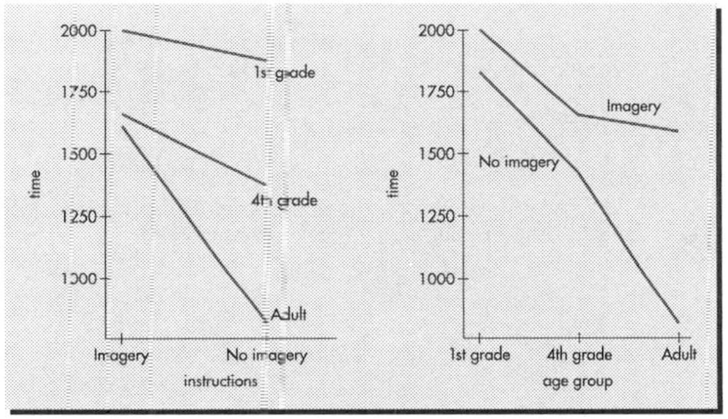

6. Describe the pattern. Write a short paragraph describing the nature of the interaction between the launch position and the launch angle. Your description should tell when you get the longer distances launching from the front position and when you get the longer distances launching from the back position.

Wrap-Up

1. Explain how Kosslyn's results tend to support his theory that word associations like "claws" and "cat" develop later in life than an understanding of the words themselves.

2. Find an example of interaction in a science or social science textbook. List the response and the two interacting factors. Then draw and label a graph that illustrates the interaction you have found.

Extensions

1. Is there another bear-launching interaction? The following diagram shows three possible positions for the fulcrum of your bear launcher.

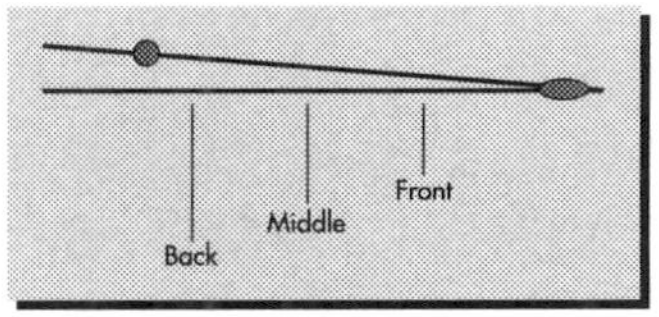

   Design and carry out an experiment to test whether there is an interaction between the position of the fulcrum and the launch angle (either 1 or 5 books under the back of the launch ramp). As before, launch four times for each combination of conditions, and use the average of the four distances as your response value.

2. Let's try a three-factor experiment. (This is an unstructured question, more in the nature of a small project.) Suppose your goal is to find the combination of launch angle, position on the ramp, and position for the fulcrum that will together produce the longest launches. Design an experiment to help you find the best combination of conditions.

Reference

S. M. Kosslyn (1980), *Image and Mind*, Cambridge, MA: Harvard Univ. Press, Chapter 9.

Funnel Swirling

NOTES FOR THE INSTRUCTOR

Statistical Setting

As concern with productivity and quality grows, it is becoming increasingly clear that design of experiments should play a role in statistical training. The "Funnel Swirling" activity was developed by Bert Gunter and Joe Ortiz, who give credit to W. Edwards Deming for the original idea, as a tool for teaching design of experiments to engineers.

Prerequisites for Students

This activity can be done on the first day of class, with no statistical preparation.

Materials

Each team of students will need access to the following set of materials: (1) a stopwatch, (2) a yardstick, (3) a protractor, (4) a 3/8″ ball bearing (other sizes can be used; see Extension below), and (5) a funnel apparatus (see Figure 1).

Each team of students will need access to the apparatus. One option is to make several sets of funnel set-ups, so that all teams can conduct trials at the same time. Another option is to make a single apparatus and have the teams take turns using it.

The key part of the funnel apparatus is a standard oil-change funnel, as sold in auto supply stores; this should be about 18″ long, although other sizes are possible. A stand is needed to hold the funnel in place during the experiment. A laboratory stand could be used, or one could build an inexpensive stand out of 3/4″ square molding and 1/4″ threaded rods. Three pieces of molding can serve as the base of the stand and a fourth piece can be used as a crossbar to support the funnel. Clothespins or wing nuts can be used to hold the crossbar in place, and self-adhesive velcro tape can be placed on the crossbar and on the edge of the funnel to help keep the funnel in place. In any event, *someone must hold the funnel still while the experiment is conducted*. The other

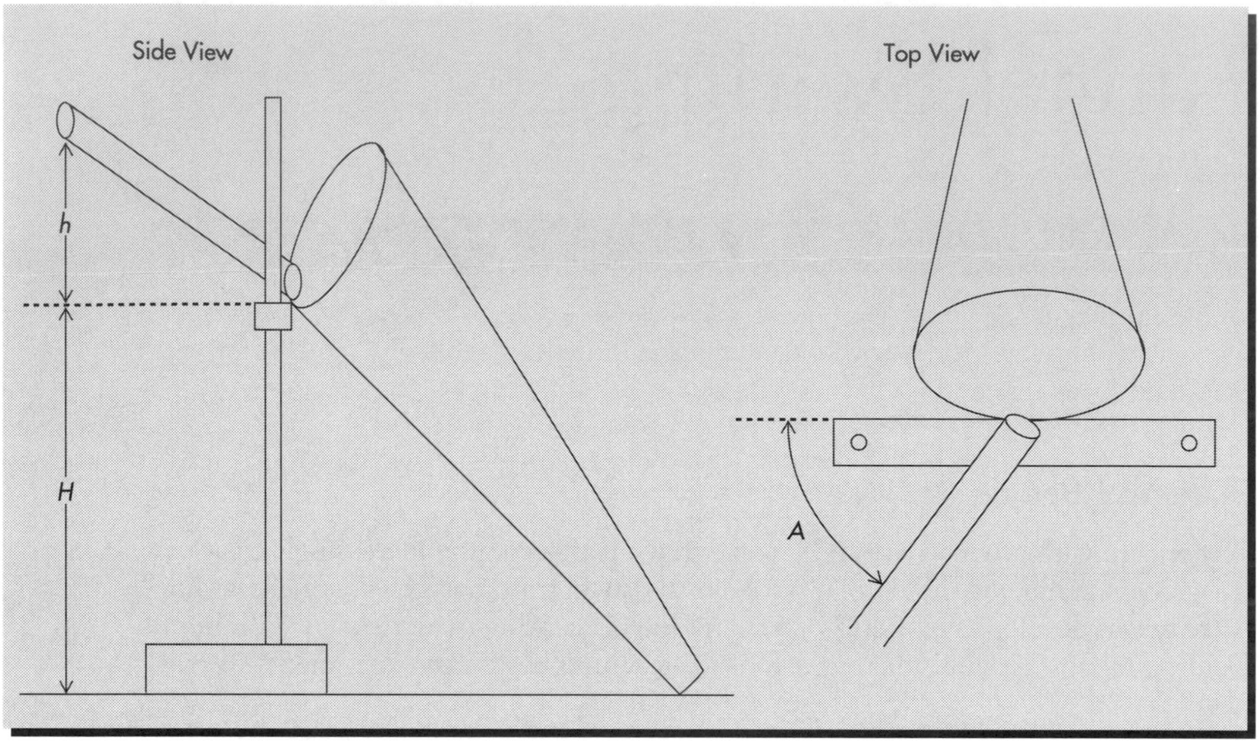

Figure 1: Funnel apparatus.

part of the funnel apparatus is a tube—a piece of 3/4″ or 1″ plastic or copper pipe; this should be about 15″ long.

Procedure

This activity should be done in teams of 3 to 5 students. Larger teams are possible. Smaller teams don't work well because it takes several persons working together just to conduct a trial.

It is a good idea to create the teams, demonstrate the activity, and allow some planning time for the teams in one class period, then have the teams conduct their trials in a later class period. This way, team members can discuss strategy together outside class. If you want to spare class time, you can have the teams conduct their trials during office hours rather than in class, so that class time is saved for the "roll-off." Conducting 8 trials takes approximately 10 to 15 minutes; planning *how* to conduct the trials takes an indefinite amount of time.

Demonstrate the process the students are to optimize by placing the crossbar at a random height, resting the end of the funnel on the crossbar, placing one end of the tube at the center of the funnel, raising the other end of the tube to a height of 4 inches, say, above the level of the crossbar, and having someone hold the funnel in place. Then drop a ball bearing through the tube and into the funnel. Tell him or her that the goal is to maximize the amount of time the ball bearing is in motion.

If you have teams conduct their trials during class time, then you should require them to submit a list of settings of the factors (height and angle) for each trial before they start their first trial, to ensure that the trials are run quickly and efficiently.

Discussion: Some teams will take a haphazard approach to the experiment, while others will try to be quite careful and systematic. However, you can expect that even the best of them will resort to some type of "one at a time" experimentation: they will hold two of the variables fixed and will vary the third. This quickly exhausts the budget and provides them with only limited information about the system (although it provides the instructor with the opportunity to show how statistical design can be very useful).

You can record each team's best time and their "optimal" settings for the variables and then discuss how they went about conducting the experiment. The roll-off introduces a design issue (blocking) that some may wish to discuss at length: Should each team use its own apparatus, or should a common apparatus be used?

During the roll-off the inherent variability in the system will cause the times obtained to differ from those the teams obtained during experimentation (i.e., repeated runs under identical settings do *not* yield identical times). Discuss with the class whether teams replicated any runs. (There are some configurations of the apparatus that lead to very unstable results, so that two runs at the same settings of the variables can produce times of 2 seconds for the first run and 16 seconds for the second run!)

Note that it is likely that something will go wrong (e.g., the stopwatch operator may fail to record the time) during experimentation. Some instructors will allow teams to repeat such trials. Others will count such mistakes against the team's limited budget. The latter approach is harsh, but it is realistic.

This activity generates data that can be used in discussing a variety of topics: variability, distributions, inference (making comparisons), etc. However, the key topic introduced is design of experiments. After teams have tried on their own to extract information about the funnel system, they are poised to learn about statistical design. For example, using the variables of funnel height and angle, one can construct a 2^2 factorial design with replication.

Extension

As noted in the Materials section, many other variables can be added to the experiment. For example, different ball bearing sizes can be used, with bearing size (say, from 1/4″ up to 5/8″) being a design variable. The height, *h*, of the top of the tube can be varied as a design factor. Another variable is the type of funnel, plastic versus metal. Indeed, one instructor used plastic funnels originally and then used metal funnels later in the course when students repeated the activity. The funnels were sufficiently different that using the metal funnels was like doing a new activity; this allowed students to use what they had learned about design of experiments in what was essentially a new setting.

The discussion of experimental design can go into many areas. Additional topics include blocking, interactions, randomization of run order, fractional factorial designs, etc.

Sample Results from Activity

Some teams will choose settings that keep the ball bearing in the funnel for only a few seconds, but times of over 20 seconds are possible. It happens that the greatest times occur with settings that are unstable—when the ball bearing just barely makes a loop when it first enters the funnel. Upon replication under these settings, the ball bearing might not make a full loop and might head straight down the funnel, giving a time of close to 0!

Sample Assessment Questions

1. Suppose your job were to find the best combination location (in direct sunlight or in shade), watering schedule (once per day or twice per day), and soil pH (low or high) for growing a plant. Describe how you would conduct an experiment to determine the optimal combination.

2. Explain what is meant in statistics by an "interaction."

Funnel Swirling

SCENARIO

Consider the following simple "game": A ball bearing will be rolled down a tube and into a funnel; it will swirl around in the funnel for a while and then roll out onto a table. The goal is to maximize the amount of time the ball bearing takes to roll down the tube and out of the funnel. You may adjust the height of the funnel and the angle of the tube as you attempt to maximize time.

Question

How good are you at finding the best setting?

Objectives

In this activity you will learn about factorial experiments and statistical design. You will also study variability and see how it is an inherent part of an experiment.

Activity

This activity should be done in teams.

1. Your instructor will divide the class into teams for the "contest" (to see which team can keep a ball bearing swirling in the funnel for the longest time). Each team will be given access to a set of materials in order to experiment and collect data.

2. Your team has a budget of $8,000. Each trial costs $1,000, so you can afford only 8 trials—use them wisely.

3. There are two variables that you can adjust during the experiment: (1) the height, *H*, of the funnel (i.e., the height of the crossbar) and (2) the angle, *A*, of the tube relative to the crossbar. The height, *h*, of the tube above the crossbar should be 4 inches for each trial. Your team may decide how to conduct your eight trials. You should decide how you will conduct your trials (i.e., the settings you will use for each variable for each trial) in advance so that the actual experimentation will go quickly.

4. During a trial one person should hold the tube and drop the ball bearing, another should hold the funnel steady (if the funnel wobbles it will absorb energy from the ball bearing, which will significantly reduce swirling time), and a third person should operate the stopwatch.

5. After you have completed your eight trials, analyze the data and determine the optimal settings of height and angle. After all teams have conducted their trials and analyzed their data a "roll-off" will be held to determine the winning team. Each team must submit its "optimal" settings for the variables at the start of the roll-off.

Wrap-Up

1. Did the results of the experiment surprise you? Which factor had the greatest effect on the ball bearing times? What do you now think are the optimal settings of the variables? How do these compare to what you expected when you started?

2. Write a brief summary of what you learned in this activity about how to design an experiment.

Extension

Many other variables can be added to the experiment. For example, different ball bearing sizes can be used, with bearing size (say, from 1/4″ up to 5/8″) being a design variable. The height, *h*, of the top of the tube can be varied as a design factor. Another variable is the type of funnel, plastic versus metal. Try repeating the activity while allowing some of these factors to vary.

Reference

1. Bert Gunter (1993), "Through a funnel slowly with ball bearings and insight to teach experimental design," *The American Statistician*, **47**:265–269.

MODELING

Models, Models, Models . . .

NOTES FOR THE INSTRUCTOR

Statistical Setting

This activity can be used together with a unit on regression and model fitting. It is an example of how real data can be modeled with fairly simple linear models. The residual plot shows a very distinct pattern and can be a way of introducing the analysis of residuals. The extension activity illustrates that a straight line is not always the best model. The data in the attachment also include time series of monthly CO_2 levels, and this data can be used for assessment or additional extension of this activity.

Prerequisites for Students

Students should know how to fit a regression line or a median fit line to data on two variables.

Materials

Access to a statistical package will greatly enhance the effectiveness of this activity.

Procedure

The focus of the activity is on finding the "best model" to fit a data set. While studying the scatter plot leads to fitting two lines to the data, another way of reaching the same conclusion could be to study the plot of the differences in the levels from year to year. If the change in levels is the same, these differences should be about the same. Figure 2 shows the differences plotted over time and suggests that the differences are larger in the latter period, which implies a steeper slope of the fitted straight line.

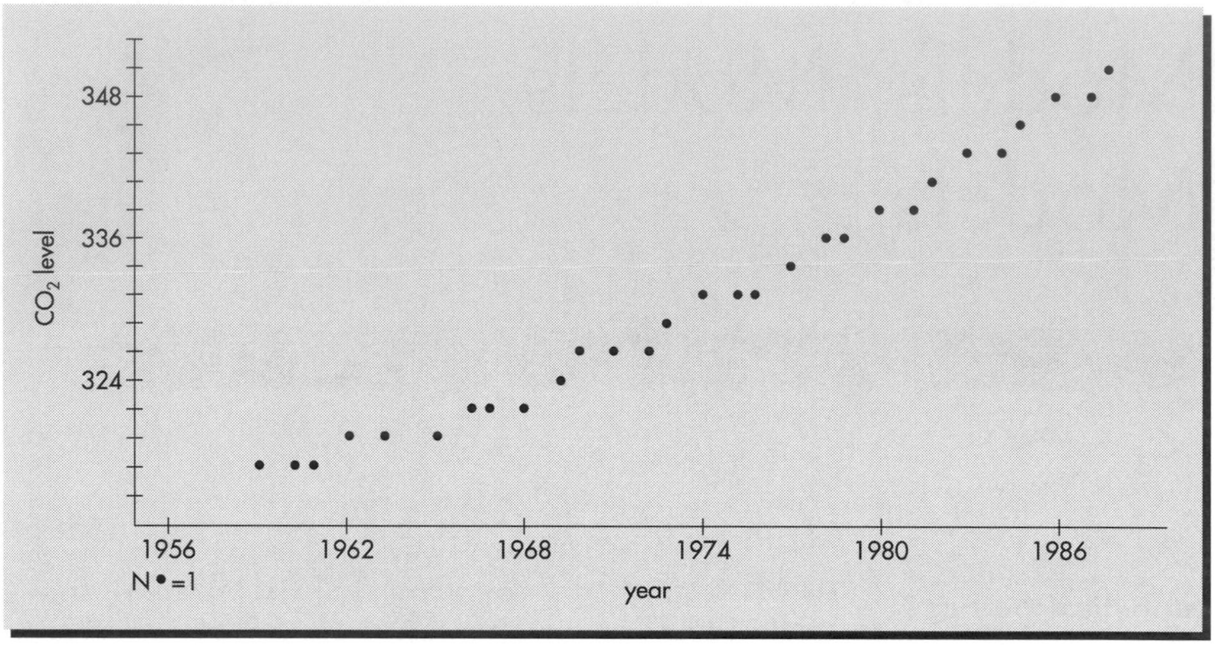

Figure 1: CO_2 levels plotted against year.

Sample Results from Activity

2. Figure 1, a scatter plot of the annual levels of CO_2 versus time, does show that the annual levels of CO_2 are steadily increasing. However, there is a slight curvature in the plot. One way of dealing with this is to fit a quadratic function, or as we suggest to fit more than one straight line, which will take care of the different rates of change.

Figure 2 shows the change in CO_2 levels from year to year. The plot indicates at least two different patterns: smaller changes earlier (1959 to 1968) and larger changes later (1969 to 1988). Thus if two lines are fit to the two time periods, one should get a better description of the data.

3. a. The output in Figure 3 is obtained from a student version of Minitab, Stat 101. The regression coefficient is 1.196, so the model for the period 1959 to 1988 indicates that the slope is 1.196, or the average annual CO_2 levels increase by 1.196 units per year. The residual plot (Figure 4) is clearly not a random scatter of the residuals and shows a distinct pattern in the values of the residuals. It therefore indicates that the model can be improved.

The model for the period 1959 to 1968 shows that the slope is .718, so that the average annual CO_2 levels increase at the rate of .718 per year during this period.

During the period 1969 to 1988, the model gave the rate of CO_2 increase to be 1.383 per year.

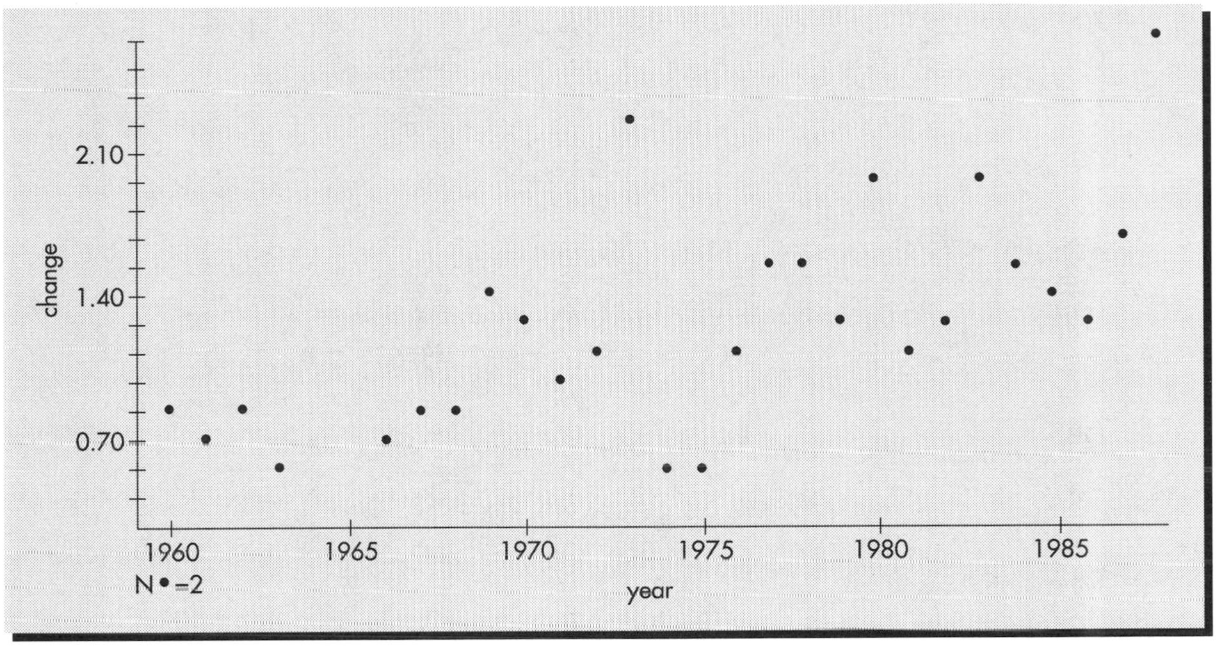

Figure 2: Increase in CO_2 levels plotted against year of increase.

The regression equation is
CO2lev = −2030 + 1.20 years

Predictor	Coef	Stdev	t-ratio	p
Constant	−2030.44	60.96	−33.31	0.000
years	1.19649	0.03088	38.74	0.000

s = 1.433 R-sq =98.2% R-sq(adj) = 98.2%

Analysis of variance

SOURCE	DF	SS	MS	F	p
Regression	1	3083.8	3083.8	1501.01	0.000
Error	27	55.5	2.1		
Total	28	3139.3			

Regression Anaylsis of Data from 1959 to 1968

The regression equation is
CO2/1 = −1090 + 0.0.718 yearl

Figure 3: Regression analysis of complete data (1959 to 1988).

Predictor	Coef	Stdev	t-ratio	p
Constant	−1089.74	28.72	−37.95	0.000
yearl	0.71770	0.01463	49.07	0.000

s = 0.1326 R-sq = 99.7% R-sq(adj) = 99.7%

Analysis of variance

SOURCE	DF	SS	MS	F	p
Regression	1	42.352	42.352	2407.93	0.000
Error	7	0.123	0.018		
Total	8	42.475			

Regression Analysis of Data from 1969 to 1988

The regression equation is
CO2/2 = −2400 + 1.38 year2

Predictor	Coef	Stdev	t-ratio	p
Constant	−2399.75	52.00	−46.14	0.000
year2	1.3801	0.02628	52.62	0.000

s = 0.6778 R-sq = 99.4% R-sq(adj) = 99.3%

Anaylsis of Variance

SOURCE	DF	SS	MS	F	p
Regression	1	1272.0	1272.0	2768.46	0.000
Error	18	8.3	0.5		
Total	19	1280.2			

Figure 3: Regression analysis of complete data (1959 to 1988). (cont.)

The difference in the rates of change in CO_2 levels does confirm that these levels are increasing faster in the more recent past. Thus breaking up the data into two parts helps us describe the data better.

b. A residual is defined as the difference between the observed value and the fitted value. Table 1 gives the residuals for the model using all the years. A positive residual implies that the model fitted value is smaller than the observed value or that the model is underpredicting. A negative value implies that the model is overpredicting.

Table 1 RESIDUALS FOR 1959 TO 1988

ROW	years	CO21ev	res
1	1959	316.1	2.62036
2	1960	317.0	2.32388
3	1961	317.7	1.82739
4	1962	318.6	1.53088
5	1963	319.1	0.83441
6	1964	*	*
7	1965	320.4	−0.25858
8	1966	321.1	−0.75507
9	1967	322.0	−1.05154
10	1968	322.8	−1.44806
11	1969	324.2	−1.24454
12	1970	325.5	−1.14102
13	1971	326.5	−1.33752
14	1972	327.6	−1.43399
15	1973	329.8	−0.43051
16	1974	330.4	−1.02698
17	1975	331.0	−1.62347
18	1976	332.1	−1.71994
19	1977	333.6	−1.41644
20	1978	335.2	−1.01294
21	1979	336.5	−0.90942
22	1980	338.4	−0.20593
23	1981	339.5	−0.30240
24	1982	340.8	−0.19891
25	1983	342.8	0.60461
26	1984	344.3	0.90811
27	1985	345.7	1.11163
28	1986	346.9	1.11514
29	1987	348.6	1.61865
30	1988	351.2	3.02219

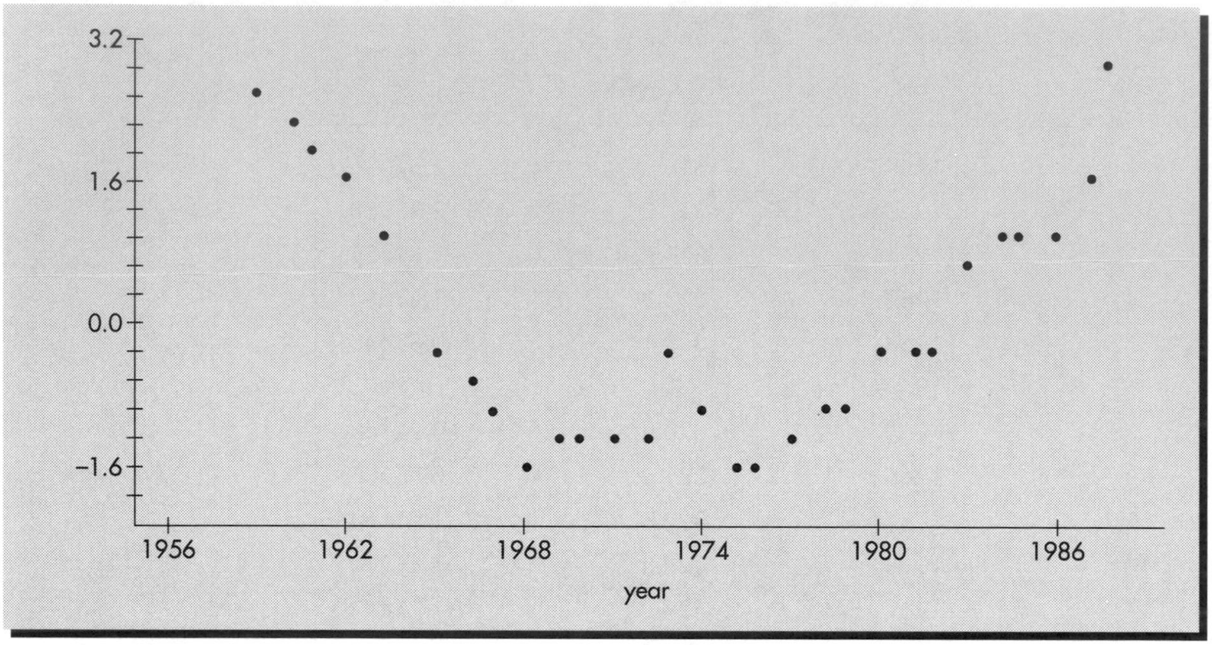

Figure 4: Plot of residuals against year.

c. The residual plot in Figure 4 shows a distinct pattern and provides us with additional proof that a single straight-line model may not be the best way to describe the data.

Models, Models, Models . . .

SCENARIO

When looking at relationships between two quantities, we are often interested in fitting a mathematical function to describe this relationship. Statisticians call these functions **models.** This activity uses data on levels of carbon dioxide in the atmosphere at a site in Hawaii to illustrate fitting a model to data. Carbon dioxide, or CO_2, is one of the gases in the environment whose levels have been increasing. Why should this be of great interest? It is so because scientists are concerned that increasing levels of complex gases, one of which is CO_2, will thicken the blanket around the Earth and prevent heat from escaping. This could result in "global warming," which could lead to disastrous coastal flooding and severe droughts. These data have been used by scientists in studies involving levels of CO_2 in the atmosphere. Reasons for the choice of the site, and additional information regarding the data, are shown after the activity. The patterns observed in this data can be considered to be typical of what could be observed globally.

Questions

How has the level of CO_2 changed over time? What mathematical function will best "model" the relationship between the two variables?

Objectives

The goals of this activity are to learn how a variable is changing with respect to another and to model the relationship between two variables. Here the two variables are the levels of CO_2 and time.

Activity

1. There is concern that levels of carbon dioxide are rising in the atmosphere. In this activity you will try to understand the rate of increase over the period 1959 to 1988 using scatter plots. You will also fit straight lines to the complete data as well as subsets of the data and compare the slopes. You will use only the annual averages that are shown as the last column in Table 2.

2. Plot the data and analyze the plots.

a. Construct a scatter plot of the data using the year as the *x*-variable and the annual CO_2 level as the *y*-variable. What does the scatter plot of the data show? Are CO_2 levels steadily increasing?

b. Again look at the scatter plot of the data, and note the pattern of CO_2 levels from 1959 to 1968 and from 1969 to 1988. From the plot, can we say that the patterns of changes in the annual CO_2 level are the same for the two periods?

3. Fit a model.

a. Fit a straight line to the data using either the median fit or the least-squares method, and note the slope of the line.

b. One way of determining if a straight line "fits" the data is to study the differences between the observed and fitted values. These differences are called the **residuals.** Calculate the residuals for the model using all the years. What does a positive residual signify? What does a negative residual signify?

c. Plot the residuals in a scatter plot against the years. If the straight line is a good fit, there should not be any visible pattern in the scatter plot and the residuals should be randomly scattered around zero. What does your residual plot show?

d. Since the CO_2 levels do not change the same way for the entire period, you will now try to describe the change with two straight lines. Fit a line using data from 1959 to 1968. Fit another line using data from 1969 to 1988.

4. Evaluate the fitted models. Look at the slopes of the fitted lines for the complete time period and for the two subsets. Complete the following sentences:

During the period 1959 to 1988, the average annual CO_2 levels __________ at the rate of __________ per year.

During the period 1959 to 1968, the average annual CO_2 levels __________ at the rate of __________ per year.

During the period 1969 to 1988, the average annual CO_2 levels __________ at the rate of __________ per year.

Did the CO_2 levels increase at the same rate in the later years as in the earlier years?

Wrap-Up

There are different ways of deciding which function can model the relation between two variables. What was the method that you used? What did you learn about the steps that one should follow when fitting a "model" to data? The data in the activity indicate that annual levels of CO_2 are increasing. There are several reasons for this increase. However, scientists agree that one of the major culprits is the increase in fossil fuel consumption. The data in Table 1 give the gross energy consumption per capita in the United States measured in millions of Btu. Gross energy includes the energy generated by primary fuels such as petroleum, natural gas, and coal, imports of their derivatives, plus hydro and nuclear power.

1. Construct a scatter plot of the data. Do you think a straight line would be a good model in this case?

2. Fit a straight line to the data, and look at the residual plot. Does the residual plot confirm your answer to 1?

3. Do you think that a nonlinear function might provide a better model? Explain why.

Table 1

Year	Gross Energy per Capita	Year	Gross Energy per Capita	Year	Gross Energy per Capita
1960	242.4	1971	326.9	1982	305.8
1961	242.1	1972	339.7	1983	301.3
1962	249.3	1973	350.5	1984	313.7
1963	255.3	1974	338.9	1985	309.6
1964	263.3	1975	326.4	1986	307.8
1965	271.2	1976	340.7	1987	311.8
1966	283.3	1977	346.5	1988	324.2
1967	289.9	1978	350.9	1989	325.3
1968	303.9	1979	349.7	1990	326.9
1969	316.7	1980	333.8	1991	334.7
1970	323.7	1981	322.7	1992	321.5

Table 2 ATMOSPHERIC CONCENTRATIONS OF CARBON DIOXIDE* ON MAUNA LOA

Year	Jan	Feb	Mar	Apr	May	Jun	Jul	Aug	Sept	Oct	Nov	Dec	Annual
1958			316.0	317.6	317.8		316.1	315.2	313.4	313.5		314.8	
1959	315.6	316.4	316.8	317.8	318.4	318.2	316.7	315.0	314.0	313.6	315.0	315.8	316.1
1960	316.5	317.1	317.8	319.2	320.1	319.7	318.3	316.0	314.2	314.1	315.1	316.2	317.0
1961	317.9	317.8	318.5	319.5	320.6	319.9	318.7	317.0	315.2	315.5	316.2	317.2	317.7
1962	318.1	318.7	319.8	320.7	321.3	320.9	319.8	317.6	316.5	315.6	316.9	317.9	318.6
1963	318.8	319.3	320.1	321.5	322.4	321.6	319.9	317.9	316.4	316.2	317.1	318.5	319.1
1964	319.6				322.2	321.9	320.4	318.6	316.7	317.2	317.9	318.9	
1965	319.7	320.8	321.2	322.5	322.6	322.4	321.6	319.2	318.2	317.8	319.4	319.5	320.4
1966	320.4	321.4	322.2	323.5	323.8	323.5	322.2	320.1	318.3	317.7	319.6	320.7	321.1
1967	322.1	322.2	322.8	324.1	324.6	323.8	322.3	320.7	319.0	319.0	320.4	321.7	322.0
1968	322.3	322.9	323.6	324.7	325.3	325.2	323.9	321.8	320.0	319.9	320.9	322.4	322.8
1969	323.6	324.2	325.3	326.3	327.0	326.2	325.4	323.2	321.9	321.3	322.3	323.7	324.2
1970	324.6	325.6	326.6	327.8	327.8	327.5	326.3	324.7	323.1	323.1	324.0	325.1	325.5
1971	326.1	326.6	327.2	327.9	329.2	328.8	327.5	325.7	323.6	323.8	325.5	326.3	326.5
1972	326.9	327.8	328.0	329.9	330.3	329.2	328.1	326.4	325.9	325.3	326.6	327.7	327.6
1973	328.7	329.7	330.5	331.7	332.7	332.2	331.0	329.4	327.6	327.3	328.3	328.8	329.8
1974	329.4	330.9	331.6	332.9	333.3	332.4	331.4	329.6	327.6	327.6	328.6	329.7	330.4
1975	330.5	331.1	331.6	332.9	333.6	333.5	331.9	330.1	328.6	328.3	329.4	330.6	331.0
1976	331.6	332.5	333.4	334.5	334.8	334.3	333.0	330.9	329.2	328.8	330.2	331.5	332.1
1977	332.8	333.2	334.5	335.8	336.5	336.0	334.7	332.4	331.3	330.7	332.1	333.5	333.6
1978	334.7	335.1	336.3	337.4	337.7	337.6	336.2	334.4	332.4	332.2	333.6	334.8	335.2

(Continued)

Table 2 ATMOSPHERIC CONCENTRATIONS OF CARBON DIOXIDE* ON MAUNA LOA

Year	Jan	Feb	Mar	Apr	May	Jun	Jul	Aug	Sept	Oct	Nov	Dec	Annual
1979	335.9	336.4	337.6	338.5	339.1	338.9	337.4	335.7	333.6	333.7	335.1	336.5	336.5
1980	337.8	338.2	339.9	340.6	341.2	340.9	339.3	337.3	335.7	335.5	336.7	337.8	338.4
1981	338.8	340.1	340.9	342.0	342.7	341.8	340.0	337.9	336.2	336.3	337.8	339.1	339.5
1982	340.2	341.1	342.2	343.0	343.6	342.9	341.7	339.5	337.8	3337.7	339.1	340.4	340.8
1983	341.3	342.5	343.1	344.9	345.8	345.3	344.0	342.4	332.9	340.0	341.2	342.9	342.8
1984	343.7	344.6	345.3	347.0	347.4	346.7	345.4	343.2	341.0	341.2	342.8	344.0	344.3
1985	344.8	345.8	347.2	348.1	348.7	347.9	346.3	344.2	342.9	342.6	344.0	345.3	345.7
1986	346.0	346.7	347.6	349.2	349.9	349.2	347.7	345.5	344.5	343.9	345.3	346.6	346.9
1987	347.7	348.1	349.1	350.6	351.5	351.0	349.2	347.7	346.2	346.2	347.4	348.7	348.6
1988	349.9	351.2	351.9	353.2	353.9	353.3	352.1	350.0	348.2	348.7	349.2	351.1	351.2
1989	352.5	352.8	353.4	355.1	355.4	354.9	353.4	351.3	349.6	349.9			

From Atmospheric CO_2 concentrations—Mauna Loa Observatory, Hawaii, 1958–1986, C. D. Keeling, Scripps Institution of Oceanography. CDIAC NDP-001/R1 (rev. 1986).
*Atmospheric CO_2 in parts per million by volume (ppmv). Annual averages based on monthly means. All numbers have been rounded to the nearest tenth.

Extension

This activity was concerned with examining annual CO_2 levels to see how they changed over time. You fit a single straight-line model to the data and evaluated the fit by studying the residuals. The patterns in the residual plot and the plot of differences suggested that two lines would fit the data better than a single line. In many situations, it may not be feasible to fit one or more straight lines to the data, and the wrap-up activity illustrated this point.

In this activity was have limited ourselves to fitting straight lines to the CO_2 data. Could we have done better if we had fitted some other function? If you have the appropriate software, fit a quadratic function to the data and study the residual plot for this fit.

References

1. Atmospheric CO_2 concentrations—Mauna Loa Observatory, Hawaii, 1958–1986, C. D. Keeling, Scripps Institution of Oceanography. CDIAC NDP-001/R1 (rev. 1986).
2. *Basic Petroleum Data Book.* Petroleum Industry Statistics, Volume XIV, Number 1, January 1994, American Petroleum Institute, Washington, D.C.

Mauna Loa

BACKGROUND
Principal Investigators
Charles D. Keeling
Timothy P. Whorf
Scripps Institution of Oceanography
University of California
La Jolla, California 92093, U.S.A.

Air sample collection—Continuous. Air samples are collected from air intakes at the top of four 7-m towers and one 27-m tower.
Four samples are collected every hour.
Details are given in Keeling et al. (1982).
Measurement apparatus—Analyses of CO_2 concentrations are made by using an Applied Physics Corporation nondispersive infrared gas analyzer with a water vapor freeze trap.
Data selection procedures—Data are selected for periods of steady hourly data to within ~0.5 ppmv; at least six consecutive hours of steady data were required to form a daily average.
Calibration gases used—CO_2-in-N_2 until December 1983 and CO_2-in-air from December 1983 to the present.
Scale of data reported—1987 WMO/Scripps mole fraction scale.
Data availability—These monthly and annual data, which are derived from daily "steady" data, are available from CDIAC. The monthly data through 1986, along with monthly data that have been adjusted to remove the seasonal effects are available in machine-readable form from CDIAC.

TREND
The Mauna Loa atmospheric CO_2 measurements constitute the longest continuous record of atmospheric CO_2 concentrations available in the world. The Mauna Loa site is considered one of the most favorable locations for measuring undisturbed air because possible local influences of vegetation or human activities on atmospheric CO_2 concentrations are considered minimal and any influences from volcanic vents may be excluded from the records. The methods and equipment used to obtain these measurements have been essentially unchanged over the 31-year record.

Because of the favorable site location, continuous monitoring, and careful selection and scrutiny of the data, the Mauna Loa record is considered to be a precise record and a reliable indicator of the regional trend in the concentrations of atmospheric CO_2 in the middle layers of the troposphere. The Mauna Loa record shows a 12 percent increase in the mean annual concentration in 31 years, from 315 parts per million by volume of dry air (ppm) in 1958 to 352 in 1989 (Keeling et al. 1989).

Monthly atmospheric CO_2 concentrations

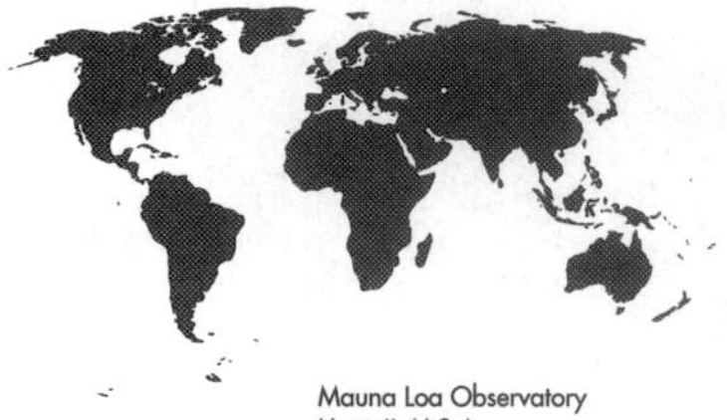

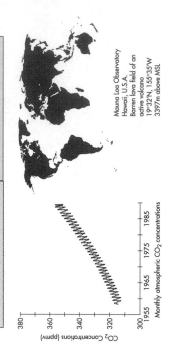

Mauna Loa Observatory
Hawaii, U.S.A.
Barren lava field of an
active volcano
19°32'N, 155°35'W
3397m above MSL

The Regression Effect

NOTES FOR THE INSTRUCTOR

Statistical Setting

This activity is best used after students have learned about correlation and regression, and it can serve as a finale to that unit. Students learn about the regression effect and are given practice identifying situations where it might occur. Students also learn to interpret the regression line as the line of the average y-values for each fixed x.

Prerequisites for Students

Students should have the ability to read and interpret scatter plots. It is preferable if they have already studied correlation and regression.

Materials

For Extension question 1, you will need to supply paired scores for two consecutive tests in a statistics class (omit names). For Sample Assessment question 2, students will need the major league baseball batting records for two consecutive years. Many almanacs have this information. Optional materials are a nickel, a \$1 bill, a \$2 bill, a \$20 bill, and photocopies of a \$500 bill and a \$1,000 bill.

Procedure

The entire class should do together question 1 through question 4. The remaining questions are best done in small groups, where students can discuss their ideas and misgivings.

After all students take the first money test, give them the answers:

I. A

II. A

III. B

This test and the make-up test are designed so that almost all students are guessing on each question. If your class is large (say, more than 75), ask all of the students who missed all three questions to stand up. With a smaller class, ask all students who missed two or three questions to stand up. With good humor, give these students some kind of "remediation"—possibly a pep talk about how you know they can do better next time or a stern lecture about paying more attention to their money. Now have all students take the money make-up test.

The answers to the money make-up test are

I. B

II. A

III. A

Finally, put the students into groups to work on the remaining questions and, preferably, one or two of the Extension questions.

Sample Results from Activity

3. The average score for the "remedial" students will most likely be higher on the money make-up test than it was on the money test.

4. The average score for the "star" students will most likely be lower on the money make-up test than it was on the money test.

5. Since the test was designed so that very few college students know the answers to any of the questions, the students who got low scores on the first test did so because they were unlucky. And the students who got high scores on the second test did so because they were lucky. It's unlikely that the entire group would be so lucky or unlucky again.

6.
   a. 71 inches
   b. 67 inches
   c. The y-intercept of the line falls at 64 inches.
   d. 69 inches (This is the $\bar{x}, \bar{y}$) point.)
   e. The sons of tall fathers tend to be shorter than their fathers, even given the fact that the group of sons averaged one inch taller than the fathers. And the sons of short fathers tend to be taller than their fathers. (This cannot be explained by the heights of the mothers. If we graph, say, the average height of the two parents on the x-axis, the same effect appears.)

Sample Results from Wrap-Up

1. If students picture a scatter plot of pairs of scores, it would look somewhat like that of the father/son heights, with a strong positive correlation between the score on the first test and the score on the second test. The correlation isn't perfect, and so the regression effect will appear.

   The supervisor selected those students with the lowest scores on an achievement test. Many of those students are actually students whose true score is somewhat better. Those students scored low because they were somewhat "unlucky" on the achievement test (meaning they were unusually tired or distracted, their guesses tended to be wrong more often than would be predicted, etc.). (Of course, some students in the low-achievers group were "lucky" and scored higher than their true ability, but this group will be smaller than the first group because there were fewer of them to begin with.) We wouldn't expect the group of unlucky students to be so unlucky again, and they weren't. Thus the relative score of the entire group went up.

   The situation with the gifted program is analogous.

2. The word "regress" means to "go back." When the correlation isn't perfect, the second score tends to go back toward the mean compared to the first score. (Galton called the phenomenon "reversion.")

Sample Assessment Questions

1. This quote is from Francis Galton's *Memories of My Life* (London: Methuen, 1908). Galton first recognized the regression effect.

   > The following question had been much in my mind. How is it possible for a population to remain alike in its features, as a whole, during many successive generations, if the average produce of each couple resemble their parents? Their children are not alike but vary: therefore some would be taller, some shorter than their average height; so among the issue of a gigantic couple there would be usually some children more gigantic still. Conversely as the very small couples. But from what I could thus far find, parents had issue less exceptional than themselves. I was very desirous of ascertaining the facts of the case.

   Translate Galton's statement into more modern English. What assumption is Galton making that isn't true? Write an explanation to Galton.

2. From almanacs for two consecutive years, get the individual batting records in either the American or National League. Do the batting records exhibit the regression effect?

3. Psychologists once recommended that instructors in a flight school should praise any exceptionally good execution of a flight maneuver by a trainee. After trying this approach for some time, the instructors reported that positive reinforcement did not work, as the trainees who had been praised for a good maneuver usually performed less well on the next try. Can you suggest a probable cause for this result? (See Kahneman, Slovic, and Tversky [Ref. 1], pp. 66–68.)

References

1. Daniel Kahneman, Paul Slovic, and Amos Tversky, eds. (1982), *Judgment Under Uncertainty: Heuristics and Biases*, Cambridge: Cambridge Univ. Press.

   Psychologists have found that people don't recognize the regression effect and that if they do notice it, they tend to feel that it needs explanation.

2. Joel R. Levin (1993), "An improved modification of a regression-toward-the-mean demonstration," *The American Statistician*, **47**(February): pp. 24–26.

   This article describes how to use two decks of cards to illustrate the regression effect. The basic idea is this. Divide one deck into a lower part and an upper part: all 1 (ace) through 6 and all 8 through 13 (king). Discard the 7's. Half of the class depict low scores and draw their "true" score from the lower half of the deck. The other half depict high scores and draw their "true" score from the top half of the deck. Each member of the class then draws his or her "error" from the second deck and adds it to his or her "true" score. On the retest, all students retain their "true" score and draw another "error" from the second deck.

3. Robert W. Mee and Tin Chiu Chua (1991), "Regression toward the mean and the paired sample *t* test," *The American Statistician*, **45**(February): pp. 39–42.

   This article shows how one should properly conduct a paired-sample comparison of means in a test-retest situation where only a subset of the population retakes the exam. For example, suppose all who fail an exam are given the opportunity to retake a similar exam. If the usual paired-sample t test is conducted from data in this setting, the "regression effect" may lead to the incorrect conclusion that some intervention (such as remedial tutoring) has been effective in raising the scores when this is not necessarily true.

The Regression Effect

SCENARIO

A mathematics supervisor in a large U.S. city got a grant to improve mathematics education. She tested all students and placed those with the lowest achievement scores in a special program. After a year, she retested them and was gratified to see that the students in the special program improved in comparison with the rest of the students.

A few years ago a school in New Jersey tested all its fourth graders to select students for a program for the gifted. Two years later the students were retested, and the school was shocked to find that the scores of the gifted group had dropped in comparison with the rest of the students.

Question

Can we conclude that the remedial program worked and that the program for the gifted was detrimental?

Objective

After completing this activity, you will understand the **regression effect** (sometimes called **regression toward the mean**) and know how to recognize where it might occur.

Activity

1. Take the following test, individually.

MONEY TEST

This is a closed wallet, multiple-choice test. If you are not sure of an answer, make your best guess. Don't leave any questions blank.

I. On the back of a nickel is
   a. Monticello
   b. the Jefferson Memorial

II. On the back of a $2 bill is
   a. signers of the declaration
   b. Independence Hall

III. On the front of a $500 bill is
   a. Madison
   b. McKinley

Your instructor will provide the answers.

2. After you have been told to do so, take the make-up test.

MONEY MAKE-UP TEST

This is a closed wallet, multiple-choice test. If you are not sure of an answer, make your best guess. Don't leave any questions blank.

I. On the front of a $20 bill is
   a. Jefferson
   b. Jackson

II. On a dollar bill, Washington is looking
   a. to his left
   b. to his right

III. On the front of a $1,000 bill is
   a. Cleveland
   b. Wilson

Your instructor will supply the answers.

3. What was the average score on the first test for the "remedial" students in your class (those students who either missed all the problems or missed most of them)? What was the average score on the second test for the "remedial" students?

4. What average score did the "star" students (those who got all three questions right on the money test) get on the make-up money test?

5. Explain why the scores of the remedial students tended to go up and the scores of the star students tended to go down. This phenomenon is called the regression effect, or regression toward the mean.

6. The following scatter plot shows the heights of 1,078 fathers and their grown sons in about 1900 (see Ref. 1). The average height of all the sons is 69 inches, and the average height of all the fathers is 68 inches. The shape of this scatter plot is typical of two positively correlated variables. Taller fathers tend to have taller sons, and shorter fathers tend to have shorter sons, but we cannot predict with certainty a son's height if we know his father's height. Even if we knew the mother's height, we still could not predict the son's height exactly. Some element of chance is involved.

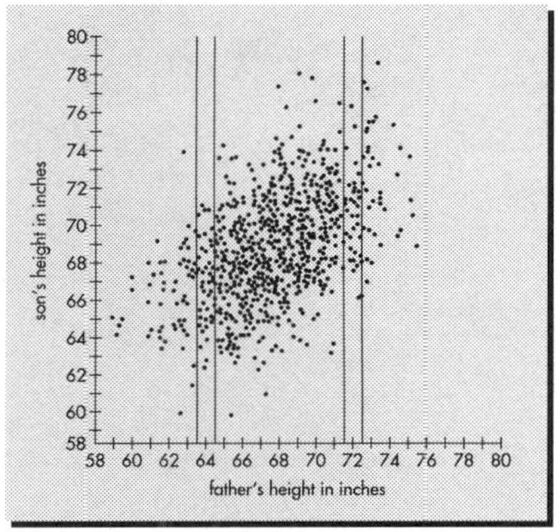

a. Use the vertical strip on the right to estimate the average height of the sons of 72-inch fathers. Place an X at this point on the graph.

b. Use the vertical strip on the left to estimate the average height of the sons of 64-inch fathers. Place an X at this point on the graph.
c. Connect the two X's with a line. This line is (approximately) the regression line.
d. Use the regression line to estimate the average height of the sons of 68-inch fathers.
e. Explain how the father and son data exhibit the regression effect.

Wrap-Up

1. How does the regression effect apply to the two situations in the Scenario?

2. Explain why "regression toward the mean" is a logical name.

Extensions

1. Your instructor will supply the scores on the first and second tests in a college statistics class. Make a scatter plot of these scores. Does this scatter plot exhibit perfect correlation? Identify the points for the students who did well on the first test and for the students who did poorly. What happened to these students on the second test compared with the other students?

2. Find some data to illustrate that stocks on the New York Stock Exchange exhibit the regression effect. Does this mean that a person should avoid buying the stocks that did well last year?

References

1. David Freedman et al. (1991), *Statistics*, second ed., New York: Norton, pp. 159–165.
2. Ann E. Watkins (1986), "The regression effect; or, I always thought that the rich get richer . . . ," *The Mathematics Teacher*, **79**:644–647.

Is Your Shirt Size Related To Your Shoe Size?

NOTES FOR THE INSTRUCTOR

Statistical Setting

Designed as an introductory activity in fitting simple linear models to data, this lesson should come early in the discussion of modeling. The data and the scatter plots are usually easy to interpret. The slopes of the regression lines are close to 1 for the relationships that work well, and the measurement scale is the same on both axes, giving students an easy slope to interpret as they are beginning their study of this important idea.

Prerequisite for Students

Students should have some familiarity with scatter plots and the equation of a straight line.

Materials

Students will need the article providing some details of Leonardo's view of the human body (attached to this activity), tape measures, and a computer or calculator capable of fitting a straight line by the least-square method.

Procedure

Students should work in small groups to decide what to measure, how to collect the data, and how to analyze the results. Then, they should combine groups for the data collection phase so that at least 15 measurements can be obtained on each pair of variables investigated.

Other issues in the design of the study should be discussed, such as the gender issue and the problem of measurement error. On the first issue, students may suggest doing separate studies on males and females, and then comparing the results. This is actually a good way to design the study. On the second issue, students might suggest making each measurement more than once on each person and using the average in the regression modeling. This is a reasonable approach if the goal is to determine whether or not Leonardo's propositions hold. It is not a good approach if measurement error is to be studied in the process of this investigation. Students might try fitting models both ways and comparing the results.

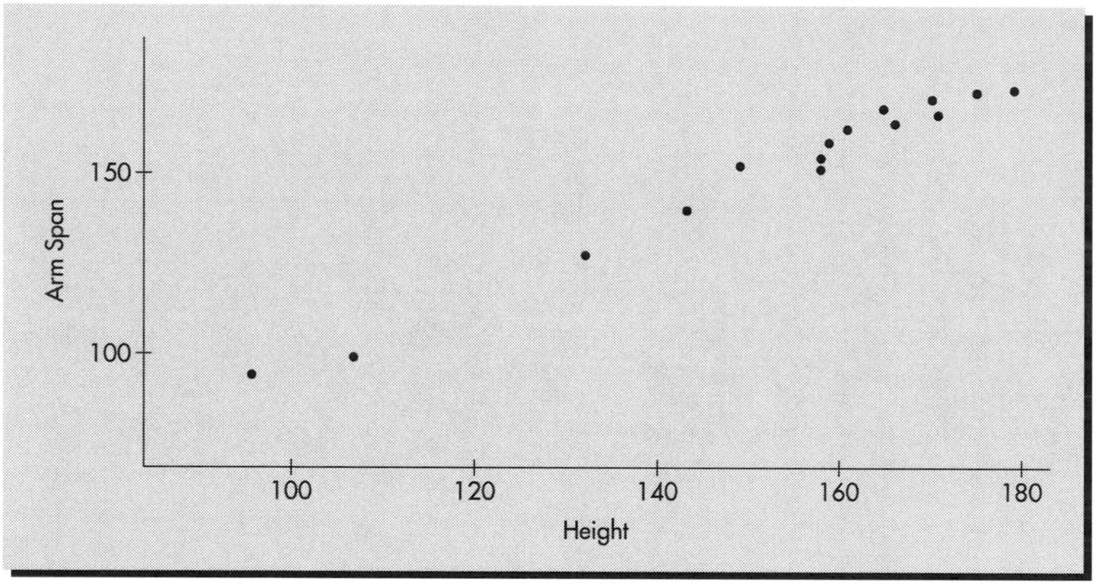

The regression equation is
ArmSpan = −5.60 + 1.03 Height

Predictor	Coef	Stdev	t-ratio	p
Constant	−5.602	5.528	−1.01	0.329
Height	1.03323	0.03582	28.84	0.000

s = 3.243 R-sq = 98.5% R-sq(adj) = 98.3%

Analysis of Variance

SOURCE	DF	SS	MS	F	p
Regression	1	8747.8	8747.8	831.99	0.000
Error	13	136.7	10.5		
Total	14	8884.5			

Figure 1

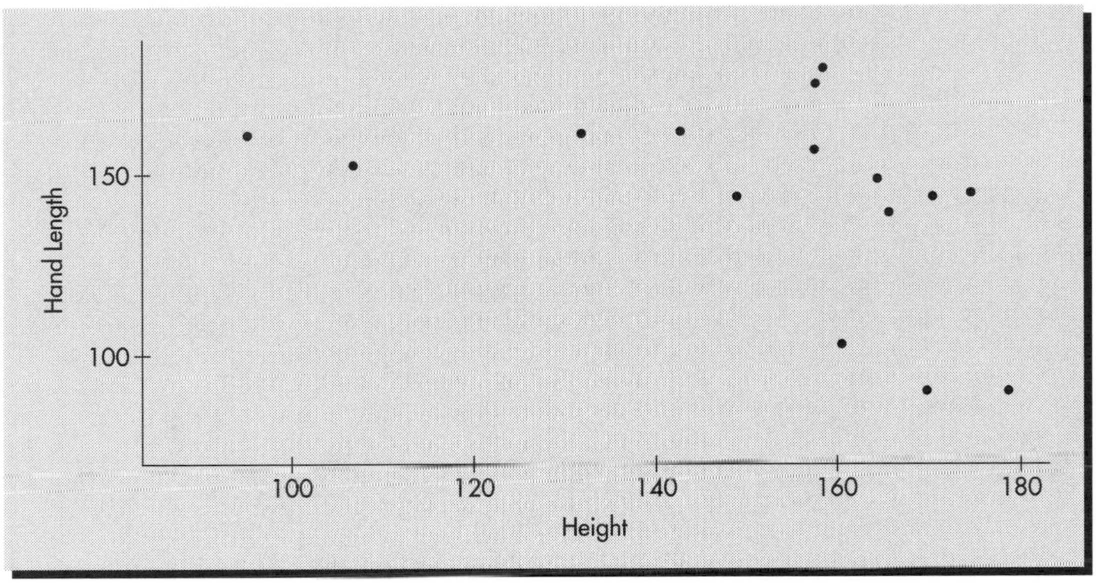

Figure 2

Sample Results from Activity

The computer printouts in Figures 1 and 2 show the results of student investigations of the arm span versus height relationship and the hand length versus height relationship, respectively. The first shows a strong, positive trend with an estimated slope close to 1. The second shows no strong linear trend, although there is a slight downward tendency. Students should discuss why the second relationship shows up as being so weak, or even misleading.

If appropriate for the level of the class, students could be asked to fit the first model without an intercept term (forcing the line to go through the origin). The estimated slope is then very close to 1.

Sample Assessment Questions

1. The plot and statistical summaries for paired data on kneeling height versus height from a sample of 15 students is shown in Figure 3.
   a. Describe the pattern in the scatter plot.
   b. Interpret the slope of the fitted regression line in terms of these measurements.
   c. Would you say that the line fits the data well? Explain.

2. Students are writing a report on the accuracy of Leonardo's conjectures. The report stresses where Leonardo seems to be on target and where he seems to be in error. Speculate as to why Leonardo's conjectures might be way off for some of the relationships investigated by these students.

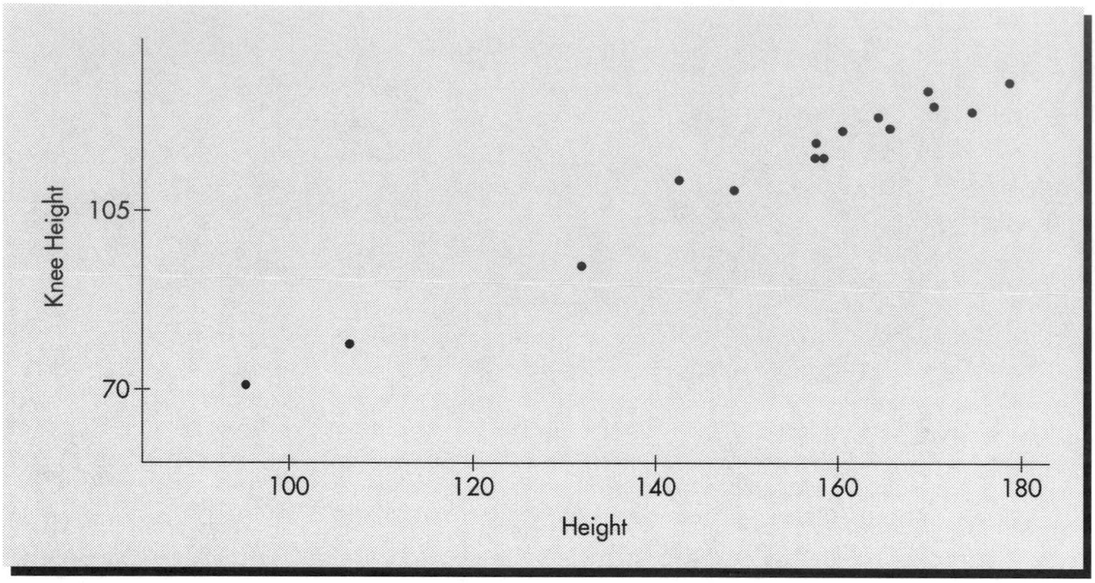

The regression equation is
KneeHt = 2.32 + 0.726 Height

Predictor	Coef	Stdev	t-ratio	p
Constant	2.325	4.493	0.52	0.614
Height	0.72624	0.02912	24.94	0.000

s = 2.636 R-sq = 98.0% R-sq(adj) = 97.8%

Analysis of Variance

SOURCE	DF	SS	MS	F	p
Regression	1	4321.8	4321.8	622.16	0.000
Error	13	90.3	6.9		
Total	14	4412.1			

Figure 3

Reference

Martin Kemp, Editor, *Leonardo on Painting*, New Haven: Yale University Press, 1989.

Is Your Shirt Size Related to Your Shoe Size?

SCENARIO

The great painter and scientist of the fifteenth century, Leonardo, studied the human body and provided detailed accounts of the relationships among various body parts. Leonardo's work on how the size of one part is related to that of another was for the purpose of instructing painters on how to paint the human body, but some of these relationships are of practical interest as well. (Who might be interested in the relationship between arm length and circumference of the neck, for example?) In this activity we will investigate the accuracy of some of Leonardo's pronouncements by modeling relationships between the sizes of various body parts.

Question

How can we represent the relationship between two variables by means of a mathematical equation?

Objective

In this lesson you will examine the relationship between two variables by fitting a least-squares regression model to paired data and interpreting the slope of the resulting straight line.

Activity

1. Read the attached article on Leonardo's view of the human body.

2. Out of the many relationships mentioned in the article, choose at least two for investigation. For example, you might study the relationship between height and arm span or between hand length and arm length, among others. Clearly state the problem under investigation.

3. List all of the variables necessary to investigate the relationships decided on in the problem statement. Decide how to measure these parts on any one person.

4. Select a group of people (at least 15) on which to make the measurements. Then, collect the necessary data.

5. Construct scatter plots to compare the variables in all of the relationships under study. For the examples cited above, one scatter plot would display height versus arm span and another would display hand length versus arm length. Describe any patterns you see in the data.

6. If appropriate, find the equation of a simple linear regression model that relates one of the paired variables to the other. Draw the line on the scatter plot, and comment on how well it fits.

7. Interpret the data by looking carefully at the two estimated parameters of each model, the slope and the intercept. How do the slopes relate to Leonardo's conjectures? What should happen to the intercepts in most of these models?

Wrap-Up

1. Once a model has been fit to data from one group of people and analysis has suggested that it fits well, verify that the model works on another small group of people.

2. Plan and carry out another investigation of this type. If the result of step 1 shows some problems with the model, plan a new investigation to see if the problems can be surmounted. If the verification shows the model to be working, specify another relationship that might be investigated.

Extensions

1. Investigate what happens to the models (linear equations) if the roles of X and Y are reversed. For example, if arm length (X) is used to predict hand length (Y), what will happen to the equation if hand length (now X) is used to predict arm

length (now Y)? Will a similar result hold when interchanging arm span and height?

2. Investigate the linear models more carefully by plotting the residuals and studying any patterns that might emerge there. Do the residuals suggest that a more complex model should be investigated?

3. Calculate correlation coefficients for each of the scatter plots investigated above. Interpret these measures of the strength of the linear relationships.

Leonardo's View of the Human Body

(Source: Martin Kemp, ed., (1989), *Leonardo on Painting*, New Haven, CT: Yale University Press.)

ON THE MEASUREMENTS OF THE HUMAN BODY

Vitruvius, the architect, has it in his work on architecture that the measurements of man are arranged by nature in the following manner: four fingers make one palm and four palms make one foot; six palms make a cubit; four cubits make a man, and four cubits make one pace; and twenty-four palms make a man; and these measures are those of his buildings.

If you open your legs so that you lower your head by one-fourteenth of your height, and open and raise your arms so that with your longest fingers you touch the level of the top of your head, you should know that the central point between the extremities of the outstretched limbs will be the navel, and the space which is described by the legs makes an equilateral triangle.

The span to which the man opens his arms is equivalent to his height.

From the start of the hair [i.e., the hairline] to the margin of the bottom of the chin is a tenth of the height of the man; from the bottom of the chin to the top of the head is an eighth of the height of the man; from the top of the breast to the top of the head is a sixth of the man; from the top of the breast to the start of the hair is a seventh part of the whole man; from the nipples to the top of the head is a quarter part of the man; the widest distance across the shoulders contains in itself a quarter part of the man; from the elbow to the tip of the hand will be a fifth part of the man; from this elbow to the edge of the shoulder is an eighth part of this man; the whole hand is a tenth part of the man.

If a man of two *braccia* is small, one of four is too large—the mean being the most praiseworthy. Half-way between two and four comes three. Therefore take a man of three *braccia* and determine his measurements with the rule I give you. And if you should say to me that you might make a mistake and judge someone to be well proportioned who is not, I reply on this point that you must look at many men of three *braccia*, of whom the great majority have limbs in conformity

with each other. From one of the most graceful of these take your measurements. The length of the hand is a third of a *braccio* and goes nine times into the man; and correspondingly the face, and from the pit of the throat to the shoulder, and from the shoulder to the nipple, and from one nipple to the other, and from each nipple to the pit of the throat.

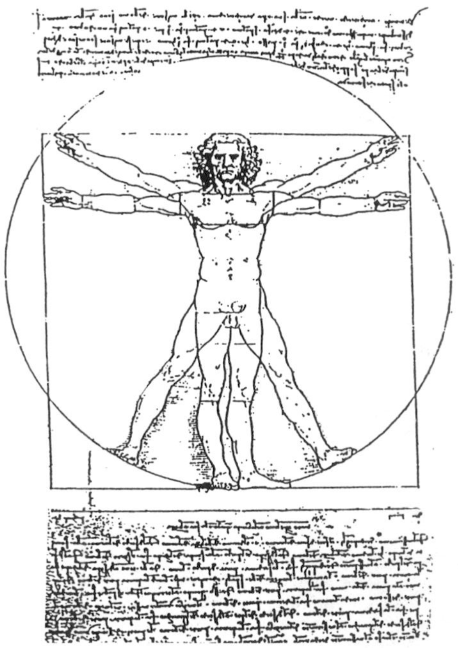

Study of human proportions in the manner of Vitruvius, Venice, Accademia.

The space between the slit of the mouth and the base of the nose is one-seventh of the face.

The space from the mouth to below the chin, *cd*, will be a quarter part of the face, and similar to the width of the mouth.

210 ACTIVITY-BASED STATISTICS GUIDE

The space between the chin and below the base of the nose, *ef*, will be a third part of the face, and similar to the nose and the forehead.

The space between the midpoint of the nose and below the chin, *gh*, will be half the face.

The space between the upper origin of the nose, where the eyebrows arise, *ik*, to below the chin will be two-thirds of the face.

The space between the slit of the mouth and above the beginning of the upper part of the chin, that is to say, where the chin ends at its boundary with the lower lip, will be a third part of the space between the parting of the lips and below the chin, and is a twelfth part of the face. From above to below the chin, *mn*, will be a sixth part of the face, and will be a fifty-fourth part of the man.

From the furthest projection of the chin to the throat, *op*, will be similar to the space between the mouth and below the chin and is a quarter part of the face.

The space from above the throat to its base below, *qr*, will be half the face and the eighteenth part of the man.

From the chin to behind the neck, *st*, is the same as the space between the mouth and the start of the hair, that is to say three-quarters of the head.

From the chin to the jawbone, *vx*, is half the head, and similar to the width of the neck from the side.

The thickness of the neck goes one and three-quarter times into the space from the eyebrows to the nape of the neck.

The distance between the centres of the pupils of the eyes is one-third of the face.

The space between the edges of the eyes towards the ears, that is to say, where the eye ends in the socket which contains it (at its outer corners), will be half the face.

The greatest width which the face has at the level of the eyes will be equivalent to that between the line of the hair at the front and the slit of the mouth.

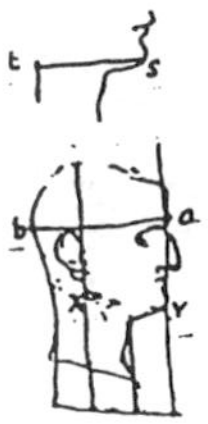

The nose will make two squares, that is to say, the width of the nose at the nostrils goes twice into the length between the tip of the nose and the start of the eyebrows; and , similarly, in profile the distance between the extreme edge of the nostril—where it joins the cheek—and the tip of the nose is the same size as between one nostril and the other seen from the front. If you divide the whole of the length of the nose into four equal parts, that is to say, from the tip to where it joins the eyebrows, you will find that one of these parts fits into the space from above the nostrils to below the tip of the nose, and the upper part fits into the space between the tear duct in the

Proportions of the face in profile, based on Windsor, RL 12304r.

IS YOUR SHIRT SIZE RELATED TO YOUR SHOE SIZE? 211

Proportions of the face from the front, based on Windsor, RL 19129r.

inner corner of the eye and the point where the eyebrows begin; and the two middle parts are of a size equivalent to the eye from the inner to the outer corner.

The hand to the point at which it joins with the bone of the arm goes four times into the space between the tip of the longest finger and the joint at the shoulder.

Ab goes four times into *ac*, and nine times into *am*. The greatest thickness of the arm between the elbow and the hand goes six times into *am*, and is similar to *rf*. The thickest part of the arm between the shoulder and the elbow goes four times into *cm*, and is similar to *hng*.

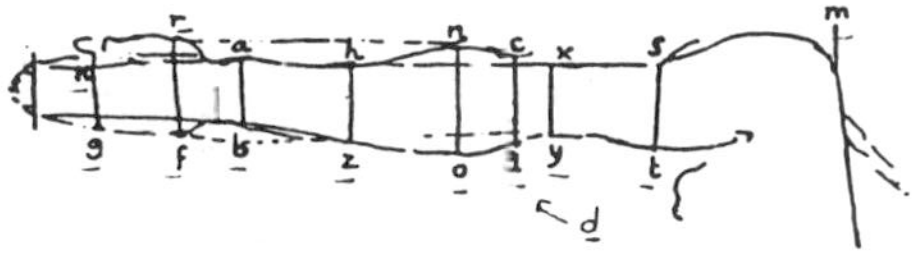

Proportions of the arm, based on Windsor, RL 19134r.

The least thick part of the arm above the elbow, *xy*, is not the base of a square, but is similar to half of the space *hz*, which is found between the inner joint of the arms and the wrist. The width of the arm at the wrist goes twelve times into the whole of the arm, that is to say, from the tip of the fingers to the shoulder joint, that is to say, three times into the hand and nine times into the arm.

If a man kneels down he will lose a quarter of his height. When a man kneels down with his hands in front of his breasts, the navel will be at the midpoint of his height and likewise the points of his elbows.

Relating to Correlation

NOTES FOR THE INSTRUCTOR

Statistical Setting

The objective of this activity is to help students understand what a correlation coefficient measures. Often sample correlation coefficients are calculated for samples as small as 5, without realizing that this estimate could be totally misleading. This activity illustrates how the sample estimate can differ from the population correlation coefficient and how this difference depends on the underlying value of the population correlation coefficient. This lesson can be used after students have been introduced to the concept of a correlation coefficient.

Prerequisite for Students

Students should have been introduced to the concept of a correlation coefficient. Since this activity is based on a computer simulation using MINITAB, students should be familiar with MINITAB, saving output and printing the output.

Procedure

The activity can be an individual or a group activity. Although it specifies the values of the sample size and the population correlation coefficient, you may let the students select their own values. You may want the students to create the MINITAB macro first and store in their accounts(disks). Students may need to be reminded that for each simulation, they have to specify the sample size n, the population correlation coefficient, ρ, and the number of simulated sample correlation coefficients. Note that in the activity, 20 sample correlation coefficients are simulated for each value of n and ρ.

Sample Results from the Activity

Two sets of dotplots, one for each sample size are attached. Each set shows 20 sample correlation coefficients for population correlation coefficients of .1, .25, .5, .75 and .9.

1. Analyzing the dotplots for variability one can see quite clearly that the smaller the underlying population correlation coefficients, the more variable the sample correlation coefficients are likely to be.
2. This variability decreases as the sample size increases.
3. A positive population correlation coefficient indicates that y tends to increase with x.
4. For samples as small as 5, we observe that the sample correlation coefficients can be negative even though the population correlation coefficient is positive when it is as small as .1.

Sample Assessment Questions

1. A newspaper reports that the correlation between two variables is .7 and that the result is a highly significant finding. It goes on to say that correlation is based on 10 data points. Do you agree with the newspaper's interpretation? Explain.

2. How would your answer to question 1 change if the sample size were known to be 500 instead of 10?

3. Is it possible to have a negative sample correlation when, in fact, two variables are known to be positively correlated? Explain.

Sample Results from the Extensions

The scatter plots below show the relationship between the paired sample observations when $\rho = .1$ and $\rho = .9$. When $\rho = .9$ the sample data is clustered around a straight line, but not as tightly clustered as many students might think. The pattern is quite disperse for the small value of ρ.

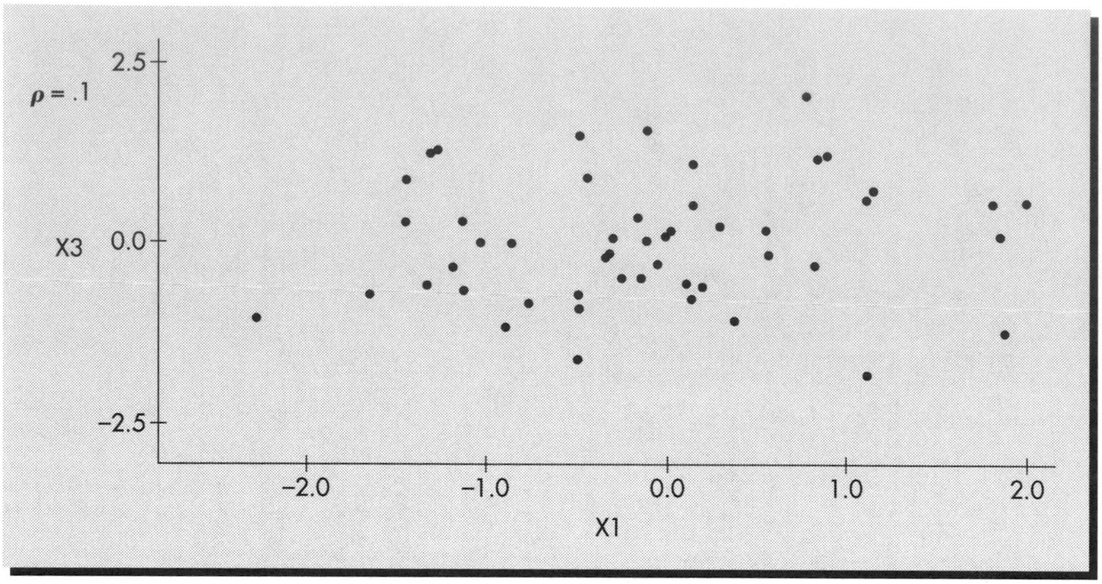
ρ = .1
2.5
X3 0.0
−2.5
−2.0
−1.0
0.0
1.0
2.0
X1

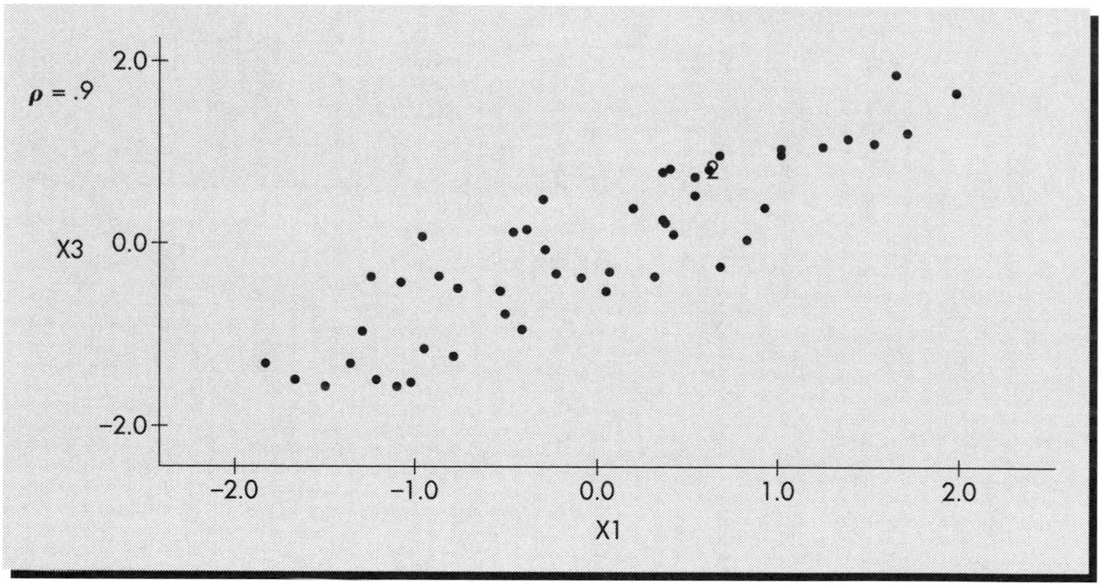
ρ = .9
2.0
X3 0.0
−2.0
−2.0
−1.0
0.0
1.0
2.0
X1

The use of SIMCOR can be extended to a study of regression coefficients. The stem plots below show distributions of regression and correlation coefficients for $\rho = 0$ and $\rho = .8$, each with sample sizes of ten.

```
Stem-and-leaf of Regress
Leaf Unit = 0.010
   2    -5 31
   5    -4 210
   8    -3 331
  14    -2 971100
  20    -1 994211
  20    -0 65
  18     0 034558
  12     1 038999
   6     2 57
   4     3 3
   3     4 28
   1     5 8
```

```
Stem-and-leaf of Regress
Leaf Unit = 0.010
   1     1 6
   1     2
   2     3 0
   3     4 4
  12     5 022357789
 (9)     6 000346789
  19     7 003359
  13     8 011247
   7     9 257
   4    10 488

   1    11 5
```

```
Stem-and-leaf of Corr
Leaf Unit = 0.010
   1    -7 4
   1    -6
   2    -5 5
   3    -4 7
   8    -3 93000
  14    -2 885443
  20    -1 998863
  20    -0 97
  18     0 1445778
  11     1 33
   9     2 45
   7     3 023
   4     4 145
   1     5
   1     6 7
```

```
Stem-and-leaf of Corr
Leaf Unit = 0.010
   1     1 4
   2     2 5
   2     3
   2     4
   8     5 347789
  13     6 23555
(11)     7 01113467889

  16     8 02234555788889
   2     9 12
```

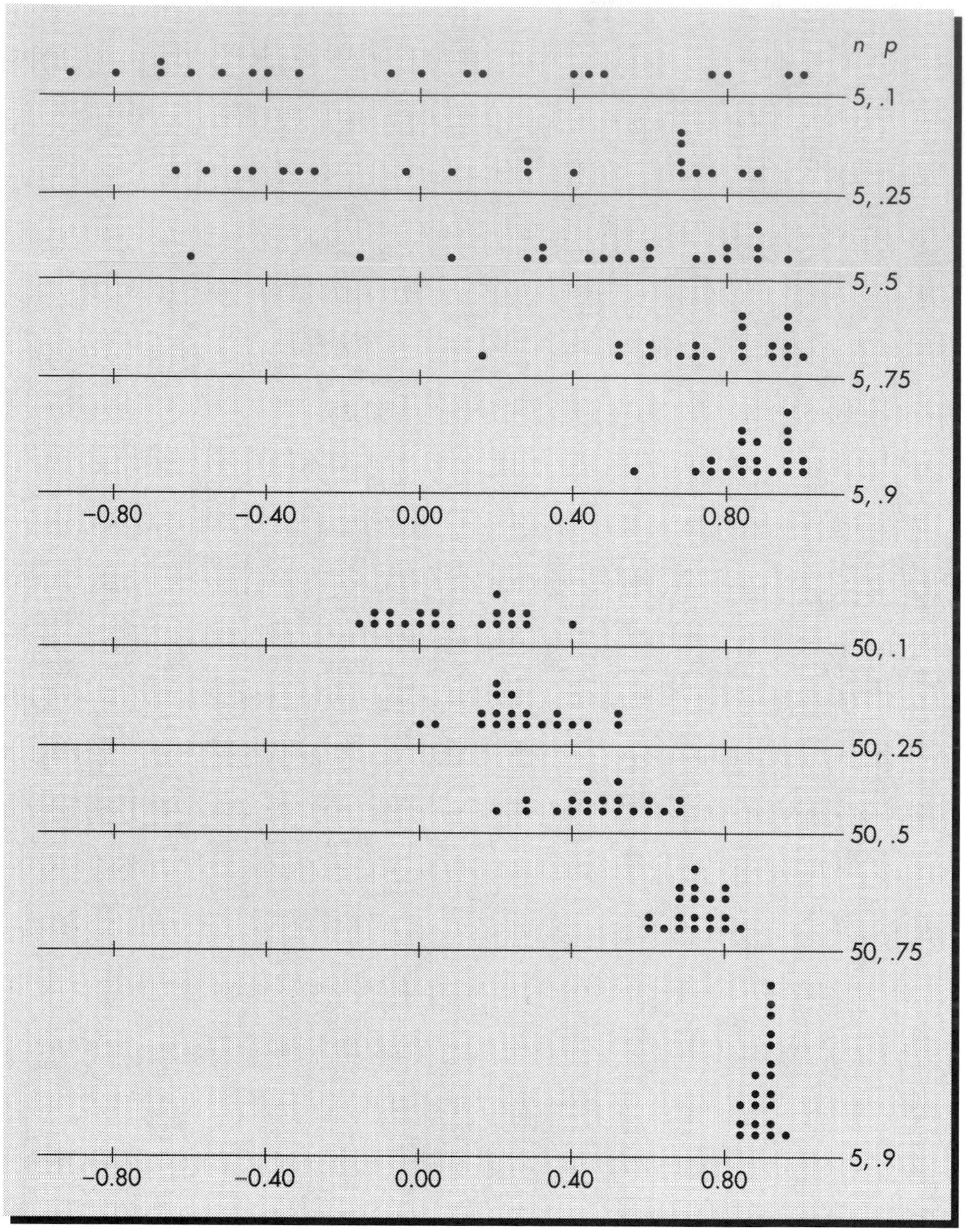

Figure 1. Dot plots of simulated sample correlation coefficients for n = 5 and n = 50.

Relating to Correlation

SCENARIO

Is your high school GPA a good indicator of your college GPA? Is a high cholesterol level related to high fat intake? Are your grades related to the number of hours you study? People are often interested in measuring the relationships between two variables. One of the most popular statistics that is used is the correlation coefficient, which measures how well the relationship between the two variables can be described by a straight line. Since a sample is used to estimate the correlation coefficient, can we find out how this estimate behaves for different values of the sample size and the population correlation coefficient?

Question

How "good" is the sample correlation coefficient as an estimate of the population correlation coefficient?

Objective

The objective of this activity is to see how the values of sample correlation coefficients depend on the sample size that is used and on the value of the population correlation coefficient.

Activity

This activity is based on a computer simulation of the sample correlation coefficients. You will generate samples of two different sizes from populations having specified correlation coefficients. The simulation is described using Minitab.

GENERATING THE RANDOM SAMPLES

1. Select the sample size, n, to be used to generate the sample correlation coefficient, r. (We suggest $n = 5$ as the starting value for n.)

2. Select the population correlation coefficient, ρ. (Use $\rho = .1$ to start the simulation.)

3. The correlation coefficient measures the linear association between two variables. So the next step is to generate the random samples of two variables, $x1$ and $x3$ that you will use to calculate the sample correlation coefficient. Your random samples will be from two normal distributions having a population correlation, ρ, that you specify. You can do that as follows:
   a. Generate two independent random samples of n observations from a **standard normal population**, and store them in two columns C1 and C2. Denote the two variables as $x1$ and $x2$.
   b. Combine the observations on $x1$ and $x2$ to get a third random variable, $x3$, such that the correlation coefficient between $x1$ and $x3$ is the prespecified ρ using the equation shown below. The variable $x3$ is stored in C3, and is given by

$$x3 = \rho \times x1 + \sqrt{1 - \rho^2} \times x2.$$

You will use observations on $x1$ and $x3$ to generate the correlation coefficient.

USING MINITAB

The following instructions show how to use a Minitab macro to do the work of generating correlation coefficients from normally distributed variables. In order to run this macro, you must store it in your computer under the name SIMCOR and learn how to execute a macro on the version of Minitab you are using. The commands for the program SIMCOR are listed below.

```
#
NOECHO
# NOTE Set k1 equal to the sample size.
# NOTE Set k2 equal to the desired correlation.
# NOTE Set k3=1.
RANDOM k1 C1 C2;
```

```
Normal 0 1.
Let C3=K2*C1+SQRT(1-K2**2)*C2
BRIEF 0
REGRESSION C3 1 C1;
COEFFICIENTS C4.
LET K4 = C4(2)
LET C5(K3)=K4
LET K5=STDEV(C1)
LET K6=STDEV(C3)
LET K7=K4*K5/K6
LET C6(K3)=K7
LET K3=K3+1
end
BRIEF 2
ECHO
```

Note that k1 is the sample size and k2 is the desired value of the correlation coefficient; both must be specified before the program will run. Note also that k3 must be set equal to 1 at the start of the session. The correlation coefficients from the random samples are stored in C6, while the regression coefficients are stored in C5.

Now, it is time to generate and study sample correlation coefficients.

1. Set up your computer to run SIMCOR.
2. Specify the sample size at $n = 5$ and the true correlation coefficient at $\rho = .1$.
3. Run the macro 20 times. This generates 20 sample correlation coefficients in C6.
4. Make a dot plot of the 20 correlations. Comment on the pattern you see.
5. Repeat steps 3 and 4 for $n = 5$ and $\rho = .25, .50, .75$ and $.90$. Construct parallel dot plots of the five sets of sample correlations. Comment on the pattern you see.
6. Repeat steps 3, 4 and 5 for $n = 50$. How does the larger sample size affect the pattern of sample correlation coefficients?

ANALYZING THE RESULTS

1. When are the sample coefficients more variable, that is, more likely to be further from the population correlation?
2. What happens to this variability when we increase the sample size?

3. What does the sign of a population correlation coefficient indicate about how y changes as x changes?
4. If the population correlation coefficient is positive, is it likely that we can get negative sample correlation coefficients? Discuss when can that happen using the dot plots.

Wrap-Up

1. The sample correlation coefficients that you generated can be considered to be estimates of the corresponding population coefficients. How "good" are these estimates for small sample sizes? Do they improve as the sample size increases?
2. Did the answer to question 1 depend on the population correlation coefficient?
3. Write a short summary of the results of your simulation, and suggest a "rule of thumb" to get a reliable estimate of ρ.

Extensions

1. Using the instructions in SIMCOR, generate random samples of the variables $x1$ and $x3$ for $n = 50$. Note that the variables $x1$ and $x3$ have a population correlation coefficient ρ, and you have to specify the value of ρ. Start with $\rho = .1$. Plot $x1$ versus $x3$ in a scatter plot using the instructions PLOT C1 C3
2. Repeat step 1 for $\rho = .5$ and for $\rho = .9$.
3. Repeat steps 1 and 2 but change the sample size, n, to 500.
4. Sketch the $y = x$ line on the plots.
5. Compare the pattern of points in the plots around the $y = x$ line as the population correlation coefficient decreased from .9 to .1.
6. Using the plots from 3, explain why you think it is possible to get a sample of five observations far from the $x = y$ line when the population correlation coefficient is small (e.g., .1).
7. Does this possibility change when the population correlation coefficient increases?

Assessment

Assessment and Activity-Based Statistics: Guidelines for Instructors

These brief notes suggest some ways instructors may integrate assessment of student learning with activities in this book. We begin by defining assessment and looking at current trends in educational assessment, and then we offer suggestions for specific assessment approaches to be used with Activity-Based Statistics (ABS).

What Is Assessment?

The term "assessment" means different things to different people. When most people hear the word "assessment," they think of grades. Assessment is most often referred to as documenting student learning and performance, usually by administering tests and assigning grades. However, assessment may also be viewed in a broader context as a way to provide students with feedback about the quality of their work so that they may improve their learning and develop standards consistent with those of the instructor, and also as a way to inform the teacher about changes needed to improve instruction.

Different forms of assessments may be used to help teachers become more aware of how well their students are learning statistics. In addition to traditional assessments like quizzes and exams, alternative assessments may be used to determine if students have mastered a topic or if more instruction is needed, or to determine if particular concepts or applications are problematic for several students who may need additional work on these topics. Instructors experimenting with new ways to teach statistics, such as using activities in this book, may be interested in using alternative forms of assessment to find out how well the group-learning format is working or whether students feel that they are learning important statistical ideas by participating in these activities. This information may be useful in helping determine which activities to use in future classes, what changes are needed in groups or group instructions, or how students are reacting to the use of ABS materials and techniques.

The educational assessment literature recommends the use of multiple assessment measures in order to gather more complete information about student learning, rather than relying exclusively on single methods such as a midterm and a final exam. Instructors using multiple methods may select from a variety of different formats in addition to using traditional methods such as quizzes and exams. These alternative methods include brief "minute" papers to find out what students understand well or find confusing, write-ups of results and conclusions from group activities, student projects or reports, journal entries, students' self-evaluations, or peer evaluations. The use of alternative assessment methods with ABS materials are detailed next.

Assessment Activities and ABS

Assessment methods may easily be incorporated into ABS in several different ways. Each activity includes suggestions for assessment activities based on that activity, either to be done in class that day, conducted outside of class, or used later on a quiz or exam. These assessments may appear in the following forms: wrap-up questions; extensions; group discussions; sample questions; projects.

WRAP-UPS

Each activity has a "Wrap-Up" section that often asks students to summarize what they have learned, sometimes in the context of explaining the main concepts to a classmate who is absent that day. Students' written responses to one or more of the Wrap-Up questions provide a quick way to focus on whether or not students have learned the main idea of the activity. As an alternative, an instructor may ask students to write a brief summary of what they learned about the main concept involved in a particular activity, such as the idea of a random sample, sampling distributions, or measurement bias.

EXTENSIONS

Each activity provides one or more "extensions" in the form of parallel or follow-up activities. These may be done in or outside of class and may include written reports of what students have learned. The instructor may quickly scan these write-ups to determine if students have successfully transferred their learning to a related but different context.

GROUP DISCUSSIONS

Instructors may choose to have small groups of students discuss their conclusions regarding a particular activity and provide a brief oral report to the class. This informal group reporting may reveal to the instructor whether or not students have achieved the desired learning goals for the activity or if more instruction is needed on that topic. Each group may instead be asked to provide a short written report, which the teacher may read for the same purpose.

SAMPLE QUESTIONS

Each activity includes one or more sample assessment questions. These questions may be used as a part of a homework assignment or as items on a quiz or exam. They may help the instructor determine if students have mastered the main goals of the activity.

PROJECTS

In addition to assessing student learning in the context of individual activities, instructors may want to use more in-depth assessments of student learning on completion of an entire unit or set of activities. Individual or group projects provide students with an authentic experience in applying their learning to real problems. Typical projects require students to think of a problem, outline their plans for collecting data to solve the problem, then complete the project, and write up the results. Projects may be used during a course (e.g., following the unit on exploring data or estimation) or at the end of a course, to document student learning of course objectives.

To ensure high-quality student projects, it is recommended that teachers monitor the project from the beginning, to make sure that the problem is feasible, to provide feedback to students as they are working on the project, to provide the scoring criteria to be used in evaluating projects, and to provide examples of exemplary projects completed by former students. Scoring rubrics may be developed to reflect what an instructor hopes to see in an excellent project. Components of a project evaluation form may include selection of appropriate statistical methods, correct analysis of data, ability to interpret and evaluate information, effective communication, etc., with a certain number of points (e.g., 0 to 3) assigned to each component.

Providing Feedback to Students

A major reason to collect assessment information is to provide evaluative feedback to students on the quality of their learning. Reading students' summaries of what they learned about the central limit theorem may reveal that students really don't understand this theorem at all and have some misunderstandings that need to be addressed. Student write-ups may also reveal errors in using statistical terms or may indicate that related concepts are not clearly understood. The instructor may provide feedback on these errors, misunderstandings, or gaps in learning to alert students to areas needing either more study or help from the teacher or teaching assistant.

HELPING STUDENTS DEVELOP APPROPRIATE STANDARDS

An important way assessment can be used to improve learning is to help students develop standards comparable to those of the instructor. It is always a joy to read and grade a high-quality student paper or report, and it is much harder to grade a report full of errors or incomplete information. In order to improve student performance on assessment tasks, students need to know what criteria are being used to evaluate their work, and they can greatly benefit from seeing examples of what an instructor considers to be an excellent product.

As activities are repeatedly used from class to class, instructors may want to collect samples of good and bad write-ups to share with future students. Seeing these examples may help students develop appropriate standards of performance so they know what is considered a high-quality summary or write-up. These samples may be kept in a notebook in class that students may refer to, with comments about what was good or bad about each student paper. (Of course, all identifying information should first be removed.)

Another effective way to help students develop standards of good performance is to collect portions of some student write-ups (e.g., summaries of what was learned); cut, paste, and copy them; and then distribute copies for students to read/discuss/critique in groups. Students may be asked to determine what is missing, what is good, or what is an error in these examples, which may help them better critique and improve their own individual or group work.

A final recommendation is to offer students opportunities to evaluate their own work, using the criteria developed by the instructor. This technique is particularly effective before submitting a report or project, where the student completes a self-assessment that is later compared to the instructor's assessment, using the same form.

INFORMALLY ASSESSING STUDENT LEARNING IN GROUPS

As students work on activities in groups, they will at different times discuss directions, the activity's outcomes, questions they have, and what they are learning. It can be quite informative for an instructor to circulate among the groups and listen for particular indicators of successful or unsuccessful student learning or interaction. Some indicators to observe include

a. the use (correct or incorrect) of the statistical terms involved,

b. indicators that the activity is progressing successfully and students are following directions,

c. indicators that students have misunderstood part of instructions or the results and therefore are deriving erroneous conclusions,

d. problems in the group dynamics, where all members are not contributing equally.

By being attuned to such possibilities, this informal assessment method may help the teacher pose questions to a group that help steer them on course, address errors before it is too late, and keep the activities running smoothly.

COURSE GRADES

Most students will engage in activities and write up their results more seriously if they know that these write-ups are evaluated and contribute to their course grade. Teachers who assign points to completed activities may evaluate student write-ups for an activity with particular objectives in mind and assign a score from 0 to 3 for each activity (where 3 means they learned it, 2 means they almost learned it, 1 means they learned a little, and 0 means they didn't learn at all).

For more complex activities with longer write-ups, checklists may be used to assign points for different parts of the activity (e.g., directions were followed correctly, graphs are correct, calculations are correct, summary is complete, etc.). This detailed feedback may help students quickly determine what they did right and what they did incorrectly.

Evaluating Student Reactions to ABS

When trying a new approach to teaching statistics, it is often helpful to solicit student reactions to the new approach. Instructors may be interested in determining how well students like using the ABS materials and how well they feel they are learning from this instructional method as opposed to traditional methods. Instructors may also be interested in obtaining feedback on the use of student learning groups or on the success of individual activities. Brief minute papers may be used to ask students questions such as the following:

> "How did you like this activity?"
>
> "What did you think you learned from this activity?"
>
> "What was most confusing?" or
>
> "How well do you think this activity helped you better understand the idea of __________?"

Students are often more forthcoming when writing an anonymous "minute" paper than if asked these same questions orally as a group. The papers take very little time to read and can be very illuminating about the success of the particular activities.

PROJECTS

Many instructors of introductory statistics have found that student projects have great educational merit and have used them quite successfully to enhance motivation, interest, and understanding in the course. Projects might be defined as activities that require planning and execution of a series of steps, over a period of time usually exceeding a month, to solve a particular problem or answer a specific question. Among the advantages of projects are the following:

- Students work on a complete problem, from formulating the question to investigating through data gathering and analysis, to reporting conclusions.
- Students bring a variety of statistical techniques to bear on one problem and thus develop connections among the topics.
- Students gain experience in using statistics in a realistic way.
- Students gain experience in group work, as projects are often conducted with partners or teams.
- Students gain experience in communicating statistical ideas.

In a one-semester or one-quarter introductory course, it is difficult to allow projects that are completely open ended. Neither the instructor nor the student may ever see the finished product. Thus, it is helpful to focus projects a little, concentrating on the main themes that should be covered in introductory statistics. These themes include data exploration, quality improvement, sample surveys, experiments, and modeling. The first two fit together quite nicely since much of quality-improvement work in industry involves careful exploration of data. Thus, four main themes for projects are suggested: exploring data and improving quality; sample survey, experi-

ment, and modeling. The following sections provide some details on the statistical ideas embedded in these themes, with real examples of the uses of these ideas in practice. The goal is to provide sound background information that students can draw upon when carrying out a project.

Theme: Exploring Data and Improving Quality

Data and Decisions

In today's information society, decisions are made on the basis of data. A student checks the calorie chart before selecting a fast-food lunch. A homeowner checks the efficiency rating before purchasing a new refrigerator. A physician checks the outcomes of recent clinical studies before prescribing a medication. An engineer tests the tensile strength of wire before it is wound into a cable. The decisions made from data—correctly or incorrectly—affect each of us every day of our lives. How to systematically study data for the purpose of making decisions is the overarching theme of modern statistics, and the basic ideas on how to make sense out of data are the subject of the first set of activities. These ideas should be studied thoroughly in an introductory statistics course because they form the building blocks for the remainder of statistics and the remainder of the activities in this program.

Quality Really Is Job One

"Quality is job one." This slogan is now synonymous with an American automobile manufacturer. But it is far more than a slogan! A commitment to quality has allowed this manufacturer to recapture a sizable share of the automobile market once lost to foreign competition. Similar stories can be told about many other firms whose products have risen in customer satisfaction since the firm began emphasizing quality within the production process (continuous process improvement). Producing a product or service of high quality is a complex matter, but virtually all of the success stories have one thing in common. Decisions were and are made on the basis of objective, quantitative information—data!

To see how the Ford Motor Company emphasizes quality, data, and statistics, consider the following statement from its manual on **continuous process control**.

> To prosper in today's economic climate, we—Ford, our suppliers and our dealer organizations—must be dedicated to never-ending improvement in quality and productivity. We must constantly seek more efficient ways to produce products and services that consistently meet customer's needs. To accomplish this, everyone in our organization must be committed to improvement and use effective methods. . . . [T]he basic concept of using statistical signals to improve performance can be applied to any area where work is done, the output exhibits variation, and there is desire for improvement. Examples range from component dimensions to bookkeeping error rates, performance characteristics of a computer information system, or transit times for incoming materials.

Ford is not alone in its emphasis on the effective use of statistics. The chairman/CEO of the Aluminum Company of America has stated

> As world competition intensifies, understanding and applying statistical concepts and tools is becoming a requirement for all employees. Those individuals who get these skills in school will have a real advantage when they apply for their first job.

Arno Penzias, vice president for Research at AT&T Bell Laboratories, has written that

> The competitive position of industry in the United States demands that we greatly increase the knowledge of statistics among our engineering graduates. Too many of today's manufacturers still rely on antiquated "quality control" methods, but economic survival in today's world of complex technology cannot be ensured without access to modern productivity tools, notably applications of statistical methods. (*Science*, 1989, **244:**1025.)

Even in our daily lives, all of us are concerned about quality. We want to purchase high-quality goods and services, from automobiles to television sets, from medical treatment to the sound at the local movie theater. Not only do we want maximum value for our dollar when we purchase goods and services, but we also want to live high-quality lives. Consequently we think seriously about the food we eat, the amount of exercise we get, and the stress that accumulate from our daily activities. How do we make the many decisions that confront us in this effort? We compare prices and value, we read food labels to determine calories and cholesterol, we ask our physician about possible side effects to a prescribed medication, and we make mental notes about our weight gains or losses from week to week. In short, we, too, are making daily decisions on the basis of objective, quantitative information—data!

Formally or informally, then, data are the basis for many of the decisions made in our world. Use of data allows decisions to be made on the basis of factual information rather than on subjective judgment, and the use of facts should produce better results. Whether or not the data lead to good results depends on how the data were produced and how the data were analyzed. For many of the simpler problems confronting us, data are readily available. That is the case, for example, when evaluating the caloric content of food or when deciding on the most efficient appliance. Thus, techniques for analyzing data will be presented first, with ideas on how to produce good data coming later.

A Model for Problem Solving

Whether we are discussing products, services, or lives, improving quality is the goal. Along the pathway to improved quality, numerous decisions must be made, and made on the basis of objective data. It is possible, however, that these decisions could affect each other, so the wise course of action is to view the set of decisions together in a wider problem-solving context.

A civil engineer is to solve the problem of hampered traffic flow through a small town with only one main street. After collecting data on the volume of traffic, the immediate decision seems easy—widen the main street from two to four lanes. Careful thought might suggest, though, that more traffic lights will be needed for the wider street so that traffic can cross it. In addition, traffic now using the side streets will begin using the main street once it is improved. A "simple" problem has become more complex once all the possible factors are brought into the picture. The whole process of traffic flow can be improved only by taking a more detailed and careful approach to the problem.

The central idea is that data analysis is to be used to improve the quality of a process, whether the "process" is traffic flow, production of automobiles, purchase of a VCR, or studying for an exam. All aspects of the process must be examined, because the goal is to improve it throughout. This may involve the study of many variables and how they interrelate, and so a systematic approach to problem solving will be essential to our making good decisions efficiently. In recent years, numerous models for solving problems have evolved within business and industry; the model outlined next contains the essential steps present in all of them.

1. State the problem of question
   This sounds like an obvious and, perhaps, easy step. But going from a loose idea or two about a problem to a clear statement of the real problem requires careful investigation of the current situation so that one can develop well-defined goals for the study. ("I am pressed to get my homework assignments done on time, and I do not seem to have adequate time to complete all my reading assignments. I still want to work out each day and to spend some time with friends. Upon review of my study habits, it seems that I study rather haphazardly and tend to procrastinate. The real problem, then, is not to find more hours for study but to develop a study that is efficient.")

2. Collect and analyze data
   Now, all factors thought to affect the problem are listed, and data are collected on all factors thought to be important. A plan is directed toward solving the specific problem addressed in the first step. Appropriate data analysis techniques are used, according to the data collected. ("The data show that I study five hours a day and work late into the evening. It also shows that the gym is crowded when I arrive, and this may slow down my workout.")

3. Interpret the data and make decisions
   After the data are analyzed and the analysis is carefully studied, potential solutions to the original problem or question may be posed. ("I observe that the studying late into the evening often is necessary because I watch television and visit with friends before studying and then am tired when I begin studying. I will set a schedule that puts my study time early in the evening, and my visiting later in the evening. Also, I will work out in the mornings rather than the afternoons, to save time.")

4. Implement and verify the decisions
   Once a solution is posed, it should be put into practice on a trial basis (if that is feasible). New data should be collected on the revised process to see if improvement is actually realized. ("I tried the earlier study time for two weeks and it worked fine. I seemed to have more time to complete my assignments, even though the data show that I was not devoting any more hours to study. The gym is just as busy in the morning, and so I realized no saving of time by the new strategy.")

5. Plan next actions
   The trial period of step 4 may show that the earlier decision solved the problem. More likely, though, the decision was only partially satisfactory. In any case, there is always another problem to tackle, and this should be planned now, while the whole process is still firmly in mind and the data are still fresh. ("I would still like to find more time for pleasure reading. I will see how I could fit that into my revised schedule.")

Application: Improved Payment Processing in a Utilities Firm

Office personnel responsible for the processing of customer payments to a large utility company noticed that they were receiving complaints both from customers (Why hasn't my check cleared?") and from the firm's accounting division ("Why did it take so long to get these receipts deposited?") The office staff decided it was time to swing their quality improvement training into action. The following is a brief summary of their quality improvement story.

1. State the problem or question
   After brainstorming on the general problem, the team collected background data on the elapsed time for processing payments and found that about 2% of the payments (representing 51,000 customers) took more than 72 hours to process. They suspected that many of these were multis (payments containing more than one check or bill) or verifys (payments in which the bill and the payment do not agree). Figure 1 is a **Pareto chart** that demonstrates the correctness of their intuition; multis account for 63% of the batches requiring more than 72 hours to process, but they comprise only 3.6% of the total payments processed. Verifys are not nearly so serious as first suspected, but they are still the second leading contributor to the problem. The problem can now be made specific; concentrating first on the multis, reduce payments requiring more than 72 hours of processing time to 1% of the total.

2. Collect and analyze data
   What factors affect the processing of multis? A **cause-and-effect** diagram, shown in Figure 2, is used to list all important factors and to demonstrate how they might relate to each other. A missing check digit on a bill leads to an incorrect account number, which, in turn, causes a non-scanable bill. Sometimes a customer keeps the bill as a receipt and, as a result, the check cannot be processed. The largest

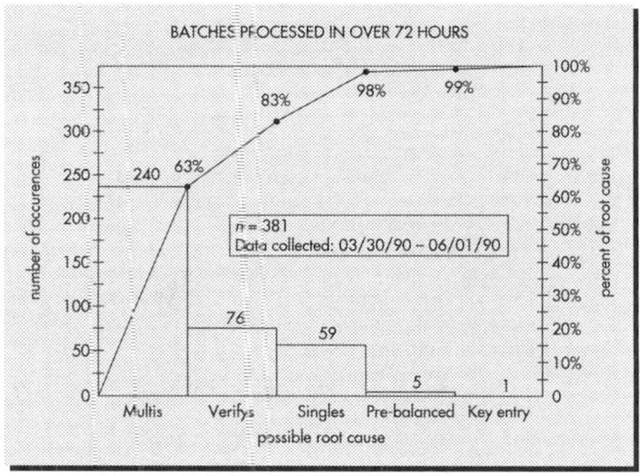

Figure 1: A Pareto Chart

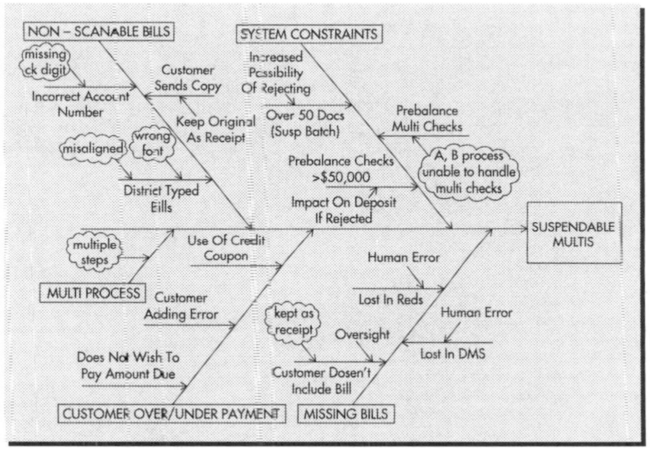

Figure 2: A Cause-and-Effect Diagram

single factor that could be corrected, in this case, was the fact that the computerized auto-balancing process was unable to handle multiple checks.

Another factor, cash carryover from the previous day, is shown to have an effect on the entire payment processing system. The effect on the elapsed time of processing payments can be seen in the **scatter plot** of Figure 3. An efficient system for handling each day's bills must reduce the carryover to the next day.

3. Interpret the data and make decisions
With data in hand that clearly show the computerized check processing system to be a major factor in the processing delays, the staff developed a plan to reprogram the machine so that it would accept multiple checks and coupons as well as both multis and verifys that were slightly out of balance.

4. Implement and verify the decisions
The reprogramming was completed, and along with a few additional changes, it produced results that were close to the goal given in the problem statement. The record of processing times during a trial period are shown in the **time series plot** of Figure 4.

5. Plan next actions
The carryover problem remains, and along the way it was discovered that the mail opening machine has an excessive amount of downtime. Resolution of these two problems led to an improvement that exceeded the goal set in the problem statement.

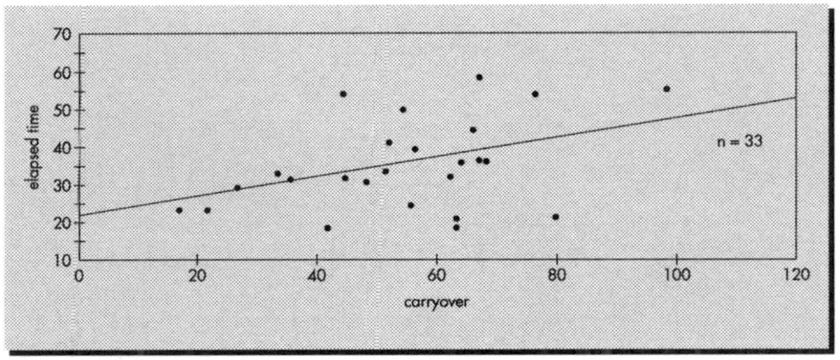

Figure 3: A Scatter plot

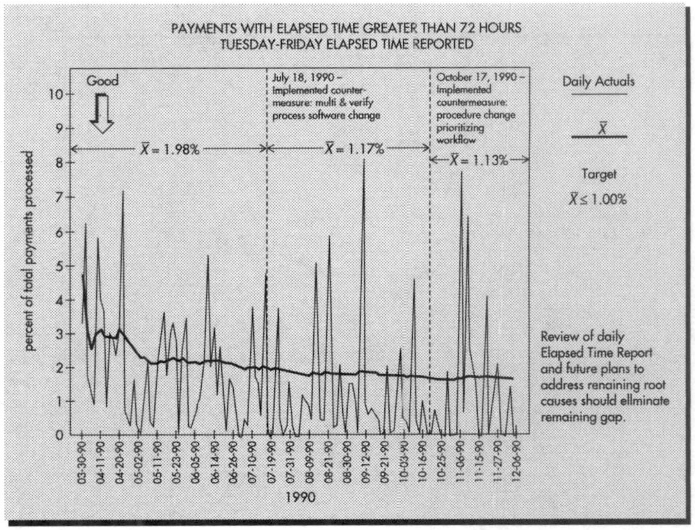

Figure 4: A Time Series Plot

On to Real Data Collection and Analysis

The heart of statistics is the collection and analysis of data. Data do not simply appear; the correct analysis is not always obvious. These two arms (collection and analysis) must work together toward the common goal of solving a problem. To achieve this goal, data must be collected according to a plan. The plan used within the quality improvement programs now meeting with great success in industry is a good one to follow in almost any setting that requires problem solving. The theme of improving the quality of a process provides a general framework for the exploration of data, giving order and direction to statistical applications at any level. The activities in the **Exploring Data and Improving Quality** section are designed to provide practical experience in using the basic statistical tools generally found essential to solving real quantitative problems.

Theme: Sample Survey

What Is a Sample Survey?

Your college administrators want to know how many students will want parking spaces for automobiles next year. How can we get reliable information on this question? One way is to ask all of the returning students, but even this procedure would be somewhat inaccurate (why?) and very time consuming. We could take the number of spaces in use this year and assume next year's needs will be about the same, but this will have inaccuracies as well. A simple technique that works very well in many cases is to select a **sample** from those students who will be attending the school next year and ask each of them if they will be requesting a parking space. From the proportion of "yes" answers, an estimate of the number of spaces required can be obtained.

The scenario outlined above has all of the elements of a typical **sample survey** problem. There is a question of "how many?" or "how much?" to be determined for a specific group of objects called a **target population**, and an approximate answer is to be derived from a **sample** of data extracted from the population of interest. Of key importance is the fact that the approximate answer will be a good approximation only if the sample truly represents the population under study. **Randomization** plays a vital role in the selection of samples that truly represent a population and, hence, produce good approximations. It is clear that we would not want to sample only our classmates or friends on the parking issue. It is less clear, but still true, that virtually any sampling scheme that depends upon subjective judgments as to who should be included will suffer from **sampling bias.**

Once we know who is to be in the sample, we still need to get the pertinent information from them. The **method of measurement,** that is, the questions or measuring devices we use to obtain the data, should be designed to produce the most accurate data possible and should be free of **measurement bias.** We could choose a number of ways to ask students about their parking needs for next year. Here are a few suggested questions:

Do you plan to drive to school next year?

Do you plan to drive to school on more than half of the school days next year?

Do you have regular access to a car for travel to school next year?

Will you drive to school next year if the cost of parking increases?

Which of these questions do you think might bias the results, and in which direction? A few moments of reflection should convince you that measurement bias could be serious in even the simplest of surveys!

Because it is so difficult to get good information in a survey, every survey should be **pretested** on a small group of subjects similar to those that could arise in the final sample. The pretest not only helps improve the questionnaire or measurement procedures, but it also helps determine a good plan for **data collection** and **data management.** Can we, for example, list a mutually exclusive and exhaustive set of meaningful options on the parking question so that responses can be easily coded for the **data analysis** phase? The data analysis should lead to clearly stated **conclusions** that relate to the original purpose of the study. The goal of the parking study is to focus on the number of spaces needed, not the types of cars students drive or the fact that auto theft may be a problem.

Key elements of any sample survey will be discussed and illustrated in this section. These elements include the following:

1. State the **objectives** clearly.
2. Define the **target population** carefully.
3. Design the **sample selection** plan using **randomization,** so as to reduce **sampling bias.**
4. Decide on a **method of measurement** that will minimize **measurement bias.**
5. Use a **pretest** to try out the plan.
6. organize the **data collection** and **data management.**
7. Plan for careful and thorough **data analysis.**
8. Write **conclusions** in the light of the original objectives.

We will now turn to some examples of popular types of sample surveys. Results from surveys like these can be found in the media on almost any day.

Application 1: A Typical Election Poll

The article reprinted here from the *Gainesville Sun* appeared close to the end of the presidential race among Bush, Clinton, and Perot. Read the article carefully before studying the material presented in the following paragraphs. As you go through this discussion, make sure all of the statements are justified, and try to answer the open questions left for the reader.

The poll attempts to answer the **question** of who would win the presidential race as of October 30, 1992. The specific question asked was, "If the presidential election were today, would you vote for George Bush, Bill Clinton, or Ross Perot?" Secondary questions dealt with subgroups of voters, such as conservatives versus liberals and Republicans versus Democrats. This study is clearly a sample survey since its goal is simply to describe the proportion of voters who plan to vote for the various candidates. (Essentially, the poll is seeking to find out "how many" people will vote for each candidate.)

The target population under study here is not the entire population of Florida or even the entire population of registered voters in Florida. It is the *population of voters most likely to go to the polls on election day.* Whether or not a voter falls into this population was determined by a series of questions, which are not presented in the article. (Speculate as to what you think these questions might have been.) The sample was selected by randomly selecting telephone numbers from different areas of the state. (How do you think this randomization might have been accomplished? Why was the state divided into areas, with randomization done within the areas?) The **sample size** was 773. (How do you think they arrived at such a strange number?)

The **measurements** (simple statements of opinion) for this study were collected by telephone conversations. The question asked was quite straightforward, so measurement bias does not appear to be a problem. The data are nicely **summarized** in proportions and **presented** on a pie chart with useful information, such as the sample size and margin of sampling error, included in the box containing the chart. Little

analysis is presented beyond this summary. The principal conclusion is that the race between Bush and Clinton is too close to call. Why would the pollsters make that conclusion when it is clear that Bush has the highest percentage of votes in the sample? To understand this, we need to delve a little more deeply into the **margin of sampling error.** Sample proportions will vary, from sample to sample, according to a definite and predictable pattern (in the long run). In fact, 95% of all possible sample proportions, for samples of size n, that could be obtained from a single population will fall within two standard deviations of the true population proportion. If a sample proportion is denoted by p, then the standard deviation of the possible sample proportions is given by

$$[p(1 - p)/n]^{1/2},$$

and 95% of the potential sample proportions will fall within

$$2[p(1 - p)/n]^{1/2}$$

of the true population proportion. This two-standard-deviation interval is called the margin of sampling error. Substituting .39 for p and 773 for n results in a calculated margin of error equal to .035. Thus, the true proportion of voters favoring Bush may well be anywhere in the interval .39 − .035 to .39 + .035, or (.355, .425). (Find the corresponding interval for Clinton. Do you see why the pollsters do not want to call this race? Find the corresponding interval for Perot. Why did the writers of this article report only one value for the margin of error?)

Notice that the sample size is in the denominator of the margin of sampling error formula. Thus, the pollster could have reduced the margin of error by increasing the sample size (why?). Now, look back at one of the surveys you have conducted earlier in this unit. Find the margin of sampling error for a proportion, and then find the sample size that would be required to cut this margin of error in half.

Let's look back at the article carefully to see what else we might like to know. Much discussion is presented on the breakdowns of the percentages among conservative, moderate, and liberal voters. Is there any way to attach a margin of sampling error to these percentages? If we could find these margins of error, would they be larger or smaller than .035? How about the breakdown between Republicans and Democrats? Can we attach a meaningful margin of error to these percentages?

It would be nice to know more about the sampling scheme. How was the randomization of telephone numbers conducted? What areas of the state were used? Why a sample size of 773? (This could indicate that some who were called refused to answer the poll. Could this cause any bias?) What were the questions that determined whether or not a response was used? (Could these questions cause any bias?)

Now that this election is over, you might want to look at the Florida vote to see if the pollsters were right in implying that the race would be close. Also, you might want to study polls of current interest in a similar fashion.

A TYPICAL ELECTION POLL
(Source: *Gainesville Sun*, October 30, 1992)

Undecided will sway Fla. vote

By BILL RUFTY
NYT Regional Newspapers

LAKELAND—The fate of Florida's 25 electoral votes apparently lies in the hands of the 3 percent of the voters who still don't know who they will vote for, according to results from the latest Florida Opinion Poll.

FLORIDA OPINION POLL

Those few voters will break the deadlock between Republican President George Bush and Democratic challenger Bill Clinton in Tuesday's presidential election.

While independent candidate Ross Perot does not have a chance to capture the state, he apparently has a tight grip on about a fifth of the vote. The poll showed that:

Amendment poll. 12A. Campaign roundup. 9A.

- 39 percent support or are leaning toward Bush.
- 37 percent support or are leaning toward Clinton.
- 21 percent support or are leaning toward Perot.
- 3 percent are undecided.

The Florida Opinion Poll, which is sponsored by The New York Times newspapers in Florida, contacted state residents in a random telephone sample.

Presidential race in Florida too close to call

If the presidential election were today would you vote for George Bush, Bill Clinton or Ross Perot?

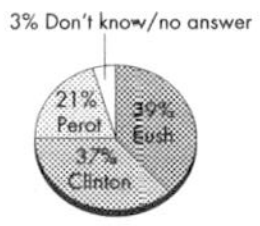

Source: The Florida Opinion Poll, Oct. 24-27, 773 registered voters who answered the question; margin of sampling error of 3.5 percentage points

Adapted from Mark Williams/NYTRENG Graphics Network

Along the political spectrum, Bush is heavily dependent on conservatives in Florida, while Clinton captures nearly half the moderate voters and nearly three-quarters of the liberals.

Perot pulls almost identical support from liberals and conservatives, while drawing a quarter of moderates.

The poll found that among conservative voters:

- 63 percent back Bush.
- 18 percent back Clinton.
- 19 percent back Perot.

Among moderate voters:

- 28 percent back Bush.
- 46 percent back Clinton.
- 26 percent back Perot.

Among liberal voters:

- 11 percent back Bush.
- 71 percent back Clinton
- 18 percent back Perot.

While Perot pulls support from across the political spectrum, he is taking away more of Bush's Republicans than Clinton's Democrats.

Among Republicans:

- 64 percent back Bush.
- 12 percent back Clinton.
- 22 percent back Perot.
- 2 percent were undecided.

But Perot takes Democratic votes from Clinton, but not as many, and there are more registered Democrats in Florida than Republicans.

Among Democrats:

- 20 percent back Bush.
- 61 percent back Clinton.
- 17 percent back Perot.
- 2 percent were undecided

Of the 6.5 million registered voters in Florida, 3.3 million, or 51 percent, are registered as Democrats, while 2.7 million voters, or 41 percent, are registered as Republicans and 550,292, or 8 percent, are registered as independents or members of third parties.

By age, candidates seem to appeal equally across all age groups, with each drawing in the 30 percent range, except for the 45- to 64-year-olds.

By gender, there appears to be no major difference in the number of women or men going for one candidate more than the others.

How polls were conducted

NYT Regional Newspapers

The latest Florida Opinion Poll was conducted by telephone from Oct. 24 to 27 with 773 voters considered most likely to go to the polls on Election Day.

The telephone numbers used in the survey were formed at random by a computer programmed to ensure that each area of the state was represented in proportion to its population.

The results based on responses from all 773 most likely voters have a margin of sampling error of 3.5 percentage points. That means if the New York Times newspapers in Florida asked every voter in the state the same questions, in most cases, the results would be within 3.5 percentage points of the results obtained by the survey.

Interviewers used a series of three questions to determine voters who were most likely to go to cast their ballots Tuesday.

In questions where only the answers of smaller groups are used, the margin of sampling error is larger.

For example, the margin of sampling error for just registered Democrats or only registered Republicans will be higher.

In addition to sampling error, the practical difficulties of conducting any poll can induce other forms of error.

Application 2: The Nielsens

Almost everyone watches television and, as a result, has some awareness of the fact that what is available to watch is determined by the Nielsen ratings. A show that does poorly in the Nielsens is not going to be on a major network very long (why?). The attached article shows the final Nielsen ratings for a week in March 1995. Read the explanation in this article before reading the discussion below.

Of the 95.4 million households in the United States, Nielsen Media Research randomly samples 4,000 on which to base their ratings. This is accomplished by randomly selecting city blocks (or equivalent units in rural areas), having an enumerator actually visit the sampled blocks to list the housing units, and then randomly selecting one housing unit per block. These sampled housing units are the basic units for all of the ratings data. After a housing unit is selected, an electronic device is attached to each TV set in the household. This device records when the set is turned on and the network to which it is tuned. Information form the network determines which show is actually playing at any point in time. This device gives information on what is happening to the TV set, but it doesn't tell who or how many people are viewing the programs. For this information, Nielsen must rely on individuals in the household recording when they personally "tune in" and "tune out."

The **rating** for a program is the percentage of the sampled households that have TV sets on and tuned to the program in question. (Note that in the estimated ratings, the denominator of the sample proportion is always 4,000.) So, a rating is an estimate of the percentage of households viewing a particular program. A **share** for a program is an estimate of the percentage of viewing households that have a TV tuned to that particular program, where a **viewing household** is one for which at least one TV set is turned on. (Will the denominator of the estimated shares be 4,000, greater than 4,000, or less than 4,000?)

In reality, the ratings and shares are slightly more complicated than explained above. A rating for any program is taken minute by minute and then averaged over the length of the program. This attempts to adjust for the fact that not all viewers watch all of a program. Thus, the final rating for "Seinfeld" would be the average of all ratings taken over the half-hour duration of the show, while the final rating for a basketball game would be the average of ratings taken over the entire time (perhaps a couple of hours) that the game was on the air.

Now, look over the attached article once again. Discuss any points that are misleading or unclear. Specifically, discuss the following:

1. Why are the shares always greater than the ratings?

2. Are there sources of potential bias in the data collection plan?

3. Can a margin of sampling error be approximated for a rating?

4. Can a margin of sampling error be approximated for a share?

5. Read the article on Computer Vision Research and relate this to the points made above. How could a computer vision device improve the Nielsens?

NBC sitcoms still dominate Thursday night

SCOTT WILLIAMS
The Associated Press

NEW YORK—The Peacock can strut again.

For the fifth consecutive week, NBC won the prime time ratings crown behind top-rated "Seinfeld" and Top 10 performances from four other shows in its Thursday lineup.

For the week, NBC averaged an 11.5 rating and a 19 percent audience share.

ABC, the season-to-date front-runner, finished second with an 11.1 rating, 19 share. CBS was third, with a 9.2 rating, 16 share.

Top 20 listings include the week's ranking, with rating for the week, season-to-date rankings in parentheses, and total homes.

An "X" in parentheses denotes one-time-only presentation. A rating measures the percentage of the nation's 95.4 million TV homes. Each ratings point represents 954,000 households, as estimated by Nielsen Media Research.

The Nielsens

1. (1) **"Seinfeld,"** NBC, 21.4, 20.4 million homes
2. (2) **"Home Improvement,"** ABC, 20.5, 19.6 million homes
3. (3) **"E.R.,"** NBC, 19.8, 18.9 million homes
3. (11) **"Friends,"** NBC, 19.8, 18.9 million homes
5. (4) **"Grace Under Fire,"** ABC, 19.6, 18.7 million homes
6. (7) **"NYPD Blue,"** ABC, 16.9, 16.1 million homes
6. (5) **"60 minutes,"** CBS, 16.9, 16.1 million homes
8. (12) **"Mad About You,"** NBC, 15.3, 14.6 million homes
9. (10) **"Hope & Gloria,"** NBC, 14.9, 14.2 million homes
10. (16) **"Murphy Brown,"** CBS, 14.8, 14.1 million homes
11. (9) **"Murder She Wrote,"** CBS, 14.3, 13.6 million homes
11. (17) **"20-20,"** ABC, 14.3, 13.6 million homes
13. **"Mcbains' 87th Precinct—NBC Sunday Movie,"** 13.8, 13.2 million homes
14. **"Chicago Hope,"** CBS, 13.7, 13.1 million homes
15. (14) **"Ellen,"** ABC, 13.5, 12.9 million homes
16. (20) **"Dave's World,"** CBS, 13.4, 12.8 million homes
16. (17) **"Awake To Danger—NBC Monday Movies,"** 13.4, 12.8 million homes
18. (8) **"Roseanne,"** ABC, 13.3, 12.7 million homes
19. (22) **"Betrayed–ABC Sunday Movie,"** 13.2, 12.6 million homes
20. **"Cybill,"** CBS, 13.1, 12.5 million homes
21. **"Thunder Alley,"** ABC, 12.7
22. **"Step by Step,"** ABC, 12.5
23. **"Primetime Live,"** ABC, 12.4
24. **"Frasier,"** NBC, 12.3
24. **"Full House,"** ABC, 12.3
26. **"Family Matters,"** ABC, 12.1
27. **"Boy Meets World,"** ABC, 12.0
28. **"America's Funniest Home Video,"** ABC, 11.8
29. **"Law and Order,"** NBC, 11.7
29. **"Nanny,"** CBS, 11.7
31. **"Dateline NBC,"** (Tuesday) 11.5
31. **"Dateline NBC,"** (Wednesday) 11.5
33. **"On Our Own,"** ABC, 11.3
34. **"Wings,"** NBC, 11.1
34. **"Beverly Hills, 90210,"** Fox, 11.1
36. **"Return-TV Censored Bloopers,"** NBC, 10.9
37. **"Lois & Clark,"** ABC, 10.8
38. **"John Larroquette Show,"** NBC, 10.6
39. **"Far and Away—ABC Monday Movie,"** 10.4
40. **"Something Wilder,"** NBC, 10.2
41. **"Melrose Place,"** Fox, 10.0
42. **"Dateline NBC,"** (Friday) 9.9
43. **"Fresh Prince of Bel Air,"** NBC, 9.7
44. **"Coach,"** ABC, 9.6
44. **"Unsolved Mysteries,"** NBC, 9.6
46. **"Walker, Texas Ranger,"** CBS, 9.5
47. **"America's Funniest Home Videos,"** ABC 9.4
48. **"All American Girl,"** ABC, 9.3
48. **"Cosby Mysteries,"** NBC, 9.3
48. **"Peter Jennings Reporting,"** ABC, 9.3
51. **"Under One Roof,"** CBS, 9.2
52. **"Simpsons,"** Fox, 9.1
53. **"Blossom,"** NBC, 9.0
53. **"Sister, Sister,"** ABC, 9.0
55. **"Greatest Commercials,"** CBS, 8.8
56. **"George Wendt Show,"** CBS, 8.7
56. **"Living Single,"** Fox, 8.7
58. **"Northern Exposure,"** CBS, 8.4
58. (X) **"Mommies,"** NBC, 8.4
60. **"Empty Nest,"** NBC, 8.3
60. **"Married . . . With Children,"** Fox 8.3
60. **"X-Files,"** Fox, 8.3
63. **"The Dead Pool—CBS Tuesday Movie,"** 8.2
64. **"America's Most Wanted,"** Fox, 8.1
64. **"Seaquest DSV,"** NBC, 8.1
66. **"Extreme,"** ABC, 8.0
66. **"Mommies,"** NBC, 8.0

(Gainesville Sun 3/26/95)

UF Computer Vision Researchers Develop "Smart" Technology For Nielsen's People-Watching TV

The next time you sit down to watch television, don't be surprised to find that your television is watching you.

At least that's what A.C. Nielsen Media Research is planning.

Transferring cutting-edge "smart" technology developed at UF's Computer Vision research Center, Nielsen hopes to employ a passive "peoplemeter" to silently and automatically record who is watching what, when and where in 4,000 Nielsen homes across the country.

"This passive peoplemeter means that, within the next two years, the demographic information concerning television viewing that we pass on to our clients will be much more accurate," said Jo LaVerde, Nielsen's director of communications.

The device Nielsen now uses requires volunteers to identify themselves with the push of a button. But many viewers, especially children, forget to log on. As a result, researchers claim that only about 50 percent of data collected now is usable.

The peoplemeter uses computer image recognition, an emerging technology that uses lasers to follow moving images and a computer that recognizes those images—whether they are people, pets or enemy tanks.

"The applications for this kind of technology are limitless, for both the military and civilian sectors," said UF computer and information sciences Professor Gerhard Ritter, who is heading the research at the Computer Vision Research Center. "We're talking about very Tom Clancy stuff here."

For Nielsen, more accurate demographic information means greater profits for their clients, including those in the advertising industry, which alone spends more than $30 billion a year on television advertising.

But computer recognition systems, like the human eye, can be easily fooled by camouflage or disguise.

"If it's easy to fool a human, just imagine how easy it is to fool a machine," said Ritter. "We've only just discovered the tip of the iceberg with this technology."

Ritter hopes to eventually create an image recognition system that acts like the human eye and responds to visual stimuli the same way the brain does, making it as foolproof as possible.

Ritter's computer vision research is supported by more than $3 million in National Science Foundation and Department of Defense grants. Another $100,000 comes from A.C. Nielsen. Ritter expects to finish a prototype passive peoplemeter in two years.

Alligator, 11/25/92 — The University of Florida is an Equal Opportunity/Affirmative Action Institution

Theme: Experiment

What Is an Experiment?

How many times has a parent or a teacher admonished a student to turn off the radio while doing homework? Does listening to music while doing homework help or hinder? To answer specific questions like this, we must conduct carefully planned experiments. Let's suppose we have a history lesson to study for tomorrow. We could have some students study with the radio on and some study with the radio off. But, the time of day that the studying takes place could affect the outcomes as well. So, we have some study in the afternoon with the radio on and some study in the afternoon with the radio off. Other students study in the evening, some with the radio on, and some with the radio off. The measurements on which the issue will be decided (for now) are the scores on tomorrow's quiz.

Since males might produce different results from females, perhaps we should **control** for sex by making sure that both males and females are selected for each of the four **treatment** slots. On thinking about this **design** for a moment, we conclude that the native ability of the students might have some affect on the outcome as well. All of the students are from an honors history course, so it's difficult to differentiate on ability. Therefore, we will **randomly** assign students (of similar ability) to the four treatment groups in the hope that any undetected differences in ability will balance out in the long run.

The above outline of a study has most of the key elements of a **designed experiment.** The goal of an experiment is to measure the effect of one or more **treatments** on **experimental units** appropriate to the question at hand. Here, there are two main treatments, the radio and the time of day that study occurs. Another **variable** of interest is the sex of the student (the experimental unit in this case), but this variable is directly **controlled** in the design by making sure we have data from both sexes for all treatments. The variability "ability" cannot be controlled as easily, so we **ran-**

domize the assignment of students to treatments to reduce the possible biasing effect of ability on the response comparisons.

Key elements of any experiment will be discussed and illustrated in this section. These key elements include the following:

1. Clearly define the **question** to be investigated.
2. Identify the key variables to be used as **treatments.**
3. Identify other important variables that can be **controlled.**
4. Identify important background (lurking) variables that cannot be controlled but should be balanced by **randomization.**
5. Randomly assign treatments to the **experimental units.**
6. Decide on a **method of measurement** that will minimize **measurement bias.**
7. Organize the **data collection** and **data management.**
8. Plan for careful and thorough **data analysis.**
9. Write **conclusions** in the light of the original question.
10. Plan a **follow-up** study to answer the question more completely or to answer the next logical question on the issue at hand.

We will now see how these steps are followed in a real experiment of practical significance.

Application: Does Aspirin Help Prevent Heart Attacks? The Physician's Health Study*

During the 1980s, approximately 22,000 physicians over the age of 40 agreed to participate in a long-term health study for which one important question was to determine whether or not aspiring helps to lower the rate of heart attacks (myocardial infarctions). The **treatment** for this part of the study was aspirin, and the **control** was a placebo. Physicians were randomly assigned to one treatment or the other as they entered the study so as to minimize bias caused by uncontrolled factors. The method of assignment was equivalent to tossing a coin and sending the physician to the aspirin arm of the study if a head appeared on the coin? After the assignment, neither the participating physicians nor the medical personnel who treated them knew who was taking aspirin and who was taking placebo. This is called a double-blind experiment. (Why is the double blinding important in a study such as this?) The method of measurement was to observe the physicians carefully for an extended period of time and record all heart attacks, as well as other problems, that might occur.

Other than aspirin, there are many variables that could have an effect on the rate of heart attacks for the two groups of physicians. For example, the amount of exercise they get and whether or not they smoke are two prime examples of variables that should be controlled in the study so that the true effect of aspirin can be measured. The tables below show how the subjects eventually divided according to exercise and to cigarette smoking. (See the complete tables at the end of this section for more details.) Do you think the randomization scheme did a good job in controlling these variables? Would you be concerned about the results for aspirin being unduly influenced by the fact that most of the aspirin takers were also nonsmokers? Would you be concerned about the placebo group possibly having too many who do not exercise?

*Source: "The final report on the aspirin component of the ongoing physicians' health study," *The New England Journal of Medicine*, **231**(3) 1989, pp. 129–135.

	Aspirin	Placebo
Exercise Vigorously		
yes	7,910	7,861
no	2,997	3,060
Cigarette smoking		
never	5,431	5,488
past	4,373	4,301
current	1,213	1,225

The data analysis for this study reports that 139 heart attacks developed among the aspirin users and 239 heart attacks developed in the placebo group. This was said to be a significant result in favor of aspirin as a possible preventative for heart attacks. To see why this was so, work through the following steps.

1. Given that there were approximately 11,000 participants in each arm of the study, calculate the proportion of heart attacks among those taking aspirin. Calculate the proportion of heart attacks among those taking placebo.

2. Calculate the standard deviations for each of these proportions and use them to form confidence intervals for the true proportions of heart attacks to be expected among aspirin users and among nonaspirin users in the population from which these sampling units were selected.

3. Looking at the two confidence intervals, can you see why the researchers in this study declared that aspirin had a significant effect in reducing heart attacks? Explain.

However, heart attacks aren't the only cause for concern. Another is that too much aspirin can cause an increase in strokes. Among the aspirin users on the study, 119 had strokes during the observation period. Within the placebo group, only 98 had strokes. Although the number of strokes is higher than the researchers would have liked, the difference between the two numbers was no cause for alarm. That is, there did not appear to be a significant increase in the number of strokes for the aspirin group. Follow the three steps listed above for constructing and observing confidence intervals to see why the researchers were not overly concerned about the difference in numbers of strokes between the two arms of the study.

Much more data relating to this study are provided in the tables here. This should lead to other questions of interest regarding the relationships among aspirin use, heart

attacks, and other factors. (For example, does age play a role in the effectiveness of aspirin? How about cholesterol level?)

Table 1 CONFIRMED CARDIOVASCULAR END POINTS IN THE ASPIRIN COMPONENT OF THE PHYSICIANS' HEALTH STUDY, ACCORDING TO TREATMENT GROUP*

End Point	Aspirin Group	Placebo Group	Relative Risk	95% Confidence Interval	P Value
Myocardial infarction					
Fatal	10	26	0.34	0.15–0.75	0.007
Nonfatal	129	213	0.59	0.47–0.74	<0.00001
Total	139	239	0.56	0.45–0.70	<0.00001
Person-years of observation	54,560.0	54,355.7	—	—	—
Stroke					
Fatal	9	6	1.51	0.54–4.28	0.43
Nonfatal	110	92	1.20	0.91–1.59	0.20
Total	119	98	1.22	0.93–1.60	0.15

Table 2 CONFIRMED DEATHS, ACCORDING TO TREATMENT GROUP

Cause*	Aspirin Group	Placebo Group	Relative Risk	95% Confidence Interval	p-Value
Total cardiovascular deaths†	81	83	0.96	0.60–1.54	0.87
Acute myocardial infarction (410)	10	28	0.31	0.14–0.68	0.004
Other ischemic heart disease (411–414)	24	25	0.97	0.60–1.55	0.89
Sudden death (798)	22	12	1.96	0.91–4.22	0.09
Stroke (430, 431, 434, 436)‡	10	7	1.44	0.54–3.88	0.47
Other cardiovascular (402, 421, 424, 425, 428, 429, 437, 440, 441)	15	11	1.38	0.62–3.05	0.43
Total noncardiovascular deaths	124§	133	0.93	0.72–1.20	0.59
Total deaths with confirmed cause	205	216	0.95	0.79–1.15	0.60
Total deaths¶	217	227	0.96	0.80–1.14	0.64

*Numbers are code numbers of the *International Classification of Diseases*, ninth revision.
†All fatal cardiovascular events are included, regardless of previous nonfatal events.
‡This category includes ischemic (3 in the aspirin group and 3 in the placebo group), hemorrhagic (7 aspirin and 2 placebo), and unknown cause (0 aspirin and 2 placebo).
§This category includes one death due to gastrointestinal hemorrhage.
¶Additional events that could not be confirmed because records were not available included 23 deaths (12 aspirin and 11 placebo), of which 11 were suspected to be cardiovascular (7 aspirin and 4 placebo) and 12 noncardiovascular (5 aspirin and 7 placebo).

Table 3 RISK OF TOTAL MYOCARDIAL INFARCTION ASSOCIATED WITH ASPIRIN USE, ACCORDING TO LEVEL OF CORONARY RISK FACTORS

	Aspirin Group	Placebo Group	Relative Risk	p-Value of Trend in Relative Risk
	no. of myocardial infarctions/total no. (%)			
Age (yr)				
40–49	27/4527 (0.6)	24/4524 (0.5)	1.12	
50–59	51/3725 (1.4)	87/3725 (2.3)	0.58	0.02
60–69	39/2045 (1.9)	84/2045 (4.1)	0.46	
70–84	22/740 (3.0)	44/740 (6.0)	0.49	
Cigarette smoking				
Never	55/5431 (1.0)	96/5488 (1.8)	0.58	
Past	63/4373 (1.4)	105/4301 (2.4)	0.59	0.99
Current	21/1213 (1.7)	37/1225 (3.0)	0.57	
Diabetes mellitus				
Yes	11/275 (4.0)	26/258 (10.1)	0.39	0.22
No	128/10,750 (1.2)	213/10,763 (2.0)	0.60	
Parental history of myocardial infarction				
Yes	23/1420 (1.6)	39/1432 (2.7)	0.59	0.97
No	112/9505 (1.2)	192/9481 (2.0)	0.58	
Cholesterol level (mg per 100 ml)*				
<159	2/382 (0.5)	9/406 (2.2)	0.23	
160–209	12/1587 (0.8)	37/1511 (2.5)	0.29	0.04
210–259	26/1435 (1.8)	43/1444 (3.0)	0.61	
≥260	14/582 (2.4)	23/570 (4.0)	0.59	
Diastolic blood pressure (mm Hg)				
≤69	2/583 (0.3)	9/562 (1.6)	0.21	
70–79	24/2999 (0.8)	40/3076 (1.3)	0.61	0.88
80–89	71/5061 (1.4)	128/5083 (2.5)	0.55	
≥90	26/1037 (2.5)	43/970 (4.4)	0.56	
Systolic blood pressure (mm Hg)				
<109	1/330 (0.3)	4/296 (1.4)	0.22	
110–129	40/5072 (0.8)	75/5129 (1.5)	0.52	0.48
130–149	63/3829 (1.7)	115/3861 (3.0)	0.55	
[≥150	19/454 (4.2)	26/412 (6.3)	0.65	

Table 3 RISK OF TOTAL MYOCARDIAL INFARCTION ASSOCIATED WITH ASPIRIN USE, ACCORDING TO LEVEL OF CORONARY RISK FACTORS

	Aspirin Group	Placebo Group	Relative Risk	p-Value of Trend in Relative Risk
	no. of myocardial infarctions/total no. (%)			
Alcohol use				
Daily	26/2718 (1.0)	55/2727 (2.0)	0.45	
Weekly	70/5419 (1.3)	112/5313 (2.1)	0.61	0.26
Rarely	40/2802 (1.4)	65/2897 (2.2)	0.63	
Vigorous exercise at least once a week				
Yes	91/7910 (1.2)	140/7861 (1.8)	0.65	0.21
No	45/2997 (1.5)	92/3060 (3.0)	0.49	
Body-mass index†				
≤23.0126	26/2872 (0.9)	41/2807 (1.5)	0.61	
23.0127–24.4075	32/2700 (1.2)	46/2627 (1.8)	0.63	0.90
24.4076–26.3865	32/2713 (1.2)	75/2823 (2.7)	0.44	
≥26.3866	49/2750 (1.8)	76/2776 (2.7)	0.65	

*To convert cholesterol value to millimoles per liter, multiply by 0.02586.
†Body-mass index is the weight (in kilograms) times the height (in meters) squared.

Theme: Modeling

What Is a Model?

Traffic safety depends, in part, on the ability of drivers to be able to anticipate their stopping distance as related to the speed at which they are traveling. In fact, handbooks for drivers have tables of typical stopping distances as a function of speed. How do we know the relationship between the speed of a car and stopping distance? Why does the handbook have just one table rather than a different table for each type of car? How were the data in the table obtained? These are all questions that relate to the notion of **modeling** the relationship between speed and stopping distance for cars.

In this discussion, a **model** refers to a mathematical relationship (often expressed in a formula) among two or more variables. At this introductory level, most examples will involve the study of relationships between only two variables. For example, the area of a circle is related to the radius of the circle by a very specific (and very accurate) formula, which could be called a model for this well-known relationship. Once the radius of a circle is known, the area can be predicted from the model; the area need not be calculated directly. Charts in a pediatrician's office show the relationship between the weight and age of growing children. These relationships also come about through modeling weight as a function of age, but these models are much less accurate than the one involving the radius and area of a circle (why?).

Models, then, are used widely in common situations. The reason for constructing a model is usually to allow prediction of a value not readily observed or to study the possible causal link between variables. The driver (or, at least, the police officer) wants to predict stopping distance by knowing only the speed of the car and to see how increased speed will cause the results to change; the driver does not want to conduct his or her own experiments to find out how these two variables are related. The mother and father want to know if their child is experiencing normal weight gain and to predict the weight that the child might attain in six more months; they do not want to measure other children, on their own, to establish these facts.

Constructing a Model

The steps in constructing a model are the same as those introduced under the general problem-solving format in the earlier section on quality. First, a **clear statement of the problem** is essential. Establishing a relationship between speed and stopping distance for cars is too general a question. Driver reaction time plays a key role in making quick stops—is the model concerned with distance from the point at which the driver sees an emergency situation or from the point at which brakes are applied? (What other conditions would have to be considered and qualified?)

Under the conditions set in the statement of the problem, **data are collected and analyzed.** Data for establishing a model for stopping distance versus speed come from highly controlled experiments on test tracks, using vehicles of a variety of sizes. Thus, the model reported is a sort of average for various vehicles, but it is usually stated that the figures are for dry pavement only (why?).

The **interpretation of the data** often involves fitting a variety of models to the experimental data and choosing the one that seems to provide the best explanation. Very complicated models could be produced to explain the nature of stopping distance as it relates to speed, with adjustments for road conditions, weight of vehicle, etc., but a model that is simple to interpret and use may work almost as well. So, the "best" model is difficult to define and often depends on the intended uses of the model.

Verifying that the selected model actually works is the next step. Once a model for predicting stopping distance is obtained from experimental data, it should be tried in a new experiment to see if the results are **reproducible.**

Planning the next actions is something to be considered carefully in modeling problems because, in general, all models can be improved or adjusted to specific cases. A model is an approximation to reality and, hence, should never be considered as the final word.

Application: Body Composition*

Health and physical fitness are of vital concern to many people in this day and age. It is well documented that we can choose lifestyles that will drastically alter how we feel and how productive we are, and one of the key variables in this process is weight. (Why are there so many diet books on the market?) But to a large extent weight is determined by a person's size and body structure. A more important related variable that can be controlled more directly by diet and exercise is percent body fat. The problem is that percent body fat is somewhat difficult to measure. This leads directly to a modeling problem. The goal is to find a model that estimates percent body fat as a function of easily measured variables.

The best way to measure percent body fat is by a laboratory technique called hydrostatic weighing (underwater weighing), but this is expensive in terms of time, equipment, and training of technicians. Many anthropometric body measurements (height, weight, skinfolds fat measurements) are, however, relatively easy to obtain. So, the refined goal is to build a model to estimate percent body fat from anthropometric measurements.

Background data from a large number of people varying in age and fatness show, under careful exploration, that some of the anthropometric measurements are more highly correlated with body density (which is related to percent fat) than others (see Table 1). In particular, the skinfolds measurements show consistently high correlation for both men and women. So these measurements will be used as the basis for constructing the model. The data exploration also shows that men and women have quite different body compositions (see Table 2), and this suggests that we may need two models, one for men and one for women. Age turns out to be quite important in

*Source: A. S. Jackson and M. L. Pollock, (1985), "Practical assessment of body composition," The Physician and Sports Medicine, **13**(5) May.

APPLICATION: BODY COMPOSITION 247

Table 1 LINEAR CORRELATIONS BETWEEN BODY DENSITY AND ANTHROPOMETRIC VARIABLES FOR ADULTS

Variables	Men (n = 402)	Women (n = 283)
Height	−0.03	−0.06
Weight	−0.63	−0.63
Body mass index*	−0.69	−0.70
Skinfolds		
Chest	−0.85	−0.64
Axilla	−0.82	−0.73
Triceps	−0.79	−0.77
Subscapula	−0.77	−0.67
Abdomen	−0.83	−0.75
Suprailium	−0.76	−0.76
Thigh	−0.78	−0.74
Sum of seven	−0.88	−0.83
Circumferences		
Waist	−0.80	−0.71
Gluteal	−0.69	−0.74
Thigh	−0.64	−0.68
Biceps	−0.51	−0.63
Forearm	−0.35	−0.41

*wt/ht2, where weight is in kg and height is in meters.

the relationship between skinfold measurements and percent body fat, but weight does not. (Can you rationalize this?)

In using linear regression techniques to construct a model, it is found that the sum of three different skinfold measurements works about as well as the sum of all seven. In the spirit of keeping the model simple, so that it can be used and will be used correctly, it is decided that the three-measurement model is the "best." However, the three measurements used for men differ from the three measurements used for women.

Data analysis using regression techniques produced the estimates of percent body fat as a function of the sum of three skinfold measurements seen in Tables 3 and 4. These estimates came from models that made use of the hydrostatically produced fat percentages as "truth" and attempted to reproduce these values as functions of the

248 ACTIVITY-BASED STATISTICS GUIDE

Table 2 DESCRIPTIVE STATISTICS AND STATISTICAL DIFFERENCES FOR MEN AND WOMEN

Variables	Men (n = 402) Mean	SD	Range	Women (n = 283) Mean	SD	Range
General characteristics						
Age (yr)	32.8	11.0	18–61	31.8	11.5	18–55
Height (cm)	179.0	6.4	163–201	168.6	5.8	152–185
Weight (kg)	78.2	11.7	53–123	57.5	7.4	36–88
Body mass index (wt/ht²)	24.4	3.2	17–37	20.2	2.2	14–31
Laboratory determined						
Body density (gm/ml)	1.058	0.018	1.016–1.100	1.044	0.016	1.022–1.091
Percent fat (%)	17.9	8.0	1–37	24.4	7.2	8–44
Lean weight (kg)	63.5	7.3	47–100	43.1	4.2	30–54
Fat weight (kg)	14.6	7.9	1–42	14.3	5.7	2–35
Skinfolds (mm)						
Chest	15.2	8.0	3–41	12.6	4.8	3–26
Axilla	17.3	8.7	4–39	13.0	6.1	3–33
Triceps	14.2	6.1	3–31	18.2	5.9	5–41
Subscapula	16.0	7.0	5–45	14.2	6.4	5–41
Abdomen	25.1	10.8	5–56	24.2	9.6	4–36
Suprailium	16.2	8.9	3–53	14.0	7.1	3–40
Thigh	18.9	7.7	4–48	29.5	8.0	7–53
Sum of skinfolds (mm)						
All seven	122.9	52.0	31–272	125.6	42.0	35–266
Chest, abdomen, thigh	59.2	24.5	10–118			
Triceps, chest, subscapula	45.3	19.6	11–105			
Triceps, suprailium, thigh				61.6	19.0	16–126
Triceps, suprailium, abdomen				56.3	21.0	13–131

skinfold data. Although we do not have the original data, we can recapture the flavor of these models by fitting simple regression models to the data found in Table 3 or 4. From Table 3 (using the midpoint of ranges as the point observation), the relationship between percent body fat (Y) and the sum of three skinfolds (X) for the 22 and under age category is

$$Y = 0.013 + 0.270X,$$

Table 3 PERCENT FAT ESTIMATE FOR MEN: SUM OF CHEST, ABDOMEN, AND THIGH SKINFOLDS

Sum of Skinfolds (mm)	Age to Last Year: Under 22	23–27	28–32	33–37	38–42	43–47	48–52	53–57	Over 57
8–10	1.3	1.8	2.3	2.9	3.4	3.9	4.5	5.0	5.5
11–13	2.2	2.8	3.3	3.9	4.4	4.9	5.5	6.0	6.5
14–16	3.2	3.8	4.3	4.8	5.4	5.9	6.4	7.0	7.5
17–19	4.2	4.7	5.3	5.8	6.3	6.9	7.4	8.0	8.5
20–22	5.1	5.7	6.2	6.8	7.3	7.9	8.4	8.9	9.5
23–25	6.1	6.6	7.2	7.7	8.3	8.8	9.4	9.9	10.5
26–28	7.0	7.6	8.1	8.7	9.2	9.8	10.3	10.9	11.4
29–31	8.0	8.5	9.1	9.6	10.2	10.7	11.3	11.8	12.4
32–34	8.9	9.4	10.0	10.5	11.1	11.6	12.2	12.8	13.3
35–37	9.8	10.4	10.9	11.5	12.0	12.6	13.1	13.7	14.3
38–40	10.7	11.3	11.8	12.4	12.9	13.5	14.1	14.6	15.2
41–43	11.6	12.2	12.7	13.3	13.8	14.4	15.0	15.5	16.1
44–46	12.5	13.1	13.6	14.2	14.7	15.3	15.9	16.4	17.0
47–49	13.4	13.9	14.5	15.1	15.6	16.2	16.8	17.3	17.9
50–52	14.3	14.8	15.4	15.9	16.5	17.1	17.6	18.2	18.8
53–55	15.1	15.7	16.2	16.8	17.4	17.9	18.5	19.1	19.7
56–58	16.0	16.5	17.1	17.7	18.2	18.8	19.4	20.0	20.5
59–61	16.9	17.4	17.9	18.5	19.1	19.7	20.2	20.8	21.4
62–64	17.6	18.2	18.8	19.4	19.9	20.5	21.1	21.7	22.2
65–67	18.5	19.0	19.6	20.2	20.8	21.3	21.9	22.5	23.1
68–70	19.3	19.9	20.4	21.0	21.6	22.2	22.7	23.3	23.9
71–73	20.1	20.7	21.2	21.8	22.4	23.0	23.6	24.1	24.7
74–76	20.9	21.5	22.0	22.6	23.2	23.8	24.4	25.0	25.5
77–79	21.7	22.2	22.8	23.4	24.0	24.6	25.2	25.8	26.3
80–82	22.4	23.0	23.6	24.2	24.8	25.4	25.9	26.5	27.1
83–85	23.2	23.8	24.4	25.0	25.5	26.1	26.7	27.3	27.9
86–88	24.0	24.5	25.1	25.7	26.3	26.9	27.5	28.1	28.7
89–91	24.7	25.3	25.9	26.5	27.1	27.6	28.2	28.8	29.4
92–94	25.4	26.0	26.6	27.2	27.8	28.4	29.0	29.6	30.2

(Continued)

Table 3 PERCENT FAT ESTIMATE FOR MEN: SUM OF CHEST, ABDOMEN, AND THIGH SKINFOLDS

Sum of Skinfolds (mm)	Age to Last Year: Under 22	23–27	28–32	33–37	38–42	43–47	48–52	53–57	Over 57
92–97	26.1	26.7	27.3	27.9	28.5	29.1	29.7	30.3	30.9
98–100	26.9	27.4	28.0	28.6	29.2	29.8	30.4	31.0	31.6
101–103	27.5	28.1	28.7	29.3	29.9	30.5	31.1	31.7	32.3
104–106	28.2	28.8	29.4	30.0	30.6	31.2	31.8	32.4	33.0
107–109	28.9	29.5	30.1	30.7	31.3	31.9	32.5	33.1	33.7
110–112	29.6	30.2	30.8	31.4	32.0	32.6	33.2	33.8	34.4
113–115	30.2	30.8	31.4	32.0	32.6	33.2	33.8	34.5	35.1
116–118	30.9	31.5	32.1	32.7	33.3	33.9	34.5	35.1	35.7
119–121	31.5	32.1	32.7	33.3	33.9	34.5	35.1	35.7	36.4
122–124	32.1	32.7	33.3	33.9	34.5	35.1	35.8	36.4	37.0
125–127	32.7	33.3	33.9	34.5	35.1	35.8	36.4	37.0	37.6

and the model for the 53–57 age category is

$$Y = 3.510 + 0.282X.$$

For the women (Table 4), the corresponding models are (for 22 and under)

$$Y = 5.470 + 0.287X$$

and (for 53–57)

$$Y = 7.210 + 0.287X.$$

(Interpret these fitted models in terms of the variables under discussion. How are the models the same, and how do they differ? What is the practical significance of the similarities and differences?)

In extending the analysis provided above, one should study the residual plots for these fitted models. These plots might suggest that the simple linear models are not quite accurate, although they may be adequate over certain ranges of X.

How will the estimates of percent body fat be used? One possibility is to aid in setting weight reduction goals. If we know our current weight and percent body fat and

Table 4 PERCENT FAT ESTIMATE FOR WOMEN: SUM OF TRICEPS, SUPRAILIUM, AND THIGH SKINFOLDS

Sum of Skinfolds (mm)	Age to Last Year: Under 22	23–27	28–32	33–37	38–42	43–47	48–52	53–57	Over 57
23–25	9.7	9.9	10.2	10.4	10.7	10.9	11.2	11.4	11.7
26–28	11.0	11.2	11.5	11.7	12.0	12.3	12.5	12.7	13.0
29–31	12.3	12.5	12.8	13.0	13.3	13.5	13.8	14.0	14.3
32–34	13.6	13.8	14.0	14.3	14.5	14.8	15.0	15.3	15.5
35–37	14.8	15.0	15.3	15.5	15.8	16.0	16.3	16.5	16.8
38–40	16.0	16.3	16.5	16.7	17.0	17.2	17.5	17.7	18.0
41–43	17.2	17.4	17.7	17.9	18.2	18.4	18.7	18.9	19.2
44–46	18.3	18.6	18.8	19.1	19.3	19.6	19.8	20.1	20.3
47–49	19.5	19.7	20.0	20.2	20.5	20.7	21.0	21.2	21.5
50–52	20.6	20.8	21.1	21.3	21.6	21.8	22.1	22.3	22.6
53–55	21.7	21.9	22.1	22.4	22.6	22.9	23.1	23.4	23.6
56–58	22.7	23.0	23.2	23.4	23.7	23.9	24.2	24.4	24.7
59–61	23.7	24.0	24.2	24.5	24.7	25.0	25.2	25.5	25.7
62–64	24.7	25.0	25.2	25.5	25.7	26.0	26.7	26.4	26.7
65–67	25.7	25.9	26.2	26.4	26.7	26.9	27.2	27.4	27.7
68–70	26.6	26.9	27.1	27.4	27.6	27.9	28.1	28.4	28.6
71–73	27.5	27.8	28.0	28.3	28.5	28.8	29.0	29.3	29.5
74–76	28.4	28.7	28.9	29.2	29.4	29.7	29.9	30.2	30.4
77–79	29.3	29.5	29.8	30.0	30.3	30.5	30.8	31.0	31.3
80–82	30.1	30.4	30.6	30.9	31.1	31.4	31.6	31.9	32.1
83–85	30.9	31.2	31.4	31.7	31.9	32.2	32.4	32.7	32.9
86–88	31.7	32.0	32.2	32.5	32.7	32.9	33.2	33.4	33.7
89–91	32.5	32.7	33.0	33.2	33.5	33.7	33.9	34.2	34.4
92–94	33.2	33.4	33.7	33.9	34.2	34.4	34.7	34.9	35.2
95–97	33.9	34.1	34.4	34.6	34.9	35.1	35.4	35.6	35.9
98–100	34.6	34.8	35.1	35.3	35.5	35.8	36.0	36.3	36.5
101–103	35.3	35.4	35.7	35.9	36.2	36.4	36.7	36.9	37.2
104–106	35.8	36.1	36.3	36.6	36.8	37.1	37.3	37.5	37.8
107–109	36.4	36.7	36.9	37.1	37.4	37.6	37.9	38.1	38.4

(Continued)

Table 4 PERCENT FAT ESTIMATE FOR WOMEN: SUM OF TRICEPS, SUPRAILIUM, AND THIGH SKINFOLDS

Sum of Skinfolds (mm)	Age to Last Year: Under 22	23–27	28–32	33–37	38–42	43–47	48–52	53–57	Over 57
110–112	37.0	37.2	37.5	37.7	38.0	38.2	38.5	38.7	38.9
113–115	37.5	37.8	38.0	38.2	38.5	38.7	39.0	39.2	39.5
116–118	38.0	38.3	38.5	38.8	39.0	39.3	39.5	39.7	40.0
119–121	38.5	38.7	39.0	39.2	39.5	39.7	40.0	40.2	40.5
122–124	39.0	39.2	39.4	39.7	39.9	40.2	40.4	40.7	40.9
125–127	39.4	39.6	39.9	40.1	40.4	40.6	40.9	41.1	41.4
128–130	39.8	40.0	40.3	40.5	40.8	41.0	41.3	41.5	41.8

have a desired body fat percentage in mind, we can calculate our desired weight by this formula:

Desired weight = {weight − [weight(% fat/100)]}/{1 − [% fat desired/100]}.

Note that current weight *is* important in estimating a desired weight goal. (Can you explain why the formula works? Are any assumptions being made here?)

What is a reasonable percentage of body fat to set as a goal? Some recommend between 10% and 22% for men and between 20% and 32% for women. This, however, is not good enough for athletes, who should average around 12% if male and 18% if female.